D0848530

INTRODUCTION TO IONOMERS

INTRODUCTION TO IONOMERS

ADI EISENBERG
Department of Chemistry
McGill University
Montreal, Quebec, Canada

JOON-SEOP KIM
Department of Chemistry
McGill University
Montreal, Quebec, Canada &
Department of Polymer Science & Engineering
Chosun University
Kwangju, Korea

A Wiley-Interscience Publication
JOHN WILEY & SONS, INC.
New York / Chichester / Weinheim / Brisbane / Singapore / Toronto

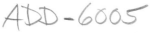
This book is printed on acid-free paper. ∞

Library of Congress Cataloging in Publication Data

Eisenberg, A. (Adi)
    Introduction to ionomers / by Adi Eisenberg and Joon-Seop Kim.
      p. cm.
    "A Wiley-Interscience publication."
    Includes index.
    ISBN 0-471-24678-6 (alk. paper)
      1. Ionomers. I. Kim, Joon-Seop. II. Title.
  QD382.I45E35 1998
  547'.84-dc21   97-31256

Printed in the United States of America.

10 9 8 7 6 5 4 3 2 1

CONTENTS

5 STYRENE IONOMERS   86

PREFACE

The field of ionomers is currently more than 30 years old. During this time, the literature in the field underwent exponential growth, with a doubling period of 6 years; and a host of new applications have appeared, ranging from membranes for the chlor-alkali industry to golf balls. Not surprisingly, our understanding of the field has also advanced, to the point at which the outcome of many experiments can be predicted, at least semiquantitatively. The ability to anticipate some experimental results is fortunate, because the field of ionomers is huge, given that many nonionic polymers can be converted into ionomers using a range of pendent ionic groups placed in various positions relative to the backbone and with greater or lesser regularity along the backbone. Presently, there are over 7500 papers and patents in the literature, which is growing at the current rate of almost 600 documents per year.

It is clearly impossible to condense this large amount of information into a book of reasonable size. A rigorous selection process had to be employed; and many topics, even important ones, had to be omitted. Naturally, any such process, especially one of the degree of selectivity involved here, is highly personal and reflects the interests and viewpoints, not to mention prejudices, of the authors. We were guided by the wish to present topics that provide insight into the relationship between the architecture of the polymer chain and the structure and properties of the polymer, primarily as they are affected by the presence of ionic groups. Thus noncrystalline materials are given greater coverage than those that are partly crystalline, and within both groups, representative rather than encyclopedic coverage is the aim.

A book such as this one would have been impossible without the generous and even enthusiastic help of friends and colleagues. It is a pleasure to express our deep gratitude to Dr. C. J. Clarke and Professor C. G. Bazuin and one of the anonymous referees for their innumerable suggestions involving both scientific and linguistic aspects. The book would undoubtedly have benefited even more if we had accepted all of them. The benefit of valuable discussions and suggestions of Mr. M. Moffitt is also gratefully acknowledged. As with all such works, and in spite of our best efforts, some mistakes undoubtedly remain, both textual and conceptual; the authors naturally assume full responsibility for those. Several colleagues were most helpful by sending us reprints and preprints of their work. In this connection, special thanks are due to doctors C. G. Bazuin, L. A. Belfiore, S. L. Cooper, B. Gabrys, J. C. Galin, M. Hara, T. Hashimoto, J. S. Higgins, R. Jérôme, C. R. Martin, R. B. Moore, M. Pineri, R. A. Register, W. M. Risen, S. Schlick, R. F. Storey, R. A. Weiss, G. L. Wilkes, C. E. Williams, S. Yano, H. Yeager.

Finally, it is a great pleasure to acknowledge our deep indebtedness to an outstanding group of graduate students, postdoctoral co-workers, visiting scientists, and colleagues, who, as a result of their inspired efforts in many capacities, have made this book possible.

LIST OF SYMBOLS AND ABBREVIATIONS

4DA	4-decylaniline
4VP	4-vinylpyridine
a	cation radius plus anion radius (distance between centers of charges at closest approach)
A,B	constants of the Halpin–Tsai equation
AMPS	[2-(acrylamido)-prop-2-yl]methanesulfonic acid
b	block
BAC	1,3-bis(aminomethyl)cyclohexane
c	ion concentration
c	volume fraction of the conducting phase
c_o	volume fraction of the conducting phase at percolation threshold
C_1, C_2	Williams-Landel-Ferry constants
CED	cohesive energy density
CP/MAS	cross-polarization/magic angle spinning
d	distance between multiplets
D	diffusion coefficient
D	self-diffusion coefficient
DB	dodecylbenzene
DEB	diethylbenzene
DMA	dimethylacetamide or dynamic mechanical analysis
DMDAAC	dimethyl diallyl ammonium chloride

DMF	dimethylformamide
DMPA	2.2-bis(hydroxymethyl)propionic acid
DMSO	dimethyl sulfoxide
DMTA	dynamic mechanical thermal analysis
DOP	dioctyl phthalate
DSC	differential scanning calorimetry
DTA	differential thermal analysis
e	electronic charge
E	Young's modulus or Young's modulus of unfilled material
$E*$	complex dynamic modulus or Young's modulus of the filled material (Pa or N/m^2)
E_a	activation energy
E'	tensile storage modulus (Pa or N/m^2)
E'_g	glassy modulus (Pa or N/m^2)
E'_{ionic}	ionic modulus (Pa or N/m^2)
E''	tensile loss modulus (Pa or N/m^2) or Young's loss modulus
EGDMA	ethylene glycol dimethacrylate
EHMA	2-ethylhexyl methacrylate
EHM model	Eisenberg–Hird–Moore model
ENB	5-ethylidene-2-norbornene
EPDM	ethylene–propylene–diene terpolymer
EPR	electron paramagnetic resonance
ESR	electron spin resonance
EW	equivalent weight
EXAFS	extended x-ray absorption fine structure
f	functionality or concentration (or volume fraction) of percolating species
f_c	critical concentration (or volume fraction) at the percolation threshold
f_v	volume fraction of an aqueous phase
f^{50}	orientation function at 50% elongation
FT-IR	Fourier transform–infrared
G	storage modulus (Shear)
G^o_{ent}	rubbery plateau shear modulus
G^o_{ionic}	ionic plateau shear modulus
G^o_N	rubbery modulus (Pa or N/m^2)

$G*$	complex dynamic shear modulus (Pa or N/m^2)
G'	shear storage modulus (Pa or N/m^2)
G''	shear loss modulus (Pa or N/m^2)
HDPE	high-density polyethylene
HSTP	halatosemitelechelic polymer
IR	infrared
k	Boltzmann constant ($= 1.38 \times 10^{-23}$ J/K)
K_i	characteristic increment of the counterion
KI	potassium iodide
LCST	lower critical solution temperature
LDPE	low-density polyethylene
LS	light scattering
m-	meta
M_1, M_2	moduli of components 1, 2 (Pa or N/m^2)
M_c	average molecular weight between ionic groups or cross-links
M_n	number-average molecular weight (g/mol)
M_w	weight-average molecular weight (g/mol)
M	modulus of the material (Pa or N/m^2)
MANa	sodium methacrylate
MDEA	N-methyl diethanolamine
MDI	4,4'-diphenyl methane diisocyanate
MST	microphase separation temperature
MW	molecular weight (g/mol)
n	critical exponent at the percolation threshold
N_A	Avogadro's number
NMR	nuclear magnetic resonance
p-	para
p_c	critical concentration at the percolation threshold
P_m	permeability coefficient
P_n	degree of polymerization
P(4VP)	poly(4-vinylpyridine)
P(E-co-MA)	poly(ethylene-co-methacrylate)
P(E-co-MAA)	poly(ethylene-co-methacrylic acid)
P(E-co-MACs)	poly(ethylene-co-cesium methacrylate)
P(E-co-MANa)	poly(ethylene-co-sodium methacrylate)
P(EA-co-4VP)	poly(ethyl acrylate-co-4-vinylpyridine)

P(EA-*co*-4VPMeI)	poly(ethyl acrylate-*co*-N-methyl-4-vinylpyridinium iodide)
P(ET-*co*-SZn)	poly(ethylene terephthalate-*co*-zinc sulfonate)
P(MMA-*co*-4VP)	poly(methyl methacrylate-*co*-4-vinylpyridine)
P(MMA-*co*-MACs)	poly(methyl methacrylate-*co*-cesium methacrylate)
P(MMA-*co*-MANa)	poly(methylmethacrylate-*co*-sodium methacrylate)
P(S-*co*-4VP)	poly(styrene-*co*-4-vinylpyridine)
P(S-*co*-4VPMeI)	poly(styrene-*co*-N-methyl-4-vinylpyridinium iodide)
P(S-*co*-AA)	poly(styrene-*co*-acrylic acid)
P(S-*co*-ANa)	poly(styrene-*co*-sodium acrylate)
P(S-*co*-MAA)	poly(styrene-*co*-methacrylic acid)
P(S-*co*-MACs)	poly(styrene-*co*-cesium methacrylate)
P(S-*co*-MANa)	poly(styrene-*co*-sodium methacrylate)
P(S-*co*-MAL)	poly(styrene-*co*-lithium methacrylate)
P(S-*co*-SC)	poly(styrene-*co*-styrenecarboxylate)
P(S-*co*-SCNa)	poly(styrene-*co*-sodium styrenecarboxylate)
P(S-*co*-SS)	poly(styrene-*co*-styrenesulfonate)
P(S-*co*-SSA)	poly(styrene-*co*-styrenesulfonic acid)
P(S-*co*-SSMn)	poly(styrene-*co*-manganese styrenesulfonate)
P(S-*co*-SSNa)	poly(styrene-*co*-sodium styrenesulfonate)
P(S-*co*-SSZn)	poly(styrene-*co*-zinc styrenesulfonate)
P(VCH-*co*-AA)	poly(vinylcyclohexane-*co*-acrylic acid)
P(VCH-*co*-ANa)	poly(vinylcyclohexane-*co*-sodium acrylate)
P(α-MS)	poly(α-methylstyrene)
PA-6	polyamide-6
PA	polyamide
PBD	polybutadiene
PDPhPhO	poly(2,6-diphenyl-1,4-phenylene oxide)
PEA	poly(ethyl acrylate)
PEEK	polyaryletheretherketone
PEO	poly(ethylene oxide)
PET	poly(ethylene terephthalate)
PFSI	perfluorosulfonate ionomer
PI	polydispersity index ($= M_w/M_n$)
PI	polyisoprene
PIB	polyisobutylene
PMMA	poly(methyl methacrylate)

PPhO	poly(phenylene oxide) or poly(2,6-dimethyl-1,4-phenylene oxide)
PPrO	poly(propylene oxide)
PS	polystyrene
PTFE	poly(tetrafluoroethylene)
PTMO	poly(tetramethylene oxide)
PVDF	poly(vinylidene fluoride)
q	charge or scattering vector (Å^{-1})
q_a	anion charge
q_c	cation charge
q_i	charge of counterion i
r	random
r	distance between the centers of charge for the anion and cation at closest approach
R	dichroic ratio ($A_\parallel / A_\perp$), the ratio of the absorbance for the electric vector parallel and perpendicular to the stretching direction or gas constant ($= 8.3145$ $\text{J·mol}^{-1}\text{·K}^{-1}$)
s	scattering vector (nm^{-1})
S-EPDM	ethylene-propylene-ethylidene norbonene sulfonate
SANS	small-angle neutron scattering
SAXS	small-angle x-ray scattering
SBS	sodium benzenesulfonate
SDBS	sodium dodecylbenzenesulfonate
SEM	scanning electron microscopy
T	absolute temperature (K) or temperature
t-	tertiary
$T_1, T_{1\rho}$	spin–lattice relaxation time
T_2	spin–spin relaxation time
T_g	glass transition temperature or glass transition
$T_{g,c}$	glass transition temperature of the cluster phase
$T_{g,m}$	glass transition temperature of the matrix phase
T_i	phase separation temperature
T_{ref}	reference temperature
tan δ	loss tangent
TBMA	t-butyl methacrylate
TDI	tolylene diisocyanate
TEM	transmission electron microscopy

TET	t-butyl methacrylate)-b-(2 ethylhexyl methacrylate-b (t-butyl methacrylate)
TGA	thermogravimetric analyzer
THF	tetrahydrofuran
UCST	upper critical solution temperature
UV/VIS	ultraviolet/visible
V_f	volume fraction of filler
VP	vinylpyridine
W_{el}	electrostatic work of removing an anion from the coordination sphere of cation
WLF	Williams–Landel–Ferry
wt	weight
X	mol fraction of crystallizable comonomer
Y	molar glass transition temperature function
Z_i	number of counterion i
Z	sum of the atomic distances of the parent polymer
Z	charge of cation
α	angle between the dipole moment vector of the vibration and the chain axis or degree of neutralization
α_g	expansion coefficient below the matrix T_g
α_L	liquid expansion coefficient
α'_L	first expansion coefficient observed in the first heating run
χ	Polymer-polymer miscibility parameter (or Flory-Huggins interaction parameter)
Δ	birefringence
ΔC_p	heat capacity change (J/K)
$\Delta \varepsilon$	dielectric increment
ΔH	enthalpy changes
ΔT_g	temperature difference between two glass transition temperatures
ΔW	association energy
ε	dielectric constant
ε''	dielectric loss
γ	shear rate
η_o	limiting viscosity at zero shear rate (Pa·s)
η_a	apparent viscosity

η_τ	melt viscosity
$/\eta*/$	complex viscosity
φ_i	volume fraction of component i
κ	specific conductivity of the membrane
κ_o	specific conductivity of conducting phase
λ	x-ray wavelength (Å) or draw ratio
ν	Poisson's ratio
θ	half the scattering angle or x-ray diffraction angle or angle between the chain axis and the stretching direction
ρ	density of polymer (g/cm^3)
σ	conductivity
υ	Poisson's ratio
ω	frequency (s^{-1})

INTRODUCTION TO IONOMERS

CHAPTER 1

INTRODUCTION

Despite the fact that the ionomer field is >30 years old (1,2), these materials remain largely underused in view of their potential for solving a wide range of industrial problems. Even the modest amounts of research carried out to date, however, suggest that the potential rewards of the exploration of ionomer applications are large. Overall, only a small number of industrial applications involve ionomers (which will be discussed in Chapter 10). However, some of those applications have turned out to be spectacularly successful in spite of the only fragmentary understanding of the fundamental aspects of ionomers at the time that the applications were conceived. This suggests that a thorough understanding of the fundamentals—notably the relationship between the chemical structure, the morphology, and the physical properties of ionomers—may well give rise to a much wider range of industrial applications of this challenging and promising class of materials. In addition to furthering academic interest in this area, this book contributes to an increase in the understanding of the fundamental aspects of ionomers, thus increasing their industrial use.

Over the years, the literature in the field has grown exponentially; a large number of conferences on the topic have been held from which proceedings have appeared (3–9), several review articles (10–28), and several edited volumes have also been published (29–33) along with one book (34). The book is now quite out of date; the proceedings of the symposia, by their very nature, deal only with highly selective aspects of this broad field; and the same is also true of many of the reviews.

In view of a number of recent significant advances in this area, the time seems right for a reassessment of the field within the framework of one single unified treatment. The present time is also appropriate because, at last, the morphology of ionomers is understood qualitatively. Furthermore, a wide range of systems has now been investigated, ranging all the way from low glass transition temperature T_g elastomers to high T_g aromatic materials, and the number of publications is extensive.

It should be stressed, however, that no attempt will be made here to provide an exhaustive review of the literature. Rather, the goal is to present the most illuminating examples, preferably for one parent polymer system, when possible, and to treat other polymer systems in terms of their similarities to or differences from this one parent polymer.

1.1. DEFINITIONS

The definition of the word *ionomer* has been the subject of some uncertainty. Historically, *ionomer* was applied to olefin-based polymers containing a relatively small percentage of ionic groups in which the "strong ionic interchain forces play the dominant role in controlling properties" (1). Over the years the definition has broadened to include other parent polymers. A difficulty with this definition arises because, under a wide range of conditions, the ionomers, especially those of higher ion contents, can behave like polyelectrolytes, particularly in solvents of relatively high dielectric constant. Thus the boundary between ionomers and polyelectrolytes is uncertain. Therefore, for the purpose of this discussion, the definition suggested by Eisenberg and Rinaudo (35) is adopted, which defines ionomers as polymers in which the bulk properties are governed by ionic interactions in discrete regions of the material (the ionic aggregates). Thus ionomer behavior is tied to properties rather than to composition, and the crucial component of ionomers is taken to be the ionic aggregate, practically paralleling the original definition. This view differentiates the materials from polyelectrolytes, which are defined as materials in which the solution properties in solvents of high dielectric constant are governed by electrostatic interactions over distances larger than typical molecular dimensions. The advantages of these definitions are that polyelectrolytes and ionomers are now clearly separated and the difficulty of delineating the borderline between them is avoided, which was necessary as long as a composition-based definition was used. The definition suggested above parallels the definitions of words such as *glass* and *rubber*, which are defined in terms of physical properties rather than in terms of composition. While the word *rubber* is associated with some compositions (e.g., butadiene, isoprene, or isobutylene), the more precise definition is given clearly in terms of material properties, e.g., extensibility. Similarly, the word *ionomer* will undoubtedly continue to be associated with materials in which the ion content is below ~15 mol % but is not necessarily tied to that composition range.

The materials that give rise to ionomeric behavior are, most frequently, copolymers of materials of low dielectric constant, such as ethylene or styrene, with ionic groups such as those based on methacrylic acid, sulfonic acid, amines or pyridines. As long as ionic aggregates, no matter how small, are present in the bulk or in polymers dissolved in media of low dielectric constant, these materials will be classified as ionomers. This definition is quite broad and allows the inclusion of all the materials commonly thought of as ionomers. At the same time, it clearly excludes some materials, such as the poly(metaphosphates) (36,37), since these show no evidence of ion aggregation, while allowing the inclusion of silicate glasses in some composition regions, since these are phase separated, with the ionic regions forming distinct domains, separated from the regions of high SiO$_2$ content.

The definition breaks down for material such as styrene-*co*-*N*-methyl-4-vinylpyridinium iodide copolymers in which no ionic aggregation has been observed because of the high glass transition of polystyrene and the weak ionic interactions. However, because ionic aggregation can still be induced by plasticization (Chapter 8) and because these materials closely resemble other ionomers, e.g., those based on poly (styrene-*co*-sodium methacrylate), the word *ionomer* is retained even for materials such as styrene-*co*-*N*-methyl-4-vinylpyridinium iodide.

1.2. NOMENCLATURE

It is clear from the above discussion that the field of ionomers is potentially enormous. In principle, one can imagine the field to be larger than that of nonionic polymeric materials, because for every conceivable nonionic polymer structure there can exist a wide range of ionomers, differing from each other only in the type, position, or concentration of the ionic groups. Practical considerations naturally limit the experimentally accessible range of materials drastically. The multiplicity of potential structures will be discussed much more extensively in Chapter 2. It is, therefore, not surprising that a uniform nomenclature scheme has not emerged for ionomers. For the random materials, it is frequently convenient to use a system like that of Navratil and Eisenberg (38), which, slightly modified, is given as:

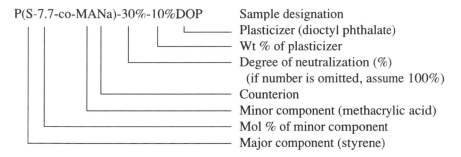

For block ionomers, the nomenclature currently employed for nonionic blocks seems appropriate:

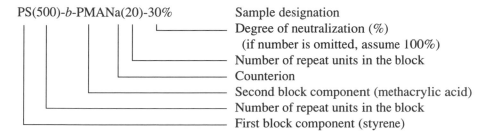

Random copolymers based on the common monomers and block copolymers comprise only a part of the family of ionomers. Because of the enormous structural variability, it is impossible to propose a nomenclature scheme that encompasses everything; and even if one could find such a scheme, undoubtedly new materials would be developed that would fall outside the suggested guidelines. Therefore, for the more complicated systems, authors usually suggest their own schemes, and those schemes are adopted here whenever possible.

1.3. HISTORICAL ASPECTS

Although the word *ionomer* dates back to 1965 (1), materials of the type now called ionomers had been synthesized and investigated long before. This brief overview of some historical aspects of ionomeric systems will extend only up until 1979, i.e., the first 15 years of intensive work on these materials. An exhaustive coverage is not attempted; this is a survey of some highlights from the author's perspectives.

One example of an early synthesis of a material of this type is the work by Littman and Marvel (39) on polymers that were later named *ionenes* (40). Shortly thereafter, a carboxylated elastomer based on butadiene and acrylic acid copolymers was vulcanized with sulfur using zinc oxide as an accelerator. This is described in a French patent dated June 21, 1933 (41). Ionic cross-links were apparently not recognized at that time. However, not long after that, it was noted that the incorporation of carboxyl groups into elastomers exerts a major influence on their properties (42).

The polymeric behavior of halatotelechelics was explored as early as 1944 (43). This was followed by a large number of subsequent papers on the topic from several laboratories. The 1950s saw the appearance of Hypalon (E. I. du Pont de Nemours & Co., Inc.), a sulfonated, chlorinated polyethylene cross-linked by both ionic and covalent bonds. Another important development of the 1950s was the appearance of an extensive and insightful review of the literature on the topic of ionic rubbers (44). Clearly, many of the effects of ionic interactions on polymer behavior were recognized at that time, and that review article is a crucial document in the early history of ionomers.

Work on styrene ionomers started in the mid 1960s, with a publication by Erdi and Morawetz (45) on the solution behavior and rheology of copolymers of styrene and neutralized methacrylic acid and with the work by Fitzgerald and Nielsen (46) on the viscoelastic properties of salts of polymeric acids. Another important event was the appearance of a patent on ionic polyurethanes (47). A crucial occurrence in 1965 was the presentation of two papers on ionomers, in one of which the word itself was proposed (1,48), and an earlier article on the ethylene-based systems which appeared in *Modern Plastics* (49). At the same time, a series of papers on ionic polyphosphates started to appear from our group (50,51). While these materials are not strictly ionomers according to the definition used here, the results illustrated clearly the large effect of ionic interactions in modifying the properties of polymers.

Two papers appeared in 1966 on the bulk properties of polymer–salt mixtures, i.e., materials like poly(ethylene oxide) (52) and poly(propylene oxide) (53), which contain a salt of low lattice energy, such as lithium perchlorate. A range of physical properties was explored, and ionic interactions were clearly recognized as being important in determining the properties of the materials. In the same year, our group (36) demonstrated that the glass transition temperatures of ionomers are linearly related to the cation charge q divided by the distance between the centers of charge a. Another event in 1966 was the appearance of a patent on ethylene ionomers (Surlyn, Du Pont) (2).

A review of ionic forces in bulk polymers was published in 1967 (32), the first since the appearance 10 years earlier of a review on ionic rubbers. This work dealt primarily with ionic interactions in phosphate polymers. Several detailed studies of ethylene ionomers appeared in 1967; of particular note are the articles of Ward and Tobolsky (54) and MacKnight et al. (55). These were the first (besides the *Polymer Preprints* abstracts mentioned above) of a long series of articles on this topic, especially from MacKnight's group. In 1967, the first pictorial model of ionomer morphology was given in the Ward and Tobolsky (54) paper, which credited Bonotto and Bonner (56). That year also saw the appearance of patents on zinc ion cross-linked floor polishes (57,58); the topic was discussed in later publication (59).

The existence of hard ionic clusters in carboxylated elastomers was proposed in 1968 by Tobolsky et al. (60). The first symposium specifically dedicated to ionomers was held that year at the American Chemical Society meeting in San Francisco; the extended abstracts are available (61). The rediscovery of ionenes by Rembaum et al. (40) also occurred in 1968. The first determination of the T_g of an ionene was published in the same year (62). These two papers were the forerunners of a long series of papers on these polymers, the early work coming primarily from Rembaum's group at the Jet Propulsion Laboratory. That same year saw the appearance of an article on an ionic homoblend, i.e., a blend of butadiene with pendent anions mixed with a butadiene with pendent cations (63). In 1969, the first study of a block ionomer was presented by Schindler and Williams (64).

The new decade saw the appearance of the first theoretical paper on ionomers, i.e., the multiplet-cluster model of Eisenberg (65). The viscoelasticity of bulk styrene ionomers was the subject of a publication from our group 2 years later (66), the same year that saw the announcement of perfluorosulfonated materials (Nafion, Du Pont). A publication on this material appeared by Grot (66), and two presentations were given at the Electrochemical Society meetings in Houston (68,69).

The second symposium on ion-containing polymers took place at the American Chemical Society meetings in 1973; extended abstracts of those meetings are available (70). Some of these abstracts were published as full papers in one volume (3); major topics are organic aspects, structure and properties of solids, solution properties, and applications. Of special interest to morphologists is the appearance in that volume of the core-shell model of ionomers by MacKnight et al. (71). Two additional models of ionomer morphology appeared in 1973, one by Marx et al. (72) and by Binsbergen and Kroon (73). In 1974, papers on the polypentenamer-based ionomers started appearing (74). This period also witnessed the early application of sophisticated spectroscopic techniques to the study of ionomers, e.g., electron spin resonance

(ESR) spectroscopy (76); the use of nuclear magnetic resonance (NMR) (75) started somewhat earlier.

From the mid 1970s to the end of the decade, the field of ionomers expanded rapidly, and only a few events can be mentioned here. Among these is the symposium on ionic polymers that was organized by Holliday at Brunel College in Uxbridge, UK, in 1975 and the appearance of the book *"Ionic Polymers"* edited by Holliday (29). A number of excellent reviews on the ionomer-like materials were made available (29), especially on the Surlyn and styrene ionomers, halatotelechelics, carboxylated elastomers, polyelectrolyte complexes, silicates, phosphates, and inorganic glasses. In the same year a paper appeared in which cross-linking of elastomers was accomplished by coordination bonding (77), and a patent by Makowski et al. (78) on the sulfonation of polystyrene using acetyl sulfate was issued. This method of sulfonation represented a most important development in that it allowed easy laboratory-scale sulfonation of polystyrene and gave rise to a host of publications on this polymer family. The first paper on micellization of ionic diblocks also appeared that year (79). A comprehensive treatment of the field appeared in 1977 (34).

The first report of an extended study of perfluorosulfonate (Nafion) membranes was published by our group in 1977 (80); several of detailed studies by many laboratories were published on this important family of materials. In the same year, a symposium was held on perfluorosulfonate membranes at the 152nd meeting of the Electrochemical Society. In 1978, a symposium on ions in polymers took place at the 176th meeting of the American Chemical Society; extended abstracts are available (81). Many of the papers were published in book form (4), including an extensive discussion of the properties of sulfonated ionomers and the use of stearates as ionic

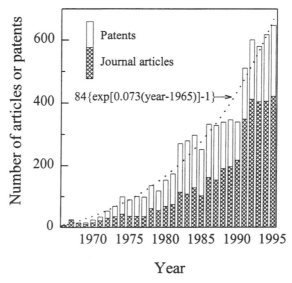

Figure 1.1. The number of journal articles (*hatched bars*) and patents (*open bars*) published between 1965 and 1995 listed under the keyword *ionomer* by the Chemical Abstracts Service.

plasticizers, work originating primarily from groups at the Exxon Research and Engineering Co. Also presented was the first neutron scattering study of coil dimensions in ionomers (82). Finally, the first Gordon Research Conference on ion-containing polymers took place in 1979 in New London, New Hampshire, with A. Eisenberg as chairman and H. S. Makowski as co-chairman. That conference marks the conclusion of the early years of ionomer research.

Since 1965, when the word *ionomer* was coined, the field has grown exponentially, as reflected for example, by the ionomer literature. A survey of chemical abstracts is, perhaps, most instructive of the growth of the field. A total of 6 papers on ionomers appeared in 1966; from 1986 to 1990, an average of 180 papers per year was published (or 340 including patents). The growth has continued so that from 1991 to 1995 the average number of papers published per year has been 400. Over the entire period since 1965, the growth has been exponential (Fig. 1.1). These numbers were obtained by searching *Chemical Abstracts* using the keyword *ionomer*. Because there are likely papers on this topic that do not use *ionomer* as a keyword, the total number of published papers in the field is larger.

1.4. REASONS FOR INTEREST IN IONOMERS

Interest in the ionomer field is continuing and growing. Ionomers offer a wide range of challenges and opportunities both to industrial and to academic workers. The industrial interest is, in part, because ionic interactions permit control of the physical properties of polymers, even at low ion concentrations. The modulus, glass transition temperature, viscosity, melt strength, fatigue, transport, and many other properties can be influenced dramatically by ion incorporation, as will be shown in the following chapters. For example, Figure 1.2**a** illustrates the behavior of the modulus as a function of temperature for seven different ion contents in the random styrene-*co*-sodium methacrylate copolymer system. While the glass transition temperature rises, the increase in the modulus over a broad range of temperatures is greater than expected from the T_g increase. The profiles of the curves change; and as can be expected, the loss tangent curves reflect this change (Figure 1.2**b**). Changes of this order of magnitude are not atypical. An even more important reason for industrial interest is the strength of coulombic interactions, which may permit the use of ionomers as additives in composites, as tie coats between two dissimilar polymer layers, in miscibility enhancement, and in the modification of mechanical properties in general and rheological properties in particular. Finally, the presence of ions enhances the conductivity of materials above the glass transition temperature. This feature is of interest in the design of solid-state electrolytes, and indeed ionomer-like materials have been used in such applications.

From the academic point of view, the relationship between the chemical structure, the morphology, and the physical properties of the materials is of primary interest. Ions tend to aggregate in media of low dielectric constant, and it is this ion aggregation that influences the properties. Understanding the nature of the aggregates has

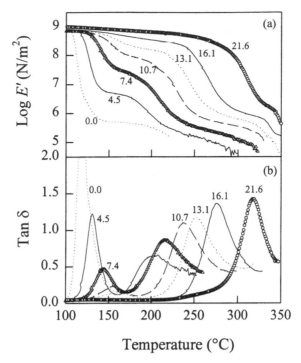

Figure 1.2. Log E' and tan δ plots (1 Hz) as a function of temperature for P(S-*co*-MANa) random copolymers. Numbers on the curves represent ion contents (mol %).

been a goal of two decades of phenomenological research, as has the relation between the morphology and the physical properties.

1.5. AIMS OF THE PRESENTATION

Since 1966, a number of conference proceedings, multiauthored compendia, and an authored book have dealt with ionomers in addition to >7500 citations (4300 journal and 3200 patents). In view of the recent efforts in modeling ionomer morphology and morphology–property relations (83), the need has arisen for a reevaluation of the sizable body of information—especially in the area of random ionomers—within the framework of this new morphological insight. One of the primary aims of this book is, therefore, the presentation of a unified picture of the relationship between the chemical structure, the morphology, and the physical properties of ionomers, wherever possible. It is hoped that, through this unified picture, coulombic interactions will gain an even wider acceptance as a means for the control of physical and physicochemical properties of polymers. Thus readers may approach new challenges with an additional tool at their disposal.

Of course, ionomers will not solve all, or even most, of the problems facing polymer scientists or engineers today; however, a tool of this versatility deserves to be considered, along with the many others available to polymer workers, such as plasticization and copolymerization. The use of ionomers as an aid to problem solving will thus always be in the background of our presentation. Ionic interactions can be used to accomplish a wide range of tasks; in some cases this route is an improvement of nonionic strategies, while in others it may provide the only solution to the problem. One unique application of ionomers that cannot be accomplished by other means is superpermselectivity as it is observed in perfluorinated ionomer membranes (5). This effect can be achieved, at present, only via ionic interactions. Furthermore, the example of phase inversion at an ionic comonomer (sodium methacrylate) content of about 6 mol % in the styrenes (84) is unique and illustrates the enormous amplification power of the effect of ionic interactions on mechanical properties. Obviously, the use of ionomers as hooks or stickers is also of great potential utility in many applications. These examples illustrate the benefit of exploring ionomers as a tool for property modification and control.

1.6. IONIC INTERACTIONS

In this introduction, we discuss several topics related to the strength of ionic interactions, the size of ionic groups, and the methods of expressing ion concentration. The strength of ionic interactions is treated first. Figure 1.3a illustrates the strength of the interactions as a function of the dielectric constant $\epsilon/\epsilon°$ for an ion pair made up of an anion with a 1-Å radius in contact with a cation with a 1-Å radius. Most common organic polymers have dielectric constants of 2–10; 2.5 was chosen as a value typical of styrenelike materials. Figure 1.3b shows the force as a function of the size of the ion pair for the same system, illustrating the effect of the distance d between centers of charge in a medium of one particular dielectric constant. Figure 1.3c shows the work necessary to separate an anion from a cation to infinity in an ion pair and the work necessary to separate one ion pair from another pair. Figure 1.3d shows work as a function of the distance between charges in a medium of a dielectric constant of 2.5.

1.7. SIZES OF IONIC GROUPS

Although ionic radii have been studied extensively, the dimensions of ionic groups are not easily accessible. Recently, an approximate method was proposed for the calculation of the effective sizes of groups such as carboxylate ions (85) and sulfonate ions (86). It is by no means simple to obtain direct information about the size of multiplets or, for that matter, about the effective sizes of ion pairs (such as sodium carboxylates). Therefore, indirect methods must be used to evaluate the sizes. One such method is described here.

If one considers the densities and the formula weights of some common inorganic compounds that resemble multiplets in their composition, one can calculate an average size per atom. The results are given in Table 1.1. The average volume per

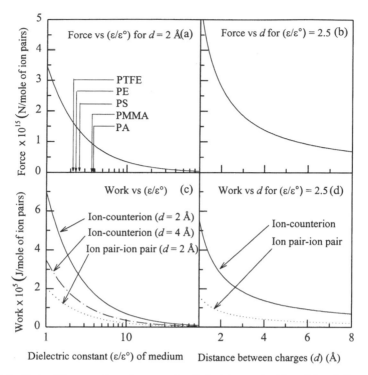

Figure 1.3. Plots of force and work versus the dielectric constant for $\epsilon/\epsilon^{\circ}$ and distance d for various ionic groups. **a,** Force versus dielectric constant for a center-to-center separations between ion pairs of 2 Å. **b,** Force versus distance between centers of charge, for a dielectric constant of 2.5. **c,** Work versus dielectric constant for a center-to-center separation between ions and ion pairs. **d,** Work versus distance between centers of charge for a dielectric constant of 2.5.

TABLE 1.1. Volume per Atom of Various Salts (87)

Salt	Formula Weight	Density, g/cm³	Number of Atoms per Equivalent	Volume per Atom, × 10² nm³
Sodium hydroxide	40.0	2.13	3	1.04
Sodium formate	68.0	1.92	5	1.18
Sodium bicarbonate	84.0	2.16	6	1.08
Sodium carbonate	106.0	2.53	6	1.16
Sodium acetate	82.0	1.53	8	1.11
Sodium oxalate	134.0	2.34	8	1.19
Mean				*1.13 ± 0.05*
Sodium sulfite	126.0	2.63	6	1.33
Sodium sulfate	142.0	2.68	7	1.26
Mean				*1.30 ± 0.04*

atom is remarkably constant; $(1.13 \pm 0.05) \times 10^{-2}$ nm^3. Thus, because a sodium carboxylate ion pair contains four atoms, the volume of the ion pair can be calculated to be ~4.5×10^{-2} nm^3. Using similar reasoning, the volume per atom in a sodium sulfonate ion pair was calculated to be 1.30×10^{-2} nm^3/atom, using physical data from sodium sulfite and sodium sulfate (87). Therefore, the volume of a sodium sulfonate group is 6.5×10^{-2} nm^3. The results of this method are approximate, but they do provide an idea of the effective sizes of multiplets, which are difficult to obtain by other methods.

1.8. METHODS OF EXPRESSING ION CONCENTRATIONS

It is useful to remind the reader at this point of the several different ways of expressing ion concentrations. As long as one deals with simple vinyl polymers, mole percent is a convenient way of expressing the concentration. Mole percent is understood to mean the ratio of the number of repeat units of the ionic moiety to the total number of repeat units ($\times 100$). In materials such as the random styrene–sodium methacrylate copolymers, this method is a simple and clear way of presenting the ion concentration. However, when dealing with polymers of different structures (e.g., the highly aromatic or rigid chain systems), one finds other units to be more convenient. For example, milliequivalents per unit weight or moles per liter can be useful. Note that the equivalent weight of a material is also a valid indication of the ion content and is favored by electrochemists. Figure 1.4 illustrates the relationship between these units for the styrene-*co*-methacrylic acid copolymers.

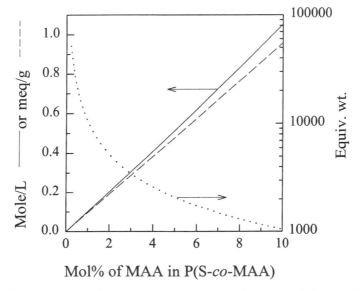

Figure 1.4. Various ways of expressing the ion content for random P(S-*co*-MAA) copolymers.

1.9. TOPICS OMITTED FROM THIS TREATMENT

Within the framework of the present review, a number of topics must be omitted. One major omission is an in-depth discussion of the synthesis of ionomers. Syntheses will be treated only briefly and in connection with the descriptions of the types of ionomers (Chapter 2). The synthesis of ionomers has become a large topic, and its detailed inclusion would increase the scope of the presentation appreciably. Another omission is the polyelectrolytes. Because polyelectrolytes are, by definition, materials in which ionic interactions extend over large dimensions in dilute solutions of high dielectric constant, these materials constitute a separate topic; and because many treatments of polyelectrolytes are already available (88–91), this topic will be completely omitted from this presentation. Dilute solution properties will also be omitted; the field is under active investigation and is in a state of flux. Nonionic water-soluble polymers are not included, because ionic interactions are not involved; chemical reactions in ionomers are treated only briefly. Electrical conductivity in ionomers is not discussed.

1.10. ORDER OF PRESENTATION

The book is presented as follows. Chapter 2 contains a brief survey of the different types of ionomers. The first part of the chapter describes architectural features of ionomers and illustrates the many different types of structures that can be achieved; the second part deals with various chemical aspects, e.g., types and placement of ions. Morphologies and coil dimensions of random ionomers are the subject of Chapter 3. This is an important section of the book, because many of the properties that make ionomers unique are related to their morphology. Extensive studies of the morphology have been performed in a number of groups; and although the survey here is brief, it gives the reader an idea of the important features that make ionomers unique. Chapter 4 deals with the glass transition in ionomers and methods of controlling it by manipulating parameters. Both matrix and cluster glass transitions are discussed.

The properties of ionomers are discussed in Chapter 5 through 7. In Chapter 5, the mechanical properties of random styrene ionomers are discussed in considerable detail, focusing first on the random styrene–sodium methacrylate copolymer, and treating other styrene ionomers in relation to this model system. The styrene ionomers were picked as a model for two reasons. First, the parent polymer is well understood, is noncrystalline, and has given rise to a wide range of ionomers because of its ease of modification. Second, the styrene ionomers are perhaps the best understood noncrystalline ionomer system, in terms of both morphology and properties. Chapter 6 deals with the mechanical properties of partly crystalline ionomers, again in relation to the styrene ionomers. The materials covered include ionomers based on ethylene, tetrafluoroethylene, and pentenamer. Chapter 7 describes other ionomers, e.g., those based on elastomers, acrylates, methacrylates, zwitterionomers, block ionomers, segmented ionenes, monochelics, telechelics, stars, and polymer–salt mixtures.

Plasticization of ionomers presents unusual challenges, because three different types of plasticizers can be considered: nonpolar, polar, and amphiphilic. Each has a different effect on the materials; Chapter 8 is devoted to this topic. Chapter 9 is focused on blends of ionomers and, specifically, on the use of ionic interactions in miscibility enhancement via ion–ion, ion pair–ion pair, ion–coordination, and ion–dipole interactions. Blending of ionomers with nonionic materials has become a major topic, but most of the literature on these materials has appeared in the form of patents; besides, the morphologic and physical property aspects have not been treated in sufficient detail in most cases. Chapter 10 deals with applications, both current and potential. A survey of the patent literature reveals that many applications are possible, and selected examples of these are described.

REFERENCES

1. Rees, R. W.; Vaughan, D. J. *Polym. Prepr. Am. Chem. Soc. Div. Polym. Chem.* **1965,** *6,* 287–295.

2. Rees, R. W. U.S. Patent 3 264 272, 1966.

3. *Ion-Containing Polymers*; Eisenberg, A., Ed.; Journal of Polymer Science, Polymer Symposium 45.

4. *Ions in Polymers*; Eisenberg, A., Ed.; Wiley, New York, 1974. Advances in Chemistry 187; American Chemical Society: Washington, DC, 1980.

5. *Perfluorinated Ionomer Membranes*; Eisenberg, A.; Yeager, H. L., Eds.; ACS Symposium Series 180, American Chemical Society: Washington, DC, 1982.

6. *Coulombic Interactions in Macromolecular Systems*; Eisenberg, A.; Bailey, F. E., Eds.; ACS Symposium Series 302; American Chemical Society: Washington, DC, 1986.

7. *Structure and Properties of Ionomers*; Pineri, M.; Eisenberg. A., Eds.; NATO ASI Series C198; Reidel: Dordrecht, The Netherlands, 1987.

8. *Multiphase Polymers: Blends and Ionomers*; Utracki, L. A.; Weiss, R. A., Eds.; ACS Symposium Series 395; American Chemical Society: Washington, DC, 1989.

9. *Properties of Ionic Polymers, Natural & Synthetic*; Salmén, L.; Htun, M., Eds.; STFI Meddelande: Stockholm, 1991.

10. Otocka, E. P. *J. Macromol. Sci. Rev. Macromol. Chem.* **1971,** *C5,* 275–294.

11. Longworth, R. *Plast. Rubber Mater. Appl.* **1978,** 75–86.

12. Bazuin, C. G.; Eisenberg, A. *Ind. Eng. Chem. Prod. Res. Dev.* **1981,** *20,* 271–286.

13. MacKnight, W. J.; Earnest, T. R. Jr. *J. Polym. Sci. Macromol. Rev.* **1981,** *16,* 41–122.

14. Lundberg, R. D. In *Encyclopedia of Chemical Technology*; Grayson, M., Ed.; Wiley: New York, 1984; 3rd ed., Vol. Suppl., pp. 546–573.

15. MacKnight, W. J., Lundberg, R. D. *Rubber Chem. Technol.* **1984,** *57,* 652–663.

16. Lundberg, R. D. In *Encyclopedia of Polymer Science and Engineering*; Mark, H. F.; Kroschwitz, J. I., Eds.; Wiley: New York, 1987; Vol. 8, pp. 393–423.

17. Tant, M. R.; Wilkes, G. L. *J. Macromol. Sci. Rev. Macromol. Chem. Phys.* **1988,** *C28,* 1–63.

18. Mauritz, K. A. *J. Macromol. Sci. Rev. Macromol. Chem. Phys.* **1988,** *C28*, 65–98.
19. Fitzgerald, J. J.; Weiss, R. A. *J. Macromol. Sci. Rev. Macromol. Chem. Phys.* **1988,** *C28*, 99–185.
20. Lantman, C. W.; MacKnight, W. J.; Lundberg, R. D. In *Comprehensive Polymer Science: The Synthesis, Characterization, Reactions & Applications of Polymers*; Booth, C.; Price, C., Eds.; Pergamon: Oxford, UK, 1989; Vol. 2, Chapter 25.
21. Jérôme, R. In *Telechelic Polymers: Synthesis and Applications*; Goethals, E., Ed.; CRC: Boca Raton, FL, 1989; Chapter 11.
22. Lundberg, R. D.; Agarwal, P. K. *Indian J. Technol.* **1993,** *31*, 400–418.
23. Hara, M.; Sauer, J. A. *J. Macromol. Sci. Rev. Macromol. Chem. Phys.* **1994,** *C34*, 325–373.
24. Eisenberg, A.; Kim, J.-S. In *Polymeric Materials Encyclopedia*; Salamone, J. C., Ed.; CRC: Boca Raton, FL, 1996; pp. 3435–3454.
25. Bazuin, C. G. In *Polymeric Materials Encyclopedia*; Salamone, J. C., Ed.; CRC: Boca Raton, FL, 1996; pp. 3454–3460.
26. Kutsumizu, S.; Yano, S. In *Polymeric Materials Encyclopedia*; Salamonte, J. C., Ed.; CRC: Boca Raton, FL, 1996; pp. 3460–3465.
27. Hara, M.; Sauer, J. A. In *Polymeric Materials Encyclopedia*; Salamone, J. C., Ed.; CRC: Boca Raton, FL, 1996; pp. 3465–3473.
28. Hara, M. In *Polymeric Materials Encyclopedia*; Salmone, J. C., Ed.; CRC: Boca Raton, FL, 1996; pp. 3473–3481.
29. *Ionic Polymers*; Holliday, L., Ed.; Applied Science: London, 1975.
30. *Developments in Ionic Polymers*—Wilson, A. D.; Prosser, H. J., Eds; Elsevier: London, 1986.
31. *Advances in Urethane Ionomers*; Xiao, H. X.; Frisch, K. C., Eds.; Technomic: Lancaster, PA, 1995.
32. *Ionomers: Characterizations, Theory, and Applications*; Schlick, S., Ed.; CRC: Boca Raton, FL, 1996.
33. *Ionomers: Synthesis, Structure, Properties, and Applications*; Tant, M. R.; Mauritz, K. A.; Wilkes, G. L., Eds.; Chapman & Hall: New York, 1997.
34. Eisenberg, A.; King, M. *Ion-Containing Polymers, Physical Properties and Structure*; Academic: New York, 1977.
35. Eisenberg, A.; Rinaudo, M. *Polymer Bull.* **1990,** *24*, 671.
36. Eisenberg, A.; Farb, H.; Cool, L. G. *J. Polym. Sci. A-2* **1966,** *4*, 855–868.
37. Eisenberg, A. *Adv. Poly. Sci.* **1967,** *5*, 59–112.
38. Navratil, M.; Eisenberg, A. *Macromolecules* **1974,** *7*, 84–89.
39. Littman, E. R.; Marvel, C. S. *J. Am. Chem. Soc.* **1930,** *52*, 287–294.
40. Rembaum, A.; Baumgartner, W.; Eisenberg, A. *J. Polym. Sci. Polym. Lett.* **1968,** *6*, 159–171.
41. Fr. Patent assigned to I. G. Fathen Ind. A. G. 701 102, 1933.
42. Bacon, R. G. R.; Farmer, E. H. *Rubber Chem. Technol.* **1939,** *12*, 200–209.
43. Cowan, J. C.; Teeter, H. M. *Ind. Eng. Chem.* **1944,** *36*, 148–152.
44. Brown, H. P. *Rubber Chem. Technol.* **1957,** *30*, 1347–1386.
45. Erdi, N. Z.; Morawetz, H. *J. Colloid Sci.* **1964,** *19*, 708–721.

46. Fitzgerald, W. E.; Nielsen, L. E. *Proc. R. Soc. Ser. A* **1964**, *282*, 137–146.

47. Dietrich, D.; Müller, E.; Bayer, O.; Peter J. (Bayer A.G.) D B Patent 1, 495 693, 1962.

48. Rees, R. W.; Vaughan, D. J. *Polym. Prepr. Am. Chem. Soc. Div. Polym. Chem.* **1965,** *6*, 296–303.

49. Rees, R. W. *Modern Plastics* **1964**, *42*, 209.

50. Eisenberg, A.; Sasada, T. In *Physics of Non Crystalline Solids*, Prins, J. A., Ed.; North Holland: Amsterdam, The Netherlands, 1965; pp. 99–116.

51. Eisenberg, A.; Sasada, T. *J. Polym. Sci.* **1968**, *C16*, 3473–3489.

52. Lundberg, R. D.; Bailey, F. E.; Collard, R. W. *J. Polym. Sci. A-1* **1966**, *4*, 1563–1577.

53. Moacanin, J.; Cuddihy, E. F. *J. Polym. Sci. C* **1966**, *14*, 313–322.

54. Ward, T. C.; Tobolsky, A. V. *J. Appl. Polym. Sci.* **1967**, *11*, 2403–2415.

55. MacKnight, W. J.; McKenna, L. W.; Read, B. E. *J. Appl. Phys.* **1967**, *38*, 4208–4212.

56. Bonotto, S.; Bonner, E. F. *Macromolecules* **1968**, *1*, 510–515.

57. Rogers, J. R.; Sesso, L. M. U.S. Patent 3 308 078, 1967.

58. Rogers, J. R.; Sesso, L. M. U.S. Patent 3 320 196, 1967.

59. Rogers, J. R.; Sesso, L. M. Soap Chem. Spec. **1968**, *44*(3)34–37.

60. Tobolsky, A. V.; Lyons, P. F.; Hata, N. *Macromolecules* **1968**, *1*, 515–519.

61. *Polymer Preprints, American Chemical Society, Division of Polymer Chemistry*, R. W. Lenz, Ed.; *Amer. Chem. Soc.* Washington DC: vol 9(1), 1968;

62. Eisenberg, A.; Yokoyama, T. *Polym. Prepr. Am. Chem. Soc. Div. Polym. Chem.* **1968,** *9*, 617–622.

63. Otocka, E. P.; Eirich, F. R. *J. Polym. Sci. A-2* **1968**, *6*, 921–932.

64. Schindler, A.; Williams, J. L. *Polym. Prepr. Am. Chem. Soc. Div. Polym. Chem.* **1968,** *10*, 832 836.

65. Eisenberg, A. *Macromolecules*, **1970**, *3*, 147–154.

66. Eisenberg, A.; Navratil, M. *J. Polym., Sci. Polym. Lett.* **1972**, *10*, 537–542.

67. Grot, W. *Chem. Ing. Technol.* **1972**, *44*, 167–169.

68. Grot, Munn, Walmsley. Presented at the 141st National Meeting of the Electrochemical Society, Houston, TX, May 1972.

69. Leitz, Accomazzo, Michalek. Presented at the 141st National Meeting of the Electrochemical Society, Houston, TX, May 1972.

70. *Polymer Preprints, American Chemical Society, Division of Polymer Chemistry*, McGrath, J. C. Ed.; *Amer. Chem. Soc.*: Washington DC, 1973. Vol 14(2).

71. MacKnight, W. J.; Taggart, W. P.; Stein, R. S. *J. Polym. Sci. Polym. Symp.* **1974**, *45*, 113–128.

72. Marx, L.; Caulfield, D. F.; Cooper, S. L. *Macromolecules* **1973**, *6*, 344–353.

73. Binsbergen, F. L.; Kroon, G. F. *Macromolecules* **1973**, *6*, 145.

74. Sanui, K.; Lenz, R. W.; MacKnight, W. J. *J. Polym. Sci. Polym. Chem. Ed.* **1974**, *12*, 1965–1981.

75. Otocka, E. P.; Davis, D. D. *Macromolecules* **1969**, *2*, 437.

76. Pineri, M.; Meyer, C.; Levelut, A. M.; Lambert, M. *J. Polym. Sci. Polym. Phys. Ed.* **1974**, *12*, 115–130.

77. Meyer, C. T.; Pineri, M. *J. Polym. Sci. Polym. Phys. Ed.* **1975**, *13*, 1057–1061.

78. Makowski, H. S.; Lundberg, R. D.; Singhal, G. L. U.S. Patent 3 870 841, 1975.

79. Selb, J.; Gallot, Y. *J. Polym. Sci. Polym. Lett. Ed.* **1975**, *13*, 615–619.

80. Yeo, S. C.; Eisenberg, A. *J. Appl. Polym. Sci.* **1977**, *21*, 875–898.

81. *Polymer Preprints, American Chemical Society, Division of Polymer Chemistry*; R. M. Ikeda, Ed.; *Amer. Chem. Soc.* Vol 19(2): pp 215–428, 1978.

82. Pineri, M.; Duplessix, R.; Gauthier, S.; Eisenberg, A. In *Ions in Polymers*; Eisenberg, A., Ed.; Advances in Chemistry 187; American Chemical Society: Washington, DC, 1980; Chapter 18.

83. Eisenberg, A.; Hird, B.; Moore, R. B. *Macromolecules* **1990**, *23*, 4098–4108.

84. Hird, B.; Eisenberg, A. *J. Polym. Sci. Part B Polym. Phys.* **1990**, *28*, 1665–1675.

85. Kim, J.-S.; Jackman, R. J.; Eisenberg, A. *Macromolecules* **1994**, *27*, 2789–2803.

86. *CRC Handbook of Chemistry and Physics*, 76th ed.; Lide, D. R., Ed.; CRC: Boca Raton, FL, 1995.

87. Nishida, M.; Eisenberg, A. *Macromolecules* **1996**, *29*, 1507–1515.

88. Rice, S. A. *Polyelectrolytes Solutions, A Theoretical Introduction*; Academic: New York, 1961.

89. Oosawa, F. *Polyelectrolytes;* Marcel Dekker: New York, 1971.

90. *Polyelectrolytes—1*; Sélégny, E.; Mandel, M.; Strauss, U. P., Eds.; Reidel: Dordrecht, The Netherlands, 1974.

91. *Polyelectrolyte Gels: Properties, Preparation, and Applications*; Harland, R. S.; Prud'-homme, R. K., Eds.; ACS Symposium Series 480, American Chemical Society: Washington, DC, 1992.

CHAPTER 2

STRUCTURAL VARIABILITY IN IONOMERS

This chapter describes the various types of ionomers. It is presented in two sections. The first is devoted to architectural features of ionomers and reviews the possible structures that can be achieved synthetically, e.g., random copolymers, block copolymers, and telechelics. It should be stressed that most of these structures can be achieved using only one type of repeat unit, e.g., styrene. While many of the styrene ionomers with these different structures have already been synthesized, some have not. The large number that have been synthesized, however, confirms the wide range of experimentally accessible architectures. The ionomers discussed in Section 2.1 are shown in figures.

The second part of the chapter deals with the different types of ions that can be incorporated into ionomers as pendent ions or counterions along with their positional variations on the repeat unit. For example, a carboxylate ion can be attached to a styrene chain right on the backbone, as in the styrene-*co*-sodium methacrylate copolymers, and in the *para* position of the benzene ring, as in *para*-carboxystyrene; or it could be separated from the main chain by short spacers. Long spacers are characteristic of the family of combs. Naturally, other ions (e.g., sulfonate or ammonium) can be treated similarly. It should be recognized that zwitterions can be used in place of the more customary pendent ions and microcounterions. The polymer backbone can, of course, also be changed. Polymer counterions are another possibility and are treated in Chapter 9, which deals with blends.

2.1. ARCHITECTURAL VARIATION

2.1.1. Monochelics, Telechelics, and Block Copolymers

2.1.1.1. Single Ions at Chain Ends. The simplest type of ionomer consists of a single ion placed at the end of a polymer chain (Fig. 2.1a). An example of this series of materials is polystyrene containing a terminal carboxylate anion. These

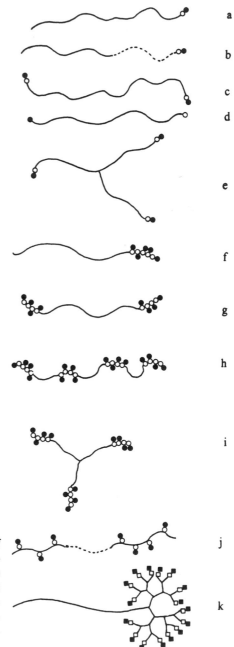

Figure 2.1. Possible molecular structures of regular ionomers. **(a)** Monochelic, **(b)** monochelic block copolymer, **(c)** telechelic, **(d)** zwitterionic telechelic, **(e)** star, **(f)** AB diblock, **(g)** ABA triblock, **(h)** $(AB)_n$ multiblock, **(i)** star block, **(j)** ABA triblock ionomer, and **(k)** ionic dendrimer attached to nonionic chain.

materials have been termed monochelics (1) and can be prepared by anionic polymerization (2–5) using, for example, alkyllithium initiators; once the polymer chain has reached the desired length and the growth is complete, the anion is terminated with CO_2 (6). The monochelics have also been called halatosemitelechelic polymers (HSTP) (7,8). 1,3-Propanesultone can be used if a sulfonate terminal group is desired (9,10). The soaps are low molecular weight examples of this series of materials; however, because of their low molecular weight, soaps and similar amphiphiles are not discussed here. Polymer chains can also be terminated with other functional groups, e.g., amines (11) and zwitterions (12). A new type of monochelic based on block copolymers of styrene and isoprene was recently prepared (13) (Fig. 2.1b).

2.1.1.2. Telechelics. The next level of complexity is encountered when one places one ion at each end of a polymer chain (Fig. 2.1c). Materials of this type, called telechelics, can be synthesized using, e.g., sodium naphthalenide as the initiator for the anionic polymerization of styrene. The polymerization is followed by termination of both ends with CO_2, propanesultone, or another group. Telechelics have been investigated widely since the mid-1940s (14), and the most extensive studies have come from Jérôme and Teyssié's group (7,8,15–17). A brief synthetic scheme for the preparation of α,ω-dicarboxy polyisoprene using a short poly(α-methyl styrene) [p(α-MS)] chain as the initiator is shown in Scheme **2.1**. The short poly(α-methyl styrene) segment in the middle of the isoprene chain is not indicated.

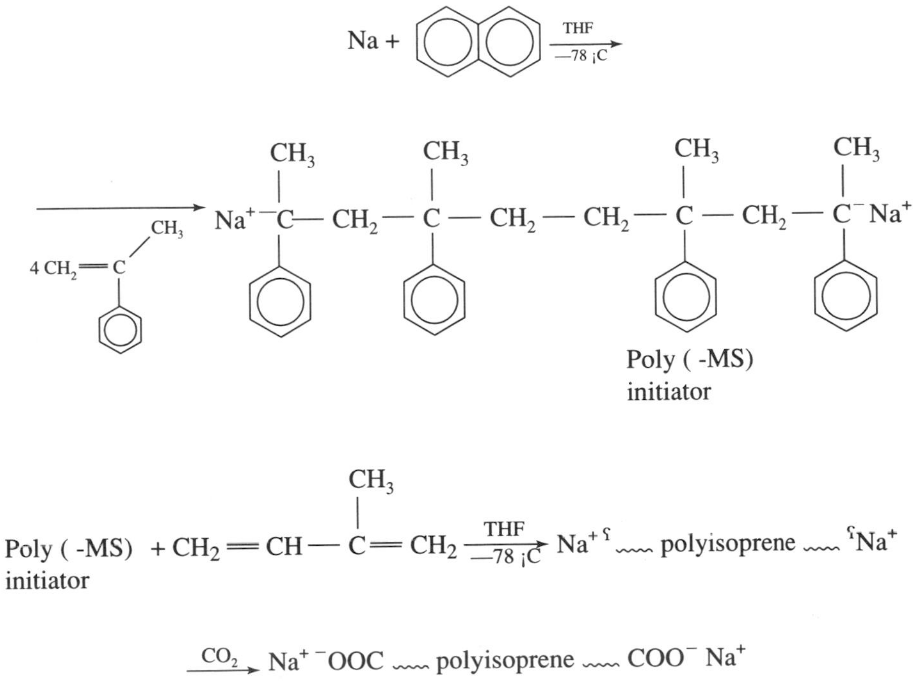

Scheme 2.1

Other ionic end groups can be used; e.g., sulfonates (16), sulfates (18,19), and borates (20). Asymmetric telechelics have also been synthesized (21). In this case, the chain (polyisoprene) contains a tertiary amine on one end of the polymer and a sulfonic acid on the other end (Fig. 2.1**d**).

The next level of complexity is reached when a three-arm star is tipped at each arm's end by an ionic group (Fig. 2.1**e**). The best known examples of this family of materials are the polyisobutylene (PIB) stars with sulfonic acids in the terminal position. They have been synthesized by Kennedy and Storey and were investigated extensively by their group (22,23). A brief synthetic route is shown in Scheme **2.2**. The PIB family of stars of this type exhibits unusual elastomeric behavior (Chapter 7).

2.1.1.3. Block Copolymers. The next family of materials to be discussed is block ionomers. The simplest block ionomers are of the AB type, in which one of the segments is nonionic (e.g., polystyrene), and the other consists of ionic repeat units (e.g., *N*-methyl-4-vinylpyridinium iodide) (24) (Fig. 2.1**f**). These materials can be prepared by anionic polymerization using alkyllithium as the initiator. ABA triblocks are a natural extension of this family (Fig. 2.1**g**). A styrene midblock capped at each end by *N*-methyl-4-vinylpyridinium iodide segments is an example of this type of material. These polymers are also synthesized by anionic polymerization; sodium naphthalenide or similar materials are used as initiators (2–5). 1,1-Diphenyl-ethylene is used as an end-capping agent to lower the activity of the polystyryl anions. A brief synthetic scheme is shown in Scheme **2.3**.

When these block ionomers are quaternized with a long-chain alkyl iodide, they are called bottle brushes (25) and represent a family of comb-type materials (see below). One can also envisage a series of bottle brushes based, e.g., on a block of methacrylate or styrenesulfonate anions in which the "hair" is provided by long-chain quaternary alkyl ammonium or alkyl pyridinium countercations. The latter are not bound to the chain by primary chemical bonds but by electrostatic interactions. In parallel with the nonionic block copolymers, it is clear that $(AB)_n$ type multiblock ionomeric materials can also be synthesized (Fig. 2.1**h**). Star blocks have also been reported (Fig. 2.1**i**) (26).

Another group of materials consists of diblocks (27), or ABA-type triblocks (28), of which one segment is completely nonionic, and the other is only partly ionic. One example is a block copolymer made of *p-tert*-butylstyrene or hydrogenated butadiene as one block and styrene as the second block, in which the styrene block is partly sulfonated (Fig. 2.1**j**). The family of *p-tert*-butylstyrene-*b*-styrene or poly(-ethylene-*r*-butadiene)-*b*-polystyrene diblocks is particularly suitable for the synthesis of this type of material, because the presence of *tert*-butyl groups on the *para* position of the benzene ring and because the saturated nature of one of the blocks prevents sulfonation from occurring at these sites (28). These systems are now receiving some attention.

Recently, another new type of block ionomer was synthesized by van Hest et al. (29). The acid form of the block copolymer consists of a block of polystyrene with an acid functionalized poly(propylene imine) dendrimer at one end. Figure 2.1**k**

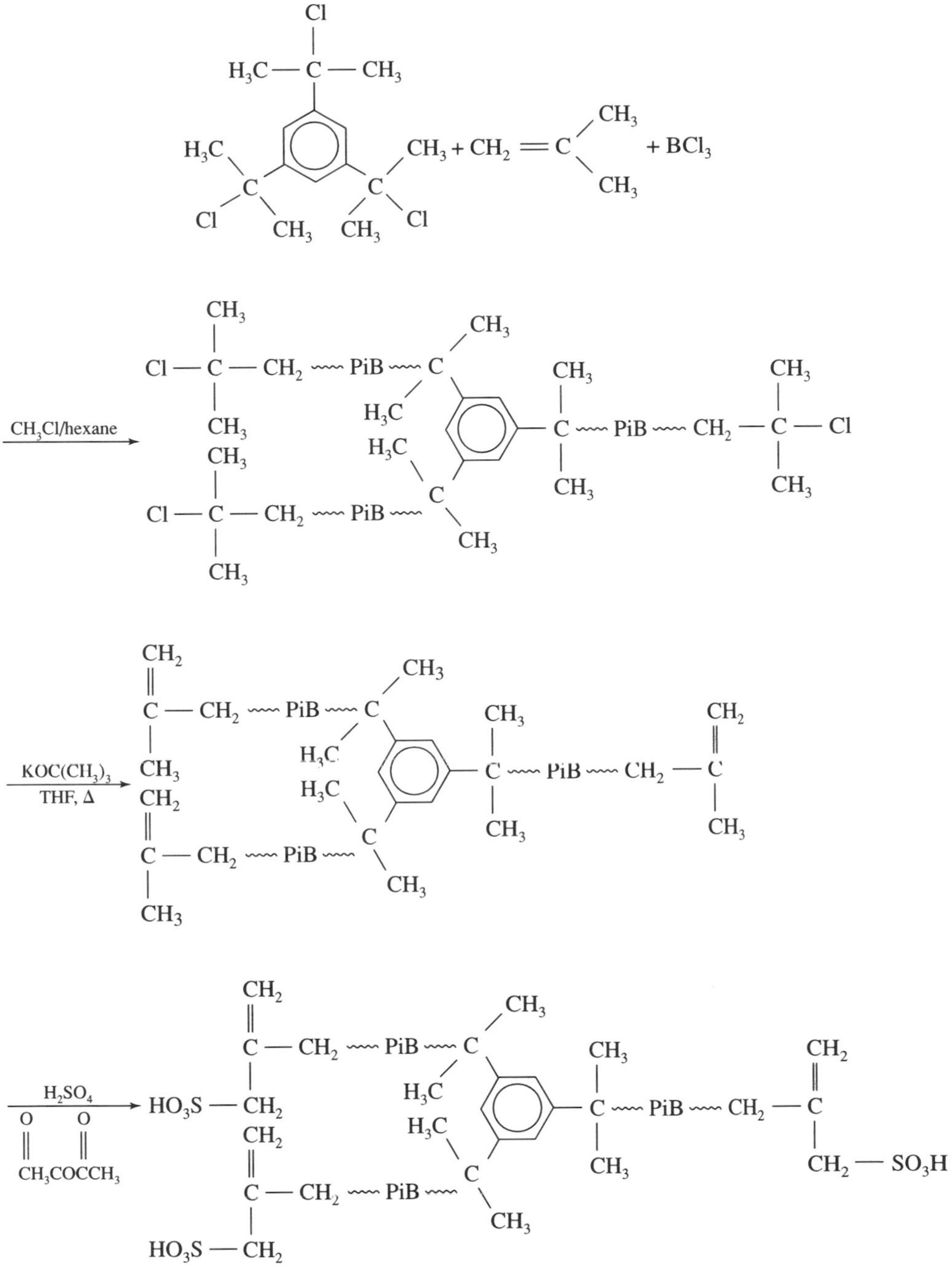

Scheme 2.2

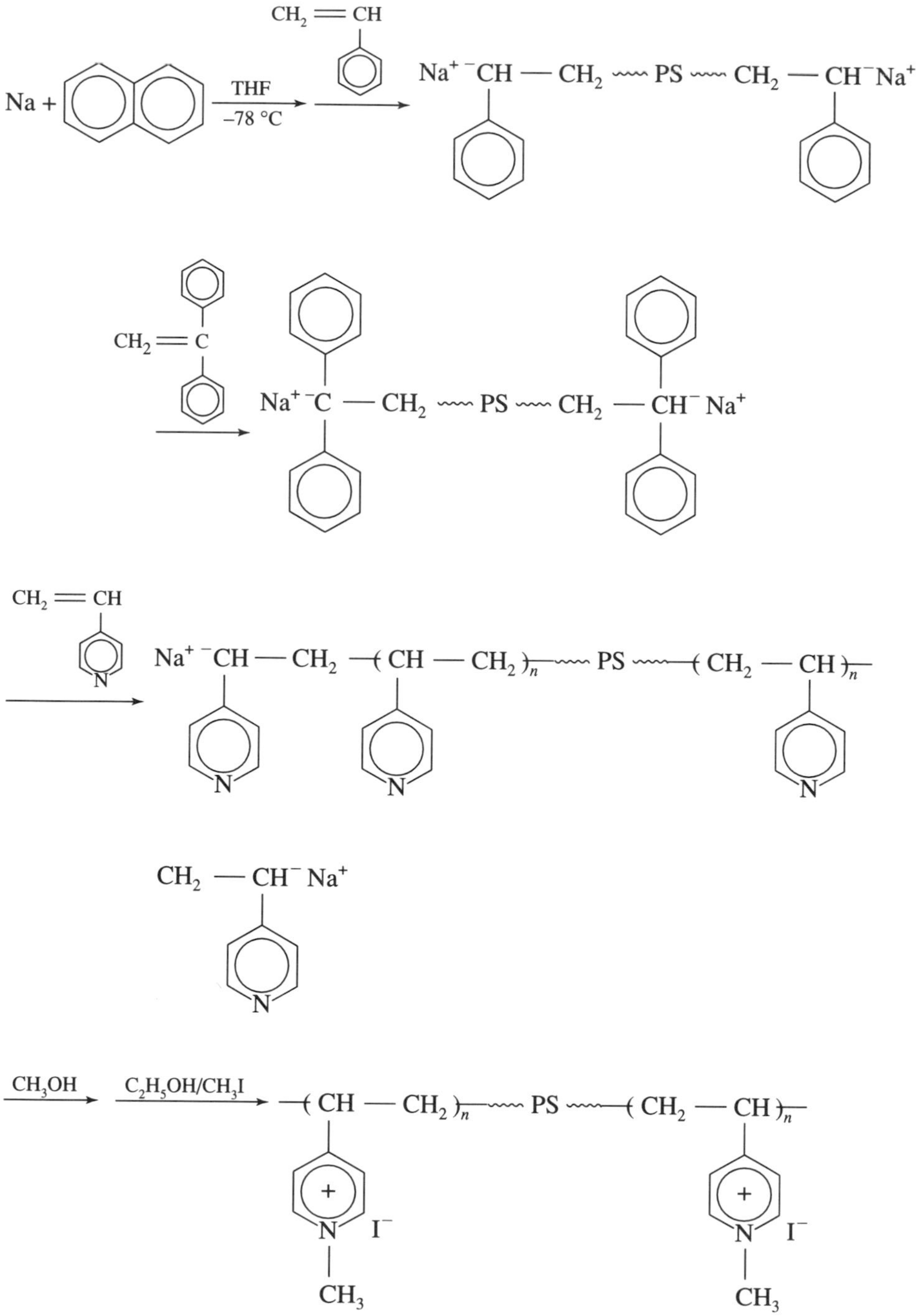

Scheme 2.3

shows the schematic structure of an ionomer containing a dendrimer of the fourth generation. Normal dendrimers in which the outermost shell is ionic are, of course, also possible (30).

2.1.2. Simple Random Copolymers

Simple random copolymers in the present context consist of a nonionic material (such as ethylene or styrene) with an ionogenic species (such as acrylic acid, methacrylic acid, or vinylpyridine) (Fig. 2.2a). Within the random ionomers, the range of materials that can be synthesized is truly enormous; thus these are the materials that have been investigated most extensively and which will be discussed in detail in Chapters 5, 6, and 7. Practically every polymer known can, in principle, be prepared as a random ionomer; and as will be seen in Section 2.2.2, the number of ionic species that can, in principle, be attached to the various backbones is also substantial.

Two other types of random ionomers need to be considered. One is the polyampholytes, i.e., polymers containing both pendent anions and pendent cations distrib-

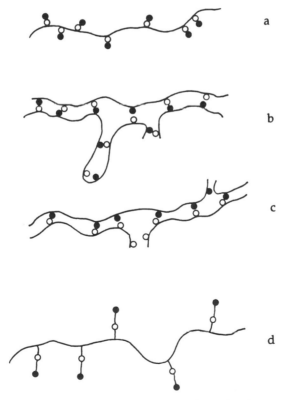

Figure 2.2. Possible molecular structures of ionomers. **(a)** Random ionomer, **(b)** polyampholyte, **(c)** homoblend, and **(d)** zwitterionomer.

uted randomly along the chain (Fig. 2.2**b**). One way to synthesize these materials involves the copolymerization of a nonionic monomer, such as styrene, with an ion pair monomer, such as vinylpyridinium–styrenesulfonate, in which the two species are joined by a strong electrostatic bond. Materials of this type are characterized, at least to some extent, by ion association, depending on the ion parameters, and exhibit many of the properties of ionomers.

Another class that must be considered here is the so-called homoblends (Fig. 2.2**c**). These materials consist of mixtures of, for example, a copolymer of polystyrene with styrenesulfonic acid and a copolymer of polystyrene with vinylpyridine (31–33). Butadiene copolymers can also be used (34). These materials exhibit some of the properties of ionomers. Note that in homoblends any one polymer chain contains only anions or only cations, whereas in polyampholytes each polymer chain contains both anions and cations. The homoblends will be discussed more extensively in Chapter 4, in the context of glass transition, and Chapter 9, in the context of mechanical properties of blends. The synthetic schemes for random copolymers are well known (35).

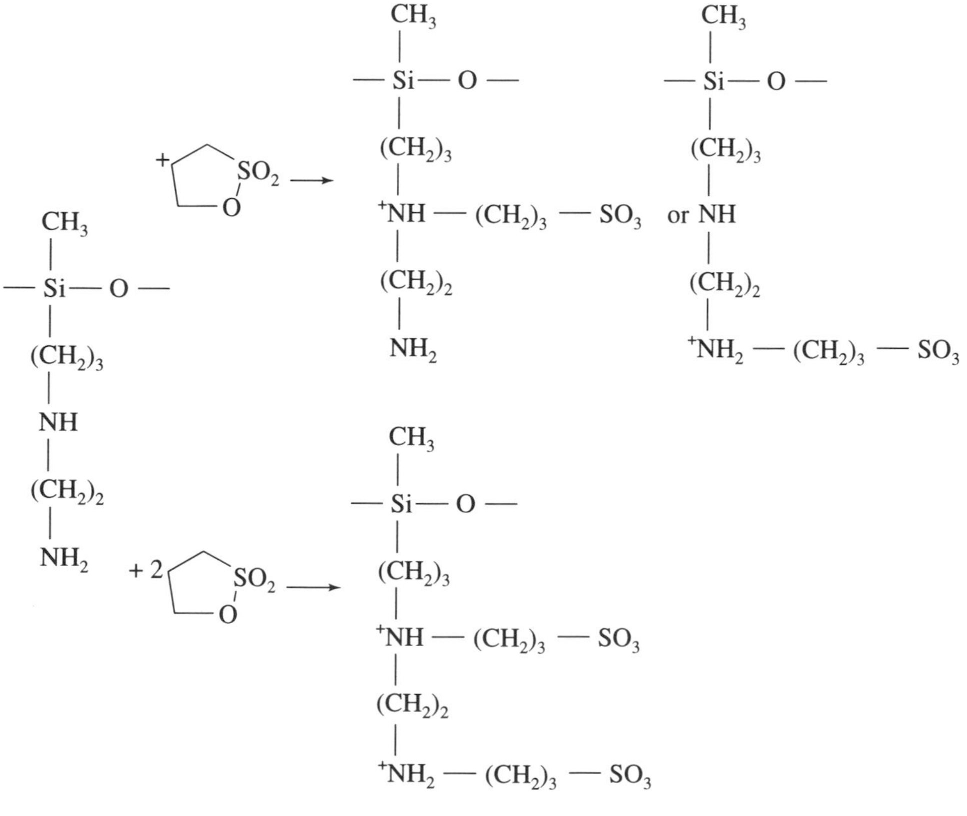

Scheme 2.4

Zwitterionomers represent a family of materials in which the side chain contains both an anion and a cation, usually in proximity (Fig. 2.2**d**) (36–39). One example of materials of this family is a copolymer of dimethylsiloxane and 4,7-diazaheptyl-4,7-di(3-propane-sulfonate) methylsiloxane (Scheme **2.4**). Another example of such materials is based on the sulfobetaines (36,38,39).

2.1.3. Other Ionomer Families

2.1.3.1. Chains with Regularly Spaced Ions.
Regular placement of ionic groups along the backbone is, in general, difficult synthetically. The one exception to this dictum is the ionenes, in which the regular placement of quaternary ammonium ions as part of the backbone can be accomplished easily by starting, for example, with tetramethyldiaminoalkanes and reacting them with dihaloalkanes (40). A chain with regularly placed ionic groups is shown in Figure 2.3**a**. Aliphatic, aromatic, and mixed varieties are possible. However, the range of backbones is somewhat limited because of the synthetic requirements. In some ways, the ionenes can be formally related to the $(AB)_n$ block copolymers; and if their block size exceeds a certain length, their behavior is quite similar to that of the block copolymers. These are treated in Chapter 7, which deals with block ionomers and other regular structures. Regular block copolymers based on ionene type coupling reactions are also possible (41). A schematic synthesis route is shown in Scheme **2.5**.

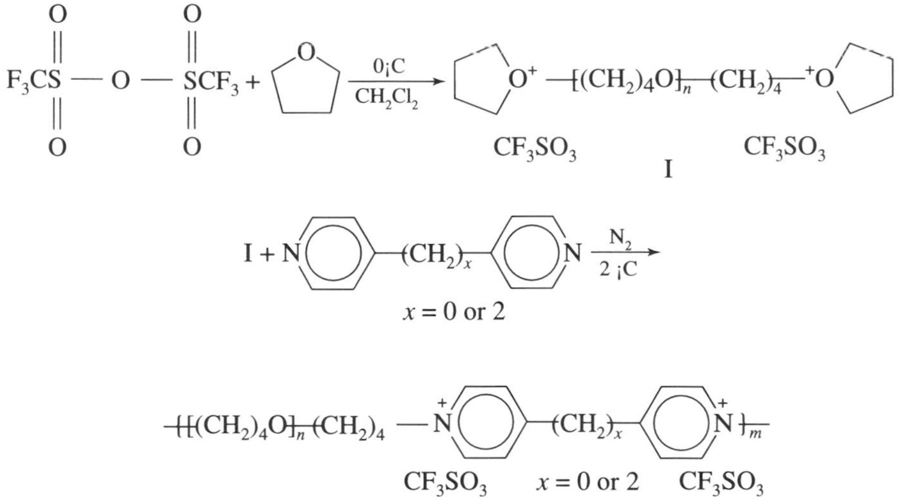

Scheme 2.5

2.1.3.2. Combs.
Combs resemble, to some extent, graft copolymers. Three different types of combs can be distinguished. In the first of these, the ion is placed close

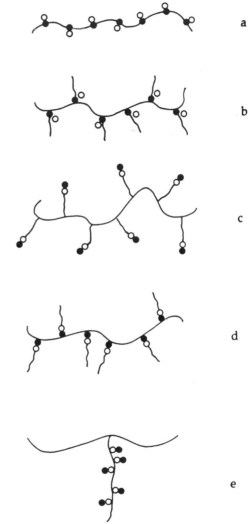

Figure 2.3. Possible molecular structures of ionomers. (**a**) Regularly spaced ionomer (ionene), (**b**) inverted comb ionomer, (**c**) comb ionomer, (**d**) counterion comb ionomer, and (**e**) graft ionomer.

to the backbone and serves as the starting point for a side chain (Fig. 2.3b). The best examples of this family of materials are random copolymers of styrene and 4-vinylpyridine quaternized with long-chain alkyl iodides (42). The side chains are monodisperse, but their placement is random. As mentioned above, block copolymers of a similar composition have been called bottle brushes. The counterion is not attached but interacts electrostatically with the pendent ion.

In the second type of comb, the ions are placed at the end of well-defined side chains (Fig. 2.3c). Several families of materials of this type have been studied, including copolymers of the alkylene oxides and undecanoic acid (by Vogl's group) and copolymers of styrene, in which the ionic groups are placed at the end of alkyl chains that are, in turn, attached at their other end to the *para* position of the benzene rings (43). The syntheses of these materials are nontrivial, and relatively little work has been done on them (Scheme **2.6**).

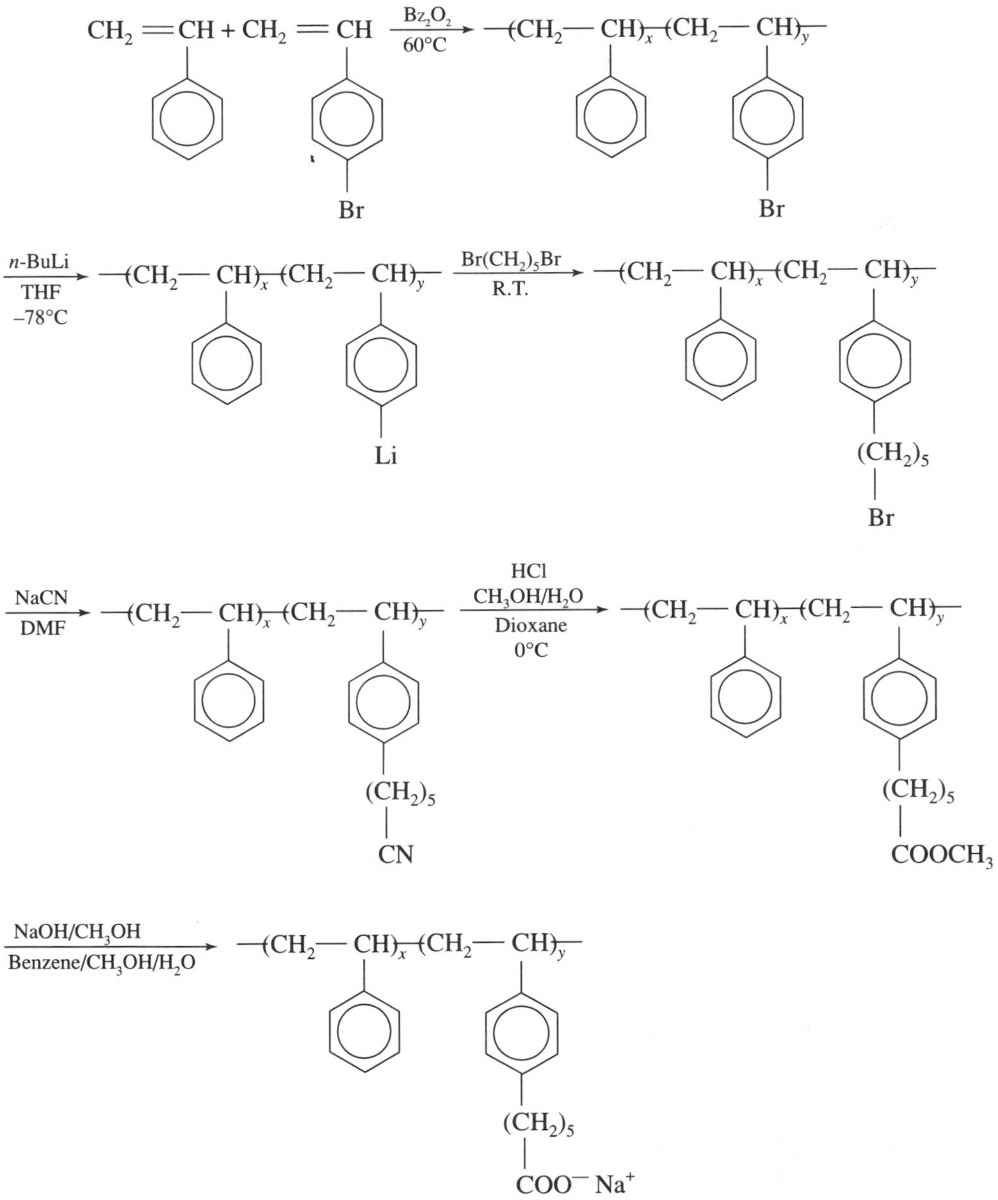

$$CH_2 = CH + CH_2 = CH \xrightarrow[60°C]{Bz_2O_2} -(CH_2-CH)_x-(CH_2-CH)_y-$$

Br

Br

$$\xrightarrow[\substack{THF \\ -78°C}]{n\text{-BuLi}} -(CH_2-CH)_x-(CH_2-CH)_y- \xrightarrow[R.T.]{Br(CH_2)_5Br} -(CH_2-CH)_x-(CH_2-CH)_y-$$

Li

$(CH_2)_5$

Br

$$\xrightarrow[DMF]{NaCN} -(CH_2-CH)_x-(CH_2-CH)_y- \xrightarrow[\substack{CH_3OH/H_2O \\ Dioxane \\ 0°C}]{HCl} -(CH_2-CH)_x-(CH_2-CH)_y-$$

$(CH_2)_5$

CN

$(CH_2)_5$

COOCH$_3$

$$\xrightarrow[Benzene/CH_3OH/H_2O]{NaOH/CH_3OH} -(CH_2-CH)_x-(CH_2-CH)_y-$$

$(CH_2)_5$

COO$^-$ Na$^+$

Scheme 2.6

In the final family of combs, the long-chain species are placed on mobile counterions (Fig. 2.3**d**). An excellent example of this type of material is styrene-*co*-vinylpyridinium copolymers, in which the counterions are alkyl sulfonates (44); another good example is the copolymers of styrene with styrene sulfonic acid, in which the counterions are long-chain amines or pyridines.

2.1.3.3. Grafts. True grafts can also be prepared (Fig. 2.3e). Examples include short poly(acrylic acid) side chains grafted onto a nonionic backbone (e.g., polystyrene) and the large family of cellulosics with grafted acrylate or methacrylate chains (45). Ionic chains with nonionic grafts are also possible.

2.1.3.4. Polymer–Salt Mixtures. The final family of materials considered here is polymer–salt mixtures. Some polymers, such as polyethers and polyamides, are known to act as solvents for salts. Polyether–salt mixtures have been investigated and show considerable resemblance to ionomers. The best explored system is based on poly(ethylene oxide) or poly(propylene oxide) with lithium perchlorate or other salts of low lattice energies. In this case, the cation associates strongly with the polyether oxygens, forming essentially a polycationic chain, whereas the perchlorate ions act as microcounterions. These materials are of particular interest in applications involving solid-state electrolytes and are briefly discussed in Chapter 7.

2.1.3.5. Concluding Comments. The structures described here represent a truly enormous range of possibilities. Many of the examples, quoted contain styrene. It would be tempting to suggest that, indeed, all of the above structures can be achieved with styrene or any other material as the parent polymer; and perhaps some day this may be possible. Currently, however, not all of these possibilities have been realized. Although styrene derivatives form the vast majority of structures outlined above, no examples of pure styrene ionenes are known in which the placement is uniform. Furthermore, the synthesis of styrene-based star polymers, while in principle possible, is extremely difficult. When one considers that it is possible, in principle, to prepare the structures shown above with a wide range of ionic groups, it becomes clear that the family of ionomers is, indeed, enormous.

2.2. CHEMICAL STRUCTURES

2.2.1. Type of Attachment of Pendent Ion

It is possible to vary the nature of the group that attaches the ion to the chain; the carboxylates serve as an excellent example of this. Researchers who wish to prepare ionomers based on polystyrene and carboxylic acids have at their disposal a wide range of unsaturated carboxylic acids. Acrylic and methacrylic acids are perhaps the best known, but butenoic acid or *p*-carboxystyrene is also possible. Furthermore, the carboxylic acid can be located on a methylene group that is attached to the phenyl ring and in a number of other positions, limited only by the imagination of the chemist. Long-chain spacers would convert the material into a comb. Pendent cations can also be located in a variety of positions. For example, the styrene-*co*-

vinylpyridine copolymers can be based on 4-vinylpyridine, 2-vinylpyridine, or 2-methyl-5-vinylpyridine. Again, the imagination of the organic chemist is the only limiting factor.

2.2.2. Nature of Pendent Ions

A relatively wide range of ionic groups is available for the preparation of ionomers, in addition to carboxylate or pyridinium. In this section, the discussion is divided into anions and cations. It is not meant to be exhaustive, but illustrative.

The most common pendent anions used in the synthesis of ionomers are the carboxylates; the sulfonates; and to a much lesser extent, the phosphonates. Carboxylate anions can be incorporated into ionomers by copolymerization with species such as acrylic acid, methacrylic acid, maleic anhydride, and a number of other unsaturated carboxylic acids that tend to polymerize or copolymerize. It should be stressed that the total number of polymerizable acids is large; such acids include maleic acid, cinnamic acid, and many more. An alternative method of preparing carboxylated ionomers is via a postpolymerization reaction. This type of reaction is particularly suitable in the case of aromatic substitutions, and a number of reactions have been developed for the carboxylation of, e.g., polystyrene (46–59).

The incorporation of sulfonic acids into polymers is also possible via copolymerization. In this case, the sulfonate group should be relatively far from the site of the double bond, as its proximity affects the reactivity ratios considerably and might even make copolymerization impossible in some cases (60,61). Furthermore, the strength of the sulfonic acid reduces the solubility of the monomers, so that common solvents are not easy to find. Nonetheless, copolymerization of sulfonated monomers with nonionic organic monomers has been carried out successfully (62–64).

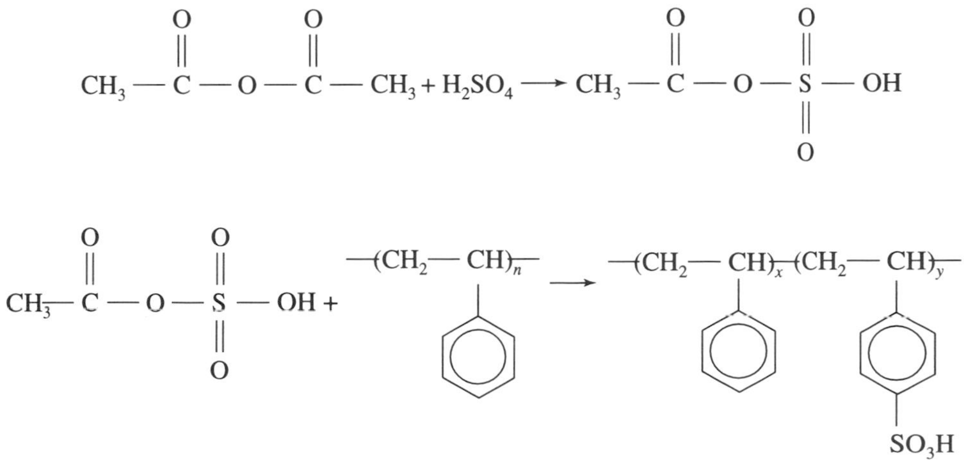

Scheme 2.7

The more common route to sulfonation is via a postpolymerization reaction of aromatic or unsaturated sites (18,28,65–76). The ease with which polysterene can be sulfonated is undoubtedly one reason for the large number of papers that have appeared in the ionomer literature on it. Makowski et al.'s method (69) is an excellent example (Scheme **2.7**).

Phosphonates have also been investigated, but to a lesser extent. Postpolymerization reactions are again involved in the preparation of these materials (77). The reaction scheme is shown in Scheme **2.8**. Figure 2.4**a** shows examples of repeat units carrying pendent anions.

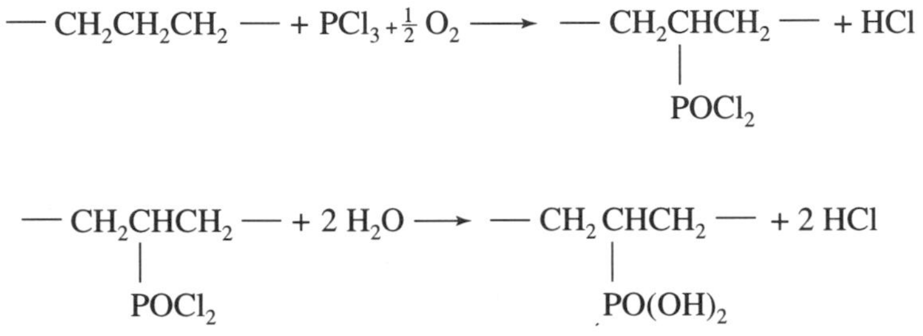

$$-CH_2CH_2CH_2 - + PCl_3 + \tfrac{1}{2}O_2 \longrightarrow -CH_2CHCH_2 - + HCl$$
$$\underset{POCl_2}{\shortmid}$$

$$-CH_2CHCH_2 - + 2\,H_2O \longrightarrow -CH_2CHCH_2 - + 2\,HCl$$
$$\underset{POCl_2}{\shortmid} \qquad\qquad\qquad \underset{PO(OH)_2}{\shortmid}$$

Scheme 2.8

Cationic polymers are also quite common in the ionomer literature (78,79). Pendent pyridine groups are among the most common examples, since the copolymerization of vinylpyridine with a range of nonionic monomers is quite straightforward. Other pendent cations are also known, such as aliphatic or aromatic amines or pyridines and their salts. Sulfonium and phosphonium ions have also been incorporated into polymeric systems. Figure 2.4**b** shows examples of repeat units carrying pendent cations.

The ionenes make up a family of materials in which a quaternary ammonium or pyridinium ion is located on the backbone. Figure 2.5 shows the structures of three ionenes, illustrating the various cations that are available for these systems.

Readers familiar with the polyelectrolyte literature will recognize that other ionic monomers are available that have not been mentioned here. These include species such as dimethyldiallyl ammonium chloride (DMDAAC) and [2-(acryloylamino)-prop-2-yl]methanesulfonic acid (APMS) (Fig. 2.6). While copolymerization of some of these monomers is possible, it is frequently difficult, so materials of this type have not been explored much as ionomers.

2.2.3. Counterions

The range of counterions that can be used with the materials mentioned above is sizable. Anionic polymers can use a wide range of metal ions, such as the alkali, alkaline earth, and transition metal or organic cations, e.g., ammonium, pyridinium (see Fig. 2.7). Multivalent cations are more difficult to incorporate than are monovalent cations, but techniques have been developed for the preparation of such materials

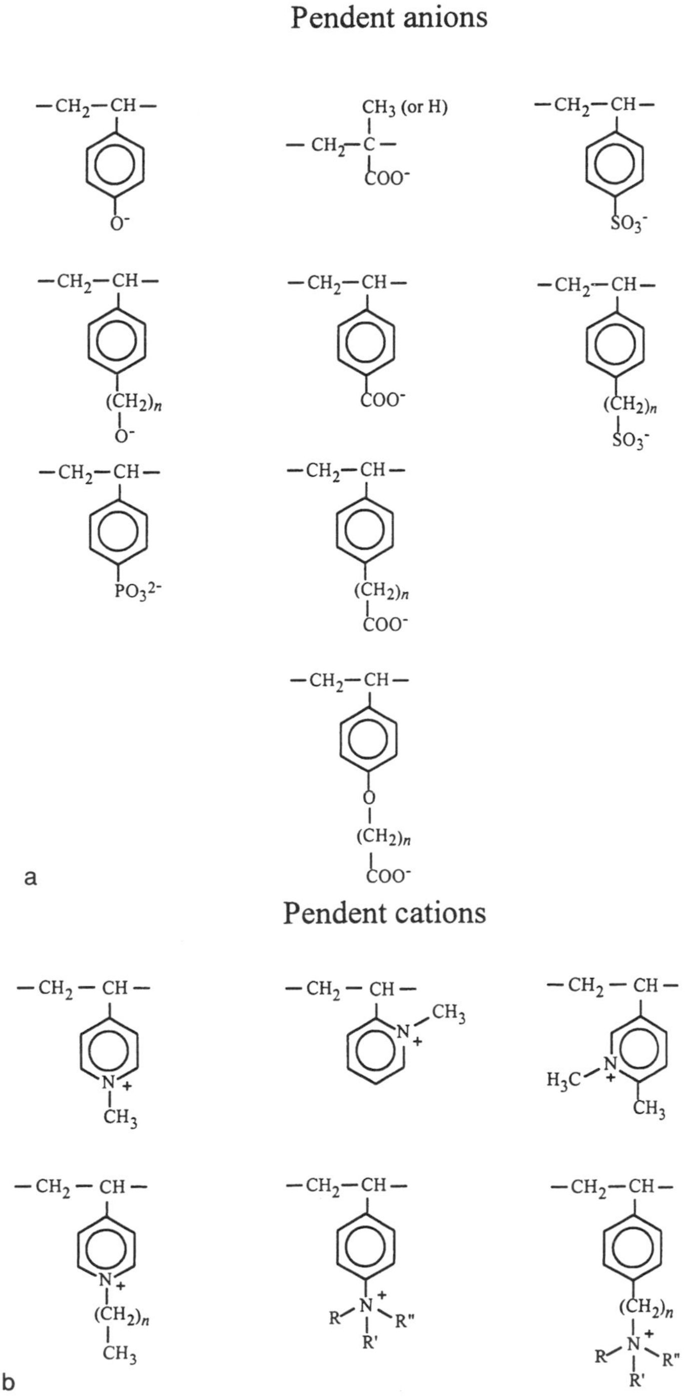

Pendent anions

a

Pendent cations

b

Figure 2.4. Examples of (a) pendent anions, mostly based on styrene, and (b) pendent cations.

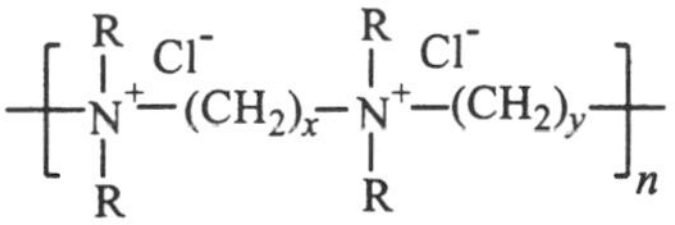

x, y ionene

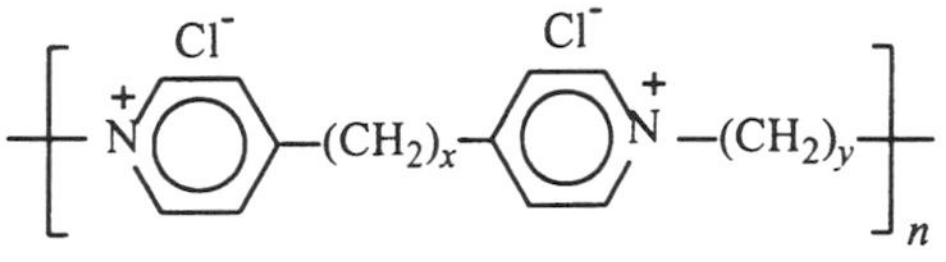

An aromatic ionene [poly(bipyridinium)]

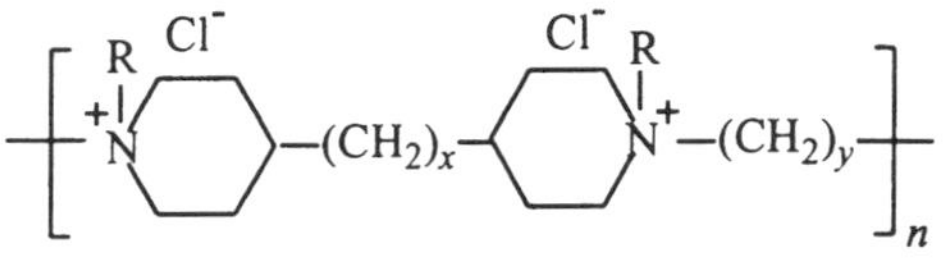

A poly(bipiperadinium)

Figure 2.5. Examples of ionenes.

(81). Cationic polymers have a somewhat smaller range of counterions available, including the halides (chloride, bromide, iodide) and organic anions (tosylate).

2.2.4. The Nonionic Host Polymer

To date, the most extensively studied ionomers have been materials based on the rubbers, ethylene, styrene, the acrylates or methacrylates, and tetrafluoroethylene, largely owing to either their commercial availability or their ease of synthesis. How-

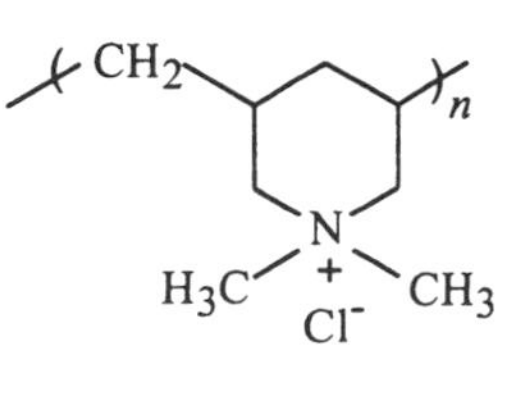

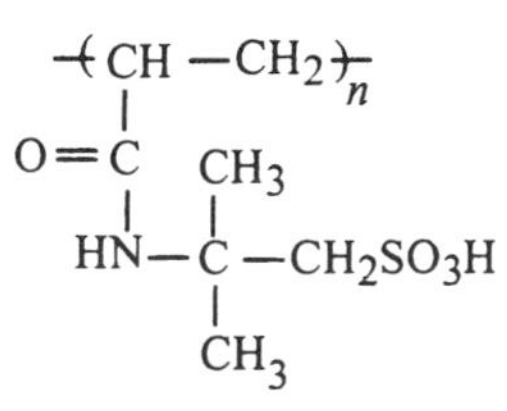

DMDAAC   APMS

Figure 2.6. Structures of DMDAAC and APMS.

Cations

Li$^+$ Na$^+$ K$^+$ Rb$^+$ Cs$^+$

Mg^{2+} Ca^{2+} Sr^{2+} Ba^{2+} Zn^{2+} Ni^{2+} Mn^{2+}

Anions

F$^-$ Cl$^-$ Br$^-$ I$^-$ OTs$^-$

Ionizable Amines

NH$_3$

R—NH$_2$ RR'—NH RR'R"—N (R, R' or R" = alkyl)

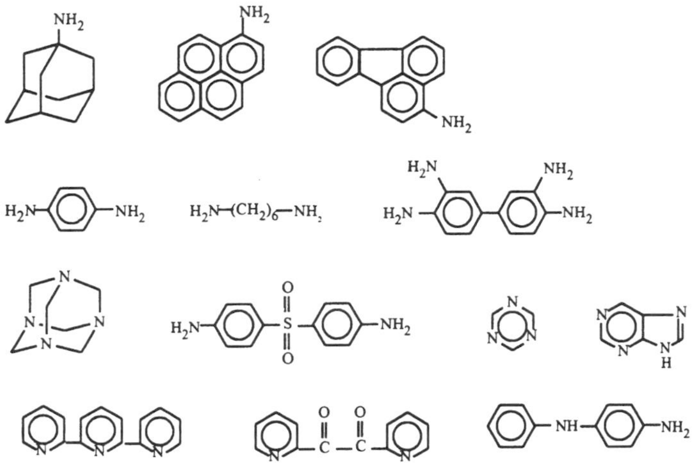

Figure 2.7. Examples of counterions. Many of the ionizable amines are described by Fan and Bazuin (80).

ever, a wide range of other monomers can be incorporated. Again, the only limitation is that of the organic chemistry required.

In addition to the extended range of monomers available for incorporation into ionomers, all the usual techniques of polymer science can also be used to modify the physical properties of ionomers. Thus copolymerization, external plasticization, internal plasticization, control over molecular weight, cross-linking, and a host of other techniques are accessible.

2.2.5. Inorganic Systems

A range of inorganic materials can also be considered to be part of the ionomer family. Ionic inorganic polymers are well known, e.g., polymethaphosphates (Na$_2$O/P$_2$O$_5 \approx 1/1$) and silicates (Na$_2$O/SiO$_2 \approx 1/1$). Most of these are not considered to be ionomers. However, at a somewhat lower ion content, the materials can be expected to exhibit some ionomer properties. For example, materials in the Na_2O/P_2O_5 system at a ratio considerably $<1/1$ or copolymers of $NaPO_3$ and HPO_3 are formally ionomers because they have a relatively small number of charges and their properties are determined by association between ionic groups. Mixed ionic inorganic and organic materials can also be envisaged. Little experimental work has been done on these systems, which are undoubtedly worthy of more extensive exploration.

2.3. CONCLUDING COMMENTS

This brief summary pointed out that an enormous range of ionomer structures is possible. While the total number of pendent ions is not large, considerable structural complexity can be introduced by varying the position of any one of the ions. A much wider range of counterions is available, especially small cations, as well as combinations of ions (as in counterion copolymers). Finally, when one considers that each of these structures can be employed, at least in principle, with any one of an extremely large number of nonionic host materials, either alone or in combination, it is clear that the ionomer field is truly gigantic. Naturally, it is not possible or even desirable to synthesize and study all of these structures. A wide enough range of experiments has now been performed to allow some modest predictions; and this base of information should decrease the need for extensive exploration when searching for particular systems to solve specific needs. The subsequent chapters describe some of these studies and serve as a guide for the selection of appropriate materials.

2.4. REFERENCES

1. Zhong, X. F.; Eisenberg, A. *Macromolecules* **1994,** *27,* 1751–1758.

2. McGrath, J. E. *Anionic Polymerization: Kinetics, Mechanisms, and Synthesis*; ACS Symposium Series 166; American Chemical Society: Washington, DC, 1981.

3. Morton, M. *Anionic Polymerization: Principles and Practice*; Academic: New York, 1983.

4. Szwarc, M.; Van Beylen, M. *Ionic Polymerization & Living Polymers*; Chapman & Hall: New York, 1993.

5. Hsieh, H. L.; Quirk, R. P. *Anionic Polymerization: Principles and Practical Applications*; Marcel Dekker: New York, 1996.

6. Horrion, J.; Jerome, R.; Teyssie, P. *J. Polym. Sci. C Polym. Lett.* **1986,** *24,* 69–76.

7. Jérôme, R. In *Telechelic Polymers: Synthesis and Applications*; Goethals, E., Ed.; CRC: Boca Raton, FL, 1989; Chapter 11.

8. Vanhoorne, P.; Jérôme, R. In *Ionomers: Characterizations, Theory, and Applications*; Schlick, S., Ed.; CRC: Boca Raton, FL, 1996, Chapter 9.

9. Möller, M.; Omeis, J.; Mühleisen, E. In *Reversible Polymeric Gels and Related Systems*;

Russo, P. S., Ed.; ACS Symposium Series 350; American Chemical Society: Washington, DC, 1987; Chapter 7.

10. Vanhoorne, P.; Van den Bossche, G.; Fontaine, F.; Sobry, R.; Jérôme, R.; Stamm, M. *Macromolecules* **1994**, *27*, 838–843.

11. Wu, J.-L.; Wang, Y.-M.; Hara, M.; Granville, M.; Jérôme, R. J. *Macromolecules* **1994**, *27*, 1195–1200.

12. Fetters, L. J.; Graessley, W. W.; Hadjichristidis, N.; Kiss, A. D.; Person, D. S.; Younghouse, L. B. *Macromolecules* **1988**, *21*, 1644–1653.

13. Pispas, S.; Hadjichristidis, N. *Macromolecules* **1994**, *27*, 1891–1896.

14. Cowan, J. C.; Teeter, H. M. *Ind. Eng. Chem.* **1994**, *36*, 148–152.

15. Broze, G.; Jérôme, R.; Teyssié, P. *Macromolecules* **1981**, *14*, 224–225.

16. Broze, G.; Jérôme, R.; Teyssié, P. *Macromolecules* **1982**, *15*, 920–927.

17. Broze, G.; Jérôme, R.; Teyssié, P.; Marco, C. *J. Polym. Sci. Polym. Phys. Ed.* **1983**, *21*, 2205–2217.

18. Misra, N.; Mandal, B. M. *Macromolecules* **1984**, *17*, 495–497.

19. Hegedus, R. D.; Lenz, R. W. *J. Polym. Sci. A Polym. Chem.* **1988**, *26*, 367–380.

20. Morrell, R. K.; Service, D. M.; Stewart, M. J.; Viguier, M.; Richards, D. H. *Br. Polym. J.* **1987**, *19*, 241–246.

21. Antonietti, M.; Heyne, J.; Sillescu, H. *Makromol. Chem.* **1991**, *192*, 3021–3034.

22. Mohajer, Y.; Tyagi, D.; Wilkes, G. L.; Storey, R. F.; Kennedy, J. P. *Polym. Bull.* **1982**, *8*, 47–54.

23. Bagrodia, S.; Mohajer, Y.; Wilkes, G. L.; Storey, R. F.; Kennedy, J. P. *Polym. Bull.* **1983**, *9*, 174–180.

24. Selb, J.; Gallot, Y. *J. Polym. Sci. Polym. Lett. Ed.* **1975**, *13*, 615–619.

25. Wollmann, D.; Williams, C. E.; Eisenberg, A. *J. Polym. Sci. Polym. Phys. Ed.* **1990**, *28*, 1979–1986.

26. Storey, R. F.; George, S. E.; Nelson, M. E. *Macromolecules* **1991**, *24*, 2920–2930.

27. Allen, R. D.; Yilgor, I.; McGrath, J. E. In *Coulombic Interactions in Macromolecular Systems*; Eisenberg, A.; Bailey, F. E., Eds.; ACS Symposium Series 302; American Chemical Society: Washington, DC, 1986; Chapter 6.

28. Weiss, R. A.; Sen, A.; Willis, C. L.; Pottick, L. A. *Polymer* **1991**, *32*, 1867–1874.

29. van Hest, J. C. M.; Baars, M. W. P. L.; Elissen-Román, C.; van Genderen, M. H. P.; Meijer, E. W. *Macromolecules* **1995**, *28*, 6689–6691.

30. Gauthier, M.; Tichagwa, L.; Downey, J. S.; Gao, S. *Macromolecules* **1996**, *29*, 519–527.

31. Smith, P. Ph.D. Dissertation, McGill University, 1985.

32. Smith, P.; Eisenberg, A. *Macromolecules* **1994**, *27*, 545–552.

33. Douglas, E. P.; Waddon, A. J.; MacKnight, W. J. *Macromolecules* **1994**, *27*, 4344–4352.

34. Otocka, E. P.; Eirich, F. R. *J. Polym. Sci. A-2* **1968**, *6*, 921–932.

35. Odian, G. G. *Principles of Polymerization*, 3rd ed.; Wiely:New York, 1991.

36. Salamone, J. C.; Volksen, W.; Israel, S. C.; Olson, A. P.; Raia, D. C. *Polymer* **1977**, *18*, 1058–1062.

37. Graiver, D.; Baer, E.; Litt, M. *J. Polym. Sci. Polym. Chem. Ed.* **1979**, *17*, 3559–3572.

38. Monroy Soto; V. M.; Galin, J. C. *Polymer* **1984**, *25*, 121–128.

39. Pujol-Fortin, M.-L.; Galin, J.-C. *Macromolecules* **1991,** *24,* 4523–4530.
40. Rembaum, A.; Baumgartner, W.; Eisenberg, A. *J. Polym. Sci. B* **1968,** *6,* 159–171.
41. Lee, B.; Wilkes, G. L.; McGrath, J. E. *Polym. Prepr. Am. Chem. Soc., Div. Polym. Chem.* **1988,** *29*(1), 138–140.
42. Wollmann, D.; Gauthier, S.; Eisenberg, A. *Polym. Eng. Sci.* **1986,** *26,* 1451–1456.
43. Gauthier, M.; Eisenberg, A. *J. Polym. Sci. Polym. Chem. Ed.* **1990,** *28,* 1549–1568.
44. Bakeev, K. N.; Shu, Y. M.; MacKnight, W. J.; Zezin, A. B.; Kabanov, V. A. *Macromolecules* **1994,** *27,* 300–302.
45. Vitta, S. B.; Stahel, E. P.; Stannett, V. T. *J. Macromol. Sci. Chem.* **1985,** *A22,* 579–590.
46. Kenyon, W. O.; Waugh, G. P. *J. Polym. Sci.* **1958,** *32,* 83–88.
47. Blanchette, J. A.; Cotman, J. D. *J. Org. Chem.* **1958,** *23,* 1117–1122.
48. Braun, D. *Makromol. Chem.* **1961,** *46,* 269–280.
49. Letsinger, R. L.; Kornet, M. J.; Mahadevan, V.; Jerina, D. M. *J. Am. Chem. Soc.* **1964,** *86,* 5163–5165.
50. Ayres, J. T.; Mann, C. K. *J. Polym. Sci. B* **1965,** *3,* 505–508.
51. Blackburn, G. M.; Brown, M. J.; Harris, M. R.; Shire, D. *J. Chem. Soc. C* **1969,** *4,* 676–683.
52. Harrison, C. R.; Hodge, P.; Kemp, J.; Perry, G. M. *Makromol. Chem.* **1975,** *176,* 267–274.
53. Farrall, M. J.; Fréchet, J. M. J. *J. Org. Chem.* **1976,** *41,* 3877–3887.
54. Farrall, M. J.; Fréchet, J. M. J. *Macromolecules* **1979,** *12,* 426 428.
55. Lundberg, R. D.; Makowski, H. S. In *Ions in Polymers,* Eisenberg, A., Ed.; Advances in Chemistry 187; American Chemical Society: Washington, DC, 1980; Chapter 2.
56. Brockman, N. L.; Eisenberg, A. *J. Polym. Sci. Polym. Chem. Ed.* **1985,** *23,* 1145–1164.
57. Argyropolous, D. S.; Bolker, H. I. *Makromol. Chem.* **1986,** *187,* 1887–1894.
58. Hird, B.; Eisenberg, A. *J. Polym. Sci. A Polym. Chem.* **1993,** *31,* 1377–1381.
59. Tomita, H.; Register, R. A. *Macromolecules* **1993,** *26,* 2791–2795.
60. Lenz, R. W. *Organic Chemistry of Synthetic High Polymers;* Interscience: New York, 1967.
61. *Polymer Handbook,* 3rd ed.; Brandrup, J.; Immergut, E. H., Eds.; Wiley: New York, 1989.
62. Weiss, R. A.; Lundberg, R. D.; Werner, A. *J. Polym. Sci. Polym. Chem. Ed.* **1980,** *18,* 3427–3439.
63. Siadat, B.; Oster, B.; Lenz, R. W. *J. Appl. Polym. Sci.* **1981,** *26,* 1027–1037.
64. Turner, S. R.; Weiss, R. A.; Lundberg, R. D. *J. Polym. Sci. Polym. Chem. Ed.* **1985,** *23,* 535–548.
65. Suter, C. M.; Evans, P. B.; Kiefer, J. M. *J. Am. Chem. Soc.* **1938,** *60,* 538–540.
66. Bordwell, F. G.; Peterson, M. L. *J. Am. Chem. Soc.* **1954,** *76,* 3952–3956.
67. Turbak, A. F. *Polym. Prepr. Am. Chem. Soc. Div. Polym. Chem* **1961,** *2*(1), 140–146.
68. O'Farrell, C. P.; Serniuk, G. E. U. S. Patent 3 836 511, 1974.
69. Makowski, H. S.; Lundberg, R. D.; Singhal, G. L., U.S. Patent 3 870 841, 1975.
70. Noshay, A.; Robeson, L. M. *J. Appl. Polym. Sci.* **1876,** *20,* 1885–1903.
71. Makowski, H. S.; Lundberg, R. D.; Westerman, L.; Bock, J. *Polym. Prepr. Am. Chem. Soc. Div. Polym. Chem.* **1978,** *19*(2), 292–297.

72. Thaler, W. A. *J. Polym. Sci. Polym. Chem. Ed.* **1982,** *20,* 875–896.
73. Rahring, D.; MacKnight, W. J.; Lenz, R. W. *Macromolecules* **1979,** *12,* 195–203.
74. Drzewinski, M.; MacKnight, W. J. *J. Appl. Polym. Sci.* **1985,** *30,* 4753–4770.
75. Bailly, C.; Williams, D. J.; Karasz, F. E.; MacKnight, W. J. *Polymer* **1987,** *28,* 1009–1016.
76. Storey, R. F.; Lee, Y. *J. Polym. Sci. Polym. Chem. Ed.* **1991,** *29,* 317–325.
77. Phillips, P. J.; MacKnight, W. J. *J. Polym. Sci. B* **1970,** *8,* 87–94.
78. Hoover, M. F. *J. Macromol. Sci. Chem.* **1970,** *A4,* 1327–1417.
79. Hoover, M. F.; Butler, G. B. *J. Polym. Sci. Polym. Symp.* **1974,** *45,* 1–37.
80. Fan X. D.; Bazuin C.G; *Macromolecules* **1995,** *28,* 8209–8215.
81. Broze, G.; Jérôme, R.; Teyssie, P. *J. Polym. Sci. Polym. Lett. Ed.* **1983,** *21,* 237–241.

CHAPTER 3

MORPHOLOGY OF RANDOM IONOMERS

The solid-state properties of the vast majority of ionomers are governed by the existence of discrete aggregates of ionic groups. As was shown in Chapter 1, in media of low dielectric constant, the driving force for ion–counterion association is high, as is the driving force for ion pair–ion pair association and even for the formation of higher aggregates. The first part of this chapter discusses the formation of primary ionic aggregates, known as multiplets. Subsequent sections are concerned with the arrangement of multiplets into more complex aggregates, including a historical overview of the types of aggregates that have been proposed for random ionomers.

3.1. MULTIPLETS

3.1.1. Size and Shape of Multiplets

Many factors enter into the energy balance that determines the final size of the multiplets, including electrostatic energies and relative sizes, which can be considered to generate a cone or solid angle between the ion pair and the polymer chain segment to which the ionic group is attached. Backbone chain rigidity and size are also of importance, as is the ion concentration.

3.1.1.1. Energetics. As pointed out in Section 1.6, the association energy of two ion pairs forming a square planar quartet, ΔW, with a center-to-center distance of 2 Å between ions in a medium of dielectric constant 2.5 (which is typical for polymers such as polystyrene, polyethylene, and polybutadiene) can be estimated to be about -40 kJ/mol. In view of the large ΔW, there is no doubt that a multiplet is the smallest aggregate that is expected at reasonable concentrations. While isolated ion pairs undoubtedly exist, especially at low ion concentrations or under circumstances in which constraints on the chain make it impossible for the segment containing the isolated ion pair to explore regions of space comparable to average intermultiplet distances, the energy differences between two ion pairs and a single multiplet are such that the majority will usually be present in the associated form.

3.1.1.2. Solid Angle and Contact Surface Area Concepts. How many ion pairs can come together to form one multiplet? An early approach to answering this question was taken by Eisenberg (1) in the original formulation of the multiplet-cluster model. One can start the consideration of multiplet formation by assuming that the multiplet is a nearly spherical liquid drop consisting of ionic groups. The hydrocarbon chain segments would be expected to be confined essentially to the surface of that liquid drop. Because the solubility characteristics of ion pairs and the polymer chains are so drastically different, there must be a fairly sharp interface, with the hydrocarbon chain outside the multiplets and the ion pairs inside. This view leads to the concept of a solid angle. You can convince yourself by looking at models of a polymer chain, e.g., of polystyrene, containing carboxylic acid groups attached to directly to the main chain; the carboxylate group is much smaller than the typical dimensions of the polymer chain segment to which it is attached. The ion pair is in the aggregate and the larger polymer chain segment is outside; thus you can think of the hydrocarbon segments as subtending a solid angle with the ion pair near the center of the sphere.

A detailed calculation of the solid angle in a situation like this still needs to be performed. However, even an approximate picture suggests that the limit of the number of ion pairs that can possibly get together if the ion pair is attached directly to a polymer chain without an intermediate spacer should be smaller than ~ 10; real multiplets in a material such as poly(styrene-*co*-sodium methacrylate) would be considerably smaller. Naturally, as the dimensions of hydrocarbon segments to which the ion pairs are attached decrease—as they do in the telechelics (2,3) and in the side chain ionomers (4)—the total number of ion pairs per multiplet can increase substantially. The approximate number of 10 units arises from highly simplified model considerations, which suggest that the solid angle subtended by a polystyrene chain attached to a single carboxylate group is on the order of 0.5π radians.

As the size of the ion pair increases and the size of the chain segment in direct contact with the ion decreases, the multiplet might become larger. The most likely geometry for multiplets would be a nearly spherical drop, because of the drastic difference in volumes. Thus one can anticipate that the smaller the ion pair, the larger the chain segment; and the closer the ion pair to the backbone (all of which influence the solid angle), the smaller the multiplet. As the size of the ion pair increases, as the segment length between the ion pair and the chain backbone increases, or as the size of the backbone chain segment connected immovably to the ion pair decreases the larger the multiplet becomes.

3.1.1.3. Shapes of Multiplets. As was expected, the shape of the multiplets in a material such as sulfonated polystyrene neutralized with Zn^{2+} was found to be spherical in a study by Li et al. (5), who used high-voltage electron microscopy to observe the ionic domains. At some point, however, other multiplet geometries might appear. The situation is probably analogous to that encountered in phase-separated block copolymers, in which different geometries appear as a function of relative block length. It should be stressed, however, that although the existence of nonspherical multiplets has been postulated (6), they have been shown to exist in random ionomers in bulk in only one system (7). The structures of multiplets in some telechelics (8), ionenes (9,10), and perfluorosulfonated ionomers in solution (11,12) were postulated to be nonspherical.

3.1.1.4. Effect of Ion Content and Dielectric Constant. Several other factors must be considered in connection with multiplet size. Some of these factors are electrostatic and involve the sizes and shapes of ions, as was implied above. Another factor is the ion content: as can be seen intuitively, at high ion contents the probability of the occurrence of nearest-neighbor ion pairs is high; this circumstance probably leads to an enlargement of the multiplet, because neighboring ion pairs can be incorporated into the same multiplet (4,13,14). Still other factors involve the nature of the polymer. For example, one can anticipate that the dielectric constant of the polymer has a major effect; if the dielectric constant is too high, as it is in the polyphosphates, multiplet formation would not be expected because the ion pairs are soluble in the polymer (15). Similar considerations might apply to polymers based on highly polar backbones or those subject to specific interactions with the ions, such as in poly(ethylene oxide) (16) and other polyethers (17).

3.1.1.5. Other Factors. Backbone flexibility is another factor; a rigid backbone would lead to smaller multiplets than a highly flexible backbone, everything else being equal, because the size of the chain segment that is rigidly attached to the ion changes. The presence of plasticizers can also be significant (Chapter 8). Other factors that must be considered are the difference in energies of one large multiplet versus two smaller multiplets and how this contributes to the total energy balance. The formation of one large multiplet involves more perturbations of the chains around the multiplets than does the formation of two or more smaller multiplets. Surface energies, however, favor larger multiplets. All these factors, naturally, contribute to the forces that result in minimum free energies, which determine the final morphology of the material. The sizes of the ions are, of course, also significant, as is their detailed structure; sodium versus zinc or carboxylate versus sulfonate ions influences the sizes of the multiplets to some extent. The detailed thermal history, as well as the method of preparation of the sample, may also be relevant.

3.1.2. Arrangements of Ion Pairs

Several attempts have been made to determine the arrangements of ion pairs within the multiplet. None is, as yet, general enough to warrant detailed discussion in this presentation. The reader is referred to the original literature. Of particular interest are the studies by extended x-ray absorption fine structure (EXAFS) of various ions in multiplets (18–29) and the suggestion by Tadano et al. (30) concerning the possible crystalline nature of the ionic inclusions in the zinc-neutralized ethylene-*co*-methacrylate ionomers in the presence of 1,3-bis(aminoethyl)cyclohexane, although this idea has not been universally accepted (31).

3.1.3. Arrangement of the Multiplets

3.1.3.1. Small-Angle Ionic Peak.
The arrangement in space of the multiplets relative to each other has been a subject of much experimental and theoretical work since the appearance of ionomers about 30 years ago. Wilson et al. (32) and Longworth and Vaughan (33) were, in 1968, the first to publish a small-angle x-ray scattering (SAXS) profile of an ethylene-based ionomer. (Fig. 3.1) Even then the authors proposed a morphological model suggesting ion aggregation. While the original interpretation of the peak as arising from ordered hydrocarbon chains between ionic aggregates is no longer accepted, the data are important because they show that in some ionomers a SAXS peak is seen at low angles; this is known as the ionic peak. A schematic presentation of ionic aggregates appeared in the same year in a paper

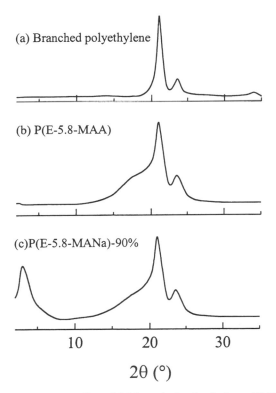

Figure 3.1. X-ray diffraction profiles of **(a)** branched polyethylene, **(b)** P(E-5.8-MAA), and **(c)** P(E-5.8-MAA) 90% neutralized with Na^+. Modified from Wilson et al. (32).

by Bonotto and Bonner (34), who suggested that in ethylene-based ionomers, the ionic aggregates contain both ionic material and acid groups (specifically for partly neutralized systems) in addition to hydrocarbon material.

3.1.3.2. Multiplet–Cluster Concept. The concept of a multiplet was proposed and developed by Eisenberg (1). Some of the features of this model were discussed above. The model considers the forces leading to the formation of aggregates (multiplets) and the subsequent clustering resulting from the rearrangement of the multiplets with increasing ion content. The details of the possible arrangements of multiplets discussed in that paper are now only of historical interest. The subsequent, more sophisticated theoretical approaches of Forsman (35,36), Dreyfus (37), and Dayte and Taylor (38) explored the balance of forces involved in ion aggregation. Although the general conclusions regarding multiplets were largely similar to those of the original paper (1), the authors provided considerably more extensive insight.

A wide range of experimental studies have been performed since 1968 to elucidate the organization of ion pairs and multiplets. The morphological studies address only the distribution of ions or ionic aggregates in the material. A successful model for the ionomer morphology must also account for the mechanical properties of the material. The absence of a relation between the morphological model and the mechanical properties is one of the shortcomings of several models that successfully treat only the SAXS profiles.

3.1.3.3. Hard-Sphere Model. In an early study, Delf and MacKnight (39) assigned the SAXS peak to scattering from the ionic aggregates. Subsequently, Marx et al. (40) suggested that the peak arises from interparticle scattering from the ionic aggregates, which were taken to be small particles located on a paracrystalline lattice. This treatment provided a quantitative fit to the experimental small-angle x-ray data. A similar model was advanced by Binsbergen and Kroon (41). In their model, the scattering moieties were points at the centers of randomly packed spheres.

Yarusso and Cooper (42) proposed a modified hard-sphere model in which the multiplets have a liquidlike order at a distance of closest approach, determined by the hydrocarbon layer attached to and surrounding each multiplet. This model was in good agreement with the experimental SAXS profiles and assumed the existence of multiplets of high electron density surrounded by a layer of hydrocarbon material of much lower electron density immersed in a medium of intermediate ion content. Furthermore, the model postulated a distance of closest approach of the multiplets, which was essentially equivalent to a hard-sphere diameter somewhat larger than that of the multiplet itself.

3.1.3.4. Core-Shell Model. Other models have also been proposed. Among those, MacKnight et al. (43) developed a model based on the radial distribution function of scattered x-rays for the partly neutralized cesium salt of poly(ethylene-*co*-methacrylic acid). This model, known as the core-shell model, assumed that the ion pairs form a core that is surrounded by a shell of material of low electron density, which is, in turn, surrounded, by another shell of somewhat higher electron density (but lower than that of the core itself). The central core is taken to have a radius of 8–10 Å and to contain ~50 ion pairs. The hydrocarbon shell surrounding this multiplet is on the order of 20 Å. The major difference between this model and the models of Marx et al. (40) and of Yarusso and Cooper (42) is the assignment of the ionic peak to intraparticle interference rather than interparticle interference. A modification was suggested by Roche et al. (6), who suggested that the geometry of the ion-rich phase is lamellar. The central lamella of high electron density material (high ion content) is sandwiched between lamellae of low electron density hydrocarbon material, which is, in turn, sandwiched between layers of intermediate electron density. Interlamellar distances are taken to be responsible for the peak.

A number of other attempts have been made to describe the morphologies of specific ionomer systems, including block ionomers (44–48), sequential ionomers (e.g., polyurethanes) (9,49), semicrystalline ionomers (33), hydrated perfluorosulfonates (50–56), and halatotelechelics (2,3). The detailed results of these treatments are not discussed here, because they do not concern dry noncrystalline random ionomers.

3.1.3.5. Dilemma of Hard-Sphere and Core-Shell Models. While the

models of Marx et al. (40) and of Yarusso and Cooper (42) were successful in modeling the SAXS profile, they, and the other morphological models, did not address themselves to the mechanical properties of the materials, which presents a problem: The distances between aggregates in these interparticle scattering models are on the order of 30 Å. On the other hand, many ionomers, especially those in which the ionic SAXS peak is seen, exhibit two glass transition temperatures T_g. The existence of two glass transition temperatures suggests that the material behaves like a phase-separated system, and because two glass transition temperatures are observed, the phase-separated regions must have dimensions of at least 50–100 Å. This size range presents a dilemma, in that one needs to pack 50–100-Å particles into a ~ 30 Å lattice.

It is also not clear how one can reconcile the core-shell model with two glass transition temperatures. The low T_g material might be either outside the outer ionic shell or between the ionic core and the outer ionic shell. We will first consider the region outside the outer ionic shell to be the low T_g region. That assumption gives a generally correct qualitative picture at low ion contents but leaves the high ion content region as a serious problem, because the packing of similarly sized spheres, even at close contact, leaves a substantial "unfilled" volume. Thus there would always be a high volume fraction (greater than about 30%) of low T_g material; however, this is not observed experimentally (see Chapter 4). For this situation, the origin of the high T_g phase remains unidentified. If the high T_g material is the nonionic shell between the core and the outer ionic shell, then we must explain how such a small region can have an independent glass transition temperature.

On the other hand, if we identify the material between the ionic core and the outer ionic shell as the low T_g phase, then the volume fraction of that phase should directly depend on the ion content, because the total number of core-shell structures increases with ion content, while the Bragg spacing shows a weak dependence on ion content. Experimentally, however, the volume fraction of the low T_g material is, at least at low ion contents, inversely related to the ion content. This problem persists even if we include the entire intrashell region. In summary, it is evident that the nature of the low and high T_g phases is not clearly defined in this model.

3.1.3.6. Morphological Continuity and Mechanical Properties.

A 1991 study strongly suggested that the SAXS peak in random ionomers is the result of interparticle interference (4). That study investigated random styrene-based ionomers in which the ionic group was placed at varying distances from the main chain via an alkyl or alkyl ether link to the benzene ring (**3.1**). It was shown that the size of

C_n alkyl side chain ionomers C_n alkyl ether side chain ionomers

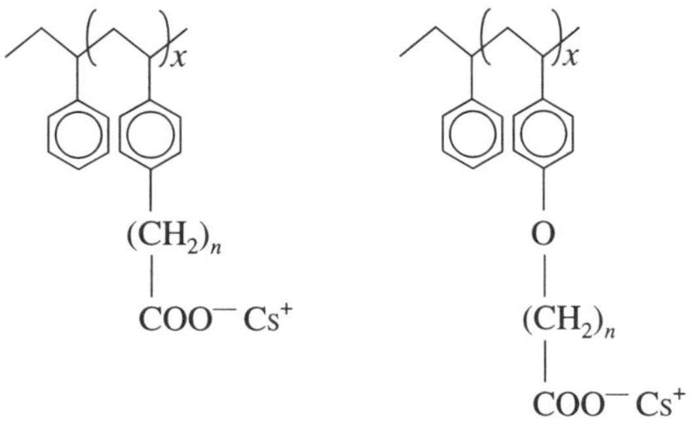

$n = 1$ (C$_2$) $n = 1$ (C$_2$)
    5 (C$_6$)   4 (C$_5$)
   10 (C$_{11}$)   10 (C$_{11}$)

Scheme 3.1

the ionic phase-separated region is a function of the side chain length; the longer the side chain length, the larger the aggregates. The relation between the radius of the aggregates and the side chain length can be extrapolated back to the styrene-*co*-cesium methacrylate [P(S-*co*-MACs)] ionomers and gives reasonable values for the aggregate size. At the other extreme, for the longer side chains, the morphology must be much like that of telechelic ionomers, which were studied previously (2,3). Thus, because the morphology of the ionomers is qualitatively independent of side chain length, it is clear that morphological continuity exists from the P(S-*co*-MACs) copolymers all the way to the cesium carboxylate terminated telechelics. From the study of the telechelics, however, it is known that only multiplets exist in that material. Therefore, because of morphological continuity, it is reasonable to suggest that P(S-*co*-MACs) copolymers also contain multiplets of the same type as are seen in the telechelics. Thus one can conclude that, at least in the styrenes and related materials, the SAXS peak is attributed to interparticle (i.e., intermultiplet) scattering and is related to the distance between scattering centers, as was suggested in a number of the earlier studies.

Mechanical properties will be discussed in detail in Chapters 5 to 7, so only a brief reference is made here to the aspects that are relevant to morphology. One mechanical property of ionomers that is difficult to explain is the level of the modulus between the two glass transition temperatures in the P(S-*co*-MANa) system. At relatively low ion content (up to ~5 mol %), the modulus is within a factor of two of that calculated on the basis of the kinetic theory of rubber elasticity: $G = \rho RT/M_c$, where G is the storage modulus, ρ is the density of polymer, R is the gas constant, T is the absolute temperature, and M_c is the average molecular weight between ionic groups or cross-links (57). Above that ion concentration, however, the value of modulus increases rapidly so that it is an order of magnitude larger than the theoretical prediction at ion concentrations of ~12 mol %. This discrepancy suggests that the multiplets do not act as simple cross-links at those concentrations. As a matter of fact, the high level of the modulus and the change in the value of that modulus with ion content suggest that phase inversion may have occurred (58) to the point at which the high T_g phase has now become not only the dominant but also probably the continuous phase.

If one looks at relative volume fractions of the two phases (discussed in detail in Chapter 5), one finds that at ~4 mol % of ions, the high and low T_g phases are present in approximately equal amounts. Because at 4 mol % the volume fraction of ionic material (i.e., ionic groups) is only ~1.2%, it is necessary to explain how that volume of ionic material can control the modulus and glass transition of 50 vol % of the polymer. Clearly, a model is needed to explain both the SAXS results and those of mechanical property studies. Such a model appeared in 1990 and is known as the Eisenberg–Hird–Moore EHM model (13).

The EHM model will be discussed in considerable detail because it helps interpret much of the subsequent material (especially the section dealing with T_g values and mechanical properties). The model is not necessarily expected to be valid for every ionomer system. For example, the lamellar zwitterionomers studied by Graiver et al. (7), the lamellar halatotelechelics studied by Broze et al. (8), the rodlike ionenes studied by Feng et al. (9), the viologen ionenes studied by Hashimoto et al. (10), and the perfluorosulfonated ionomers in solution studied by Aldebert et al. (11,12) have different morphologies. However, the model should be applicable to the vast majority of random ionomers, such as the P(S-*co*-MANa).

3.2. EHM MODEL

3.2.1. Restricted Mobility Region

The multiplet described above serves as the starting point for the new model (Fig. 3.2). One of the novel features of the model is that the mobility of the material immediately surrounding the multiplet is reduced relative to that of the bulk material (Fig. 3.3). There are several reasons for this behavior. One is a simple immobilization effect owing to the anchoring of the chain to the multiplet. The mobility of a species is a function of its molecular weight. Anchoring a chain to a multiplet increases the effective molecular weight of the segment adjacent to the multiplet, which reduces the mobility. Another factor is crowding of chains. Because the electrostatic energy released in the process of multiplet formation is considerable and because surface energies are also important, a certain amount of crowding in the immediate vicinity of the multiplet is to be expected as a result of driving forces, which tend to enlarge the multiplet (59). Finally, it is not unreasonable to expect a certain amount of chain extension in the immediate vicinity of the multiplet (35,36), which helps the multiplet accommodate a larger number of chains. All these factors contribute to a reduction in the mobility, although some are perhaps more important than others.

The thickness of the restricted mobility layer cannot be quantified at present. If the reduction of mobility is the result of immobilization of one end at the multiplet, then the persistence length (60) should be a good measure of the immobilization. If crowding is dominant, then the immobilization should have the same length scale as the crowding. Since a distribution of multiplet sizes is expected, the reduction in mobility will not be completely uniform but will depend, to some extent, on the size of the multiplet, especially if crowding is operative. This picture is also valid

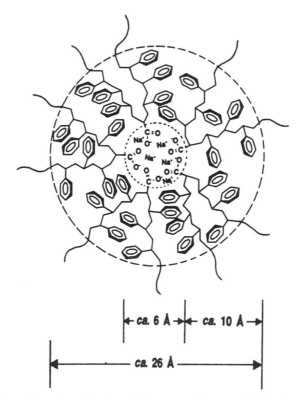

Figure 3.2. The region of restricted mobility surrounding a multiplet in a P(S-*co*-MANa) ionomer. Reprinted from Eisenberg et al. (13).

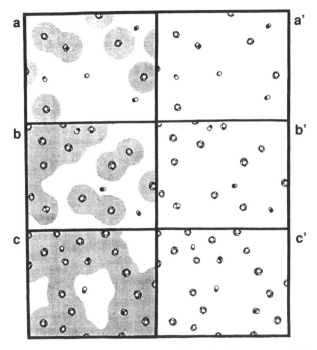

Figure 3.3. Representation of the morphologies of random ionomers at (**a**) low ion content, (**b**) intermediate ion content, and (**c**) high ion content. The *shaded areas* indicate regions of restricted mobility. **a′–c′**, the spatial arrangement of multiplets considering only electron density factors without regard to chain mobility. Reprinted from Eisenberg et al. (13).

if mobility is associated with a change of effective mass, because a smaller multiplet obviously has a much smaller core mass and therefore a higher mobility.

3.2.2. New Cluster Concept

In materials like the P(S-*co*-MANa) ionomers, the multiplet sizes are 6–8 Å (a few ion pairs), and the isolated multiplets, along with their reduced mobility layer, are expected to be too small to manifest their own T_g. The absence of a second T_g for isolated multiplets should also be seen for a wide range of polymers. Thus a lone multiplet will essentially behave as a cross-link. This behavior is indeed observed at low ion contents in most ionomers. Complications arise only when the number of multiplets increases to the point at which substantial overlap of regions of reduced mobility is encountered. These overlapping regions of restricted mobility become larger and larger as the ion content increases. Eventually, some of the regions of restricted mobility become large enough to exceed the threshold size for independent phase behavior (50–100 Å), at which point they exhibit their own T_g. The word *cluster* is suggested for these regions to reflect the clustering of multiplets in this new phase (13). Regions of restricted mobility in an isolated multiplet and in clustered

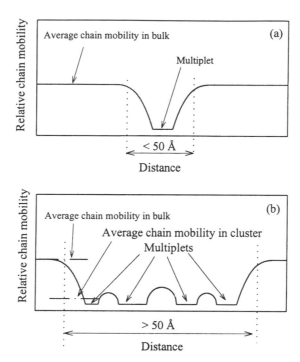

Figure 3.4. Chain mobility **(a)** in the vicinity of an isolated multiplet and **(b)** in the region of clustered multiplets. Modified from Eisenberg et al. (13).

multiplets are shown in Figure 3.3; the chain mobility in these regions is shown in Figure 3.4. The material in both the clustered and unclustered regions is polystyrene, and the total volume fraction of ion pairs is small, even in the clustered regions. At 4 mol % of ions, the volume fractions of the clustered and unclustered materials are approximately equal (this will be shown later) (61). The total volume fraction of ionic groups is on the order of 1%. Therefore, even if we assume that all of the multiplets are in the clustered regions (an overestimate), the volume fraction of ionic groups is still <2% of the volume of clusters, i.e., the cluster phase is 98% polystyrene.

It is useful to ask why we do not observe a single broad T_g but two distinct glass transitions. To answer this question, keep in mind that isolated multiplets or aggregates of multiplets that are too small to exhibit their own T_g will be perceived as cross-links. It is only when the size threshold for independent phase behavior has been exceeded that the new glass transition temperature will be seen. Once the threshold has been exceeded, the new T_g will be considerably higher than the matrix T_g, because it arises from polystyrene of reduced mobility. As shown in detail in Chapters 4 and 5, ion hopping (i.e., jumping of ion pairs between multiplets) is an essential feature of the glass transition of the clustered regions (62,63). The second glass transition requires the onset of ion hopping at an appreciable rate.

3.2.3. Prevalent Intermultiplet Distance

The question of the distance between multiplets in clusters also must be addressed. The presence of a SAXS peak suggests that there is spatial correlation between multiplets. If this were not the case, no peak would be seen, but only a small-angle upturn in intensity. In random ionomers, there is a distribution of distances between ionic groups along the chain (4); short distances along the chain result in ions being incorporated into the same multiplet, or "looping back." Beyond this loopback distance, ion pairs serve as starting points for new multiplets. Consider the ion pairs A and B in Figure 3.5, which should become incorporated into different multiplets along with C and D, such that A and C form one multiplet and B and D form another. Because A and B are separated by more than the loopback distance, the spacing of the two multiplets will be determined by the chain length A-B rather than chain length C-D.

One novel feature of the model is that the clusters are not of any particular geometric shape, but are expected to be highly irregular. As shown in Chapter 5, percolation concepts are applicable to these structures, and the clusters, with their higher T_g values, act as filler at temperatures between the two T_g values (61). The sizes of the clusters will also be broadly distributed; and in contrast to other models, no particular number of multiplets or ion pairs per cluster should be anticipated.

3.2.4. Semiquantitative Aspects

A brief discussion of the semiquantitative aspects of the model is in order. The Bragg spacing for the P(S-*co*-MACs) ionomers of intermediate ion contents (5–10 mol %) is ~23 Å (4,64). If it is assumed that (to a first approximation) all the ion pairs are incorporated into multiplets and that these are distributed on a cubic lattice, the number of ion pairs per multiplet (calculated from simple stoichiometry) is approximately five for the 7 mol % ionomer. If all the ion pairs are not in multiplets, naturally the size of the multiplets will be smaller. For 60% incorporation, the aggregation number is three. The aggregation number is taken to be four for the purpose

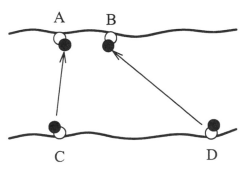

Figure 3.5. The dependence of intermultiplet distance on the spacing of ionic groups along the polymer chains. Modified from Eisenberg et al. (13).

of an illustrative calculation, and the average volume of an ion pair in the multiplet in the P(S-*co*-MANa) is taken to be ~45 $\mathring{A}^3$ (see Chapter 1). Furthermore, if the thickness of the restricted mobility region is taken as ~10 $\mathring{A}$ (65); then the volume of immobilized material per ion pair should be approximately 2500 $\mathring{A}^3$. This number does not depend strongly on the total volume/ion pair. With those numbers, approximately 80 vol % consists of material of restricted mobility at 7 mol % of ionic groups. This number is close to the experimental value of the volume fraction of material (73%) of high glass transition temperature (Chapter 5).

3.3. EXPERIMENTAL CORROBORATION

The model proposed here is in excellent agreement with an unprecedented range of data available on ionomers and can explain both the morphological and the mechanical results.

3.3.1. Small-Angle X-Ray Scattering Peak

In agreement with the proposal by Yarusso and Cooper (42), the EHM model assumes that the SAXS ionic peak arises primarily from the most common distance between multiplets (2). The model is thus based on the idea that intermultiplet (but predominantly intracluster) scattering gives rise to the SAXS peak. The breadth of the peak suggests that there is a distribution of distances between multiplets (2), and this is indeed an integral part of the EHM model. While most of the scattering is expected to be intracluster, there is no way to separate this from the scattering from neighboring multiplets in different clusters, which would be expected to be at the long end of the distribution of intermultiplet distances.

In general, the dependence of the Bragg distance calculated from the position of the peak maximum as a function of ion concentration does not follow a $1/c^{1/3}$ relation (where c is the ion concentration) but is much weaker than that (14,67). The $1/c^{1/3}$ relation would be expected from the scattering of noninteracting particles in solution. The EHM model predicts a weak dependence, because the most prevalent spacing is believed to be related to the intermultiplet distance, which is only indirectly related to ion concentration. The intermultiplet spacing is believed to be related to the loopback distance, which is a function of the nature of the polymer chain and the ion pair rather than simply ion concentration, at least to a first approximation.

The SAXS peak has been observed to persist far above the cluster T_g, as observed mechanically (67–72). This persistence is, again, in complete accord with the model, in that multiplets are not believed to disappear at the upper T_g. As shown in Chapter 4, ion hopping is intimately related to the glass transition temperature, but hopping implies only the occasional removal of an ion pair and its placement in another multiplet rather than the disappearance of multiplets. Galambos et al. (69) found that sulfonated polystyrene ionomers, cast from a tetrahydrofuran (THF)/water ratio of 90/10 at room temperature, do not exhibit a SAXS peak. However, when these materials are heated, the SAXS peak appears. This phenomenon is consistent with the model, because in solution the ions are solvated, and especially in the presence of residual water, electrostatic interactions are highly reduced. Thus, as THF evaporates, the tightly bound water molecules would be the last to leave the polymer, at a point at which the material is either close to, or even below, its glass transition temperature. Thus rearrangement on a scale that permits multiplets to be formed is obviously hindered. However, above the glass transition temperature, multiplet formation becomes possible and is observed.

3.3.2. Dynamic Mechanical Studies

Dynamic mechanical results also provide strong corroboration of the model, which explains the existence of two glass transitions. The existence of two transitions implies that each of the regions has dimensions of at least 50–100 Å, which is now reconciled with the existence of interparticle scattering at distances of only ~25–30 Å. The model postulates the existence of clustered regions in which one encounters overlapping regions of reduced mobility and regions in which the material has only a low multiplet content or is even devoid of multiplets. No intermediate regions are envisaged. If a wide range of morphologies (or mobilities) were encountered, then presumably one would see a single, broad glass transition temperature rather than the two glass transition temperatures that are observed.

The model also directly explains the increase in glass transition temperature with ion content. As the ion content increases, the average distance between multiplets decreases and the size of the multiplets increases. Both of these changes cause a reduction in the mobility, because it is expected that chains between two multiplets in a cluster would behave differently from chains at the periphery of a cluster. The larger the cluster, the larger the fraction of chains between multiplets that are embedded in the cluster as opposed to those on the surface and, therefore, the higher the glass transition temperature. With increasing ion content, the size of the intercluster regions decreases; thus the intercluster regions behave as if the cross-link density were increasing. This decrease in the size of the intercluster region leads to an increase in the matrix glass transition temperature with increasing ion content, which is observed experimentally for most ionic systems.

The discontinuities in many of the properties in the P(S-co-MANa) ionomers at ~6 mol % are also explained by the EHM model. As shown in Chapter 5, the percolation threshold for cluster behavior is reached at 5.4 mol %, which coincides, within narrow limits, with discontinuities in a large number of properties (e.g., failure of time–temperature superposition in stress–relaxation, and water uptake) (73) and parameters obtained from the dynamic mechanical property data as a function of ion content (61).

The similarity of the behavior of ionomers to that of filled systems is also completely consistent with the model (61). Below the matrix glass transition temperature, the multiplets themselves act as a filler. The volume fraction of filler, however, is extremely small, and even at 20 mol % of ionic groups in the P(S-co-MANa) system, the volume fraction of ion pairs is only about 7%. Therefore, below the matrix glass transition temperature, the modulus of material is not expected to change appreciably with ion content, and this constancy of the modulus is observed experimentally. By contrast, between the two glass transitions, it is the clustered regions that act as filler. Filler equations predict the variation of Young's modulus as a function of ion content (61) and are in excellent agreement with experiment; the volume fractions of clustered materials are taken from the areas under the loss tangent curves (see Chapter 5).

3.3.3. Volume Fraction of Clusters

The effects of changes in a wide range of molecular parameters on the mechanical properties in ionomers have been explored, and the results are uniformly in agreement with the predictions or correlations of the EHM model. If the ionic groups are located at the end of long flexible side chains, the size of multiplets is predicted to increase relative to those found in P(S-co-MANa). However, the volume fraction of clustered material at constant ion content is expected to decrease. This result is consistent with the experimental observations.

The decrease in the volume fraction of clustered material with increasing multiplet sizes at a comparable ion content is a concept that is needed to interpret results presented in several sections of the book; therefore, it is worth discussing in some detail. Let us, for the sake of simplicity, consider a sphere of radius equal to 1 cm; the surface area of this sphere would be 12.6 cm^2; and if this sphere were surrounded with material of restricted mobility of 10 Å thickness, the total volume of immobilized material would be 1.26×10^{-6} cm^3. If we now consider the same volume of material consisting of 1000 spheres of 0.1-cm radius, the total surface area would be 126 cm^2, and if the thickness of the layer of restricted mobility were kept at 10 Å, the total volume of material of restricted mobility would now be 1.26×10^{-5} cm^3. Of course, this scaling persists for all sphere sizes. Thus every time we decrease the radius of the particle by a factor of 10, the number of spheres increases by a factor of 1000 and the volume of the material of restricted mobility increases by one order of magnitude. When the radius reaches 0.01 cm, the surface area is 1260 cm^2 and the volume of material of restricted mobility is 1.26×10^{-4} cm^3. At 1 μm we have 10^{12} spheres with a total surface area of 1.26×10^5 cm^2 and a total volume of material of reduced mobility of 1.26×10^{-2} cm^3, and in the case of a 1-nm particle size (10^{21} spheres) the total surface area is 1.26×10^8 cm^2 and the total volume of material of restricted mobility is 12.6 cm^3. For a constant total volume of multiplets, it is clear that the larger the multiplet, the smaller the volume of material of restricted mobility, if the thickness of the restricted mobility layer is constant. This idea is important in the discussion of the ionic modulus for different ionomers.

3.3.4. Thickness of the Restricted Mobility Layer

Materials derived from polymers containing flexible chains, such as polypentenamer, exhibit two-phase behavior only at higher ion contents than the P(S-co-MANa) systems because the thickness of the restricted mobility layer is smaller. The same is true of poly(methyl methacrylate) (PMMA) ionomers (74). Again, correspondence with the model is clear. The thickness of the reduced mobility layer would be expected to decrease as chain flexibility increases, and this decrease in layer thickness would affect the volume fraction of reduced mobility material at any ion content. To get overlap, a higher concentration of ionic material would be needed. The persistence length of the polymer chain in PMMA is ~7 Å (74), whereas for polystyrene it is ~10 Å. It was found in the PMMA ionomer system that the phase inversion and percolation threshold occur at much higher ion contents than in styrene. Furthermore, in normal telechelics, two glass transitions are not expected, because the distances between multiplets are far bigger than twice the persistence length; and indeed two glass transitions have not been observed in telechelics, even though the multiplets in telechelics generally are much larger than the multiplets in random ionomers.

3.3.5. Unclustered Materials and Homoblends

The absence of two glass transitions in some ionomers is also consistent with the EHM model, if the multiplets either do not form or are unstable because of the large size of the ionic groups. The styrene-*co*-*N*-methyl-4-vinylpyridinium iodide ionomers are an example of this type of material. In this case, multiplets are present below the T_g, but dissociate when the T_g is reached. On the other hand, if the lifetime of multiplets at the glass transition temperature is increased by a reduction in the T_g (e.g., via plasticization), then cluster behavior is indeed observed (75).

The recently discovered mechanical cluster peak (Chapter 9) in ionomer blends of sulfonated polystyrene or its zinc salt mixed with ethyl acrylate-*co*-4-vinylpyridine copolymers is completely in accord with the model. If we take the zinc ion as the center of a one-cation "multiplet," we would expect it to be coordinated to two vinylpyridines and two sulfonate groups. This structure (Fig. 4.14) is expected to be surrounded by a layer of materials of reduced mobility, because the strong interaction between the zinc ion and the surrounding species can lead to immobilization owing to holddown (i.e., attachment to a rigid species of high mass) and possibly to crowding as well. If structures like these are close enough to each other (i.e., less than twice the persistence length), two-phase behavior will be encountered. This two-phase behavior was indeed found experimentally. Because no metal ion aggregates are present in these materials, the absence of a SAXS peak is completely consistent with the presence of two loss tangent peaks (Chapter 4). No other morphological models can explain all of these phenomena in a similarly self-consistent manner. Plasticizer effects are also explained by the model (Chapter 8).

3.3.6. Experimental Evidence for Reduced Mobility

Since the publication of the EHM model, a number of papers have appeared that directly confirm the postulated reduction in mobility. Yano et al. (76) found from dielectric studies that, in quaternized polyisoprene telechelics, the region immediately surrounding the multiplet shows lower mobility than does the bulk material.

In a subsequent NMR spectroscopic study, Gao et al. (77), using a short deuterated styrene probe located at different distances from the cores of monochelics or block ionomers, showed that the mobility decreases as one approaches the multiplet or the phase-separated block ionomer region. The probe must be at least 40 repeat units away from the block ionomer to experience unrestricted mobility, i.e., that of a chain segment not in the vicinity of a multiplet. Vanhoorne et al. (78) conducted an independent mobility study using ^{13}C NMR and again confirmed the reduction of mobility in the vicinity of the multiplet. Electron spin resonance spectroscopy was also used to study chain mobility in spin-doped P(S-*co*-MANa) random ionomers (79).

More recently, Tsagaropoulos and Eisenberg (80,81) found that filled homopolymer systems, in which the sizes of filler particles are on the order of 7 nm, also exhibit two glass transition temperatures. This finding of double glass transitions in filled systems, which had been postulated on the basis of the model (61), serves as strong confirmation of the reduced mobility concept.

Another confirmation comes from the small-angle x-ray upturn in SAXS. Several interpretations of this phenomenon have been offered, all of which agree that one is dealing with a small difference in the scattering intensity but on large dimensions. Chu et al. (82) suggested from the ultra-small-angle x-ray results that in the zinc sulfonated polystyrene system, the correlation length is ~90 nm; this value may serve as an estimate of the cluster size in that material. The existence of clustered regions of that size is, again, completely consistent with the predictions of the model.

3.4. COIL DIMENSIONS IN RANDOM IONOMERS

It is of interest to explore the short-range motions that a polymer undergoes in the process of multiplet formation, because these movements may give rise to changes in the overall chain dimensions. A number of papers appeared in the 1980s on this topic; however, there is still no generally accepted explanation.

In the first study to explore this problem, Pineri et al. (83) used deuterated styrene copolymers containing various amounts of sodium methacrylate dissolved in protonated samples of similar composition and molecular weight. For the neutralized samples, the radii of gyration of the deuterated samples were obtained at a concentration of 0.5–1% of deuterated polymer from a Zimm plot by extrapolation to 0% concentration. The results of the study suggested that the radii of gyration of the ionic samples were not appreciably different from those of the nonionic materials.

Subsequent experiments were devoted to a much more careful investigation of identical molecular weight deuterated polystyrene samples sulfonated to different extents and embedded in protonated sulfonated polystyrene (84). The results suggested that there is a reproducible coil expansion with increasing degrees of ionic substitution. For example, the radius of gyration for the unfunctionalized polystyrene sample of ~ 1000 units was 86 Å; it increased to 105 Å for the 1.9 mol % sulfonated sample, to 111 Å for the 4.2 mol % sample, and to 123 Å for the 8.5 mol % sample. These results contrast with the previous work on the P(S-*co*-MANa) copolymers and may well reflect real differences in the two systems, resulting from differences in the strength of intramultiplet interactions.

A theoretical treatment by Forsman (35) showed that associations of the type encountered in the random ionomers should lead to a coil expansion. It was also shown that the degree of expansion is related to the shape of aggregates, with expansion increasing with increasing asymmetry of the aggregates. In a subsequent paper by Forsman et al. (85), the theoretical and experimental results were compared. The paper suggests that the surface energy of aggregates (an arbitrary parameter in the theory) in the sodium sulfonated polystyrene is equal to 37.2 kJ/mol ion pairs, and that the number of ion pairs per cluster should be in the range of 15 to 12 for ion contents of 1.2–8.9 mol % sulfonation.

Squires et al. (86) suggested that aggregate formation in ionomers should not be accompanied by chain extension. This model suggests that in ionomers with a truly random distribution of aggregates linked by long subchains subject to Gaussian statistics, no chain extension should be encountered.

In more recent work on this topic, Register et al. (87) studied the chain dimensions of sodium–carboxy telechelic polystyrene ionomers and the corresponding methyl esters. They also found that the chain dimensions for the ionomer and esters are identical, in agreement with the predictions of Squires et al (86). Register et al. (88) also studied ionomer chain dimensions in sulfonated polyurethanes and found a mild chain extension. The results of Gouin et al. (44)—of the constancy of the polystyrene midblock segment in ionic triblocks—are in accord with those for the telechelics. Coil dimension in block ionomers will be discussed in Chapter 7.

Clearly, not enough work has been done to obtain a clear picture of coil dimensions in ionomers in general. Although the work on sulfonated polystyrene strongly suggests that coil expansion in that system does take place, the earlier work on carboxylated ionomers and the subsequent theoretical work of Squires et al. (86) and the experimental results of Register et al. (87) suggest that coil expansion may not be the case in all materials. There may very well be real differences between carboxylates and sulfonates, and the question still remains to be resolved for the random ionomers as a whole.

3.5. REFERENCES

1. Eisenberg, A. *Macromolecules* **1970**, *3*, 147–154.
2. Williams, C. E., Russell, T. P.; Jérôme, R., Horrion, J. *Macromolecules* **1986**, *19*, 2877–2884.
3. Williams, C. E. In *Multiphase Macromolecules Systems;* Culberston, W. M., Ed.; Plenum: New York, 1989.
4. Moore, R. B.; Bittencourt, D.; Gauthier, M.; Williams, C. E.; Eisenberg, A. *Macromolecules* **1991**, *24*, 1376–1382.
5. Li, C.; Register, R. A.; Cooper, S. L. *Polymer* **1989**, *30*, 1227–1233.
6. Roche, E. J.; Stein, R. S.; Russell, T. P.; MacKnight, W. J. *J. Polym. Sci. Polym. Phys. Ed.* **1980**, *18*, 1497–1512.

7. Graiver, D.; Litt, M.; Baer, E. *J. Polym. Sci. Polym. Chem. Ed.* **1979**, *17*, 3573–3587.

8. Broze, G.; Jérôme, R.; Teyssié, P. *J. Polym. Sci. Polym. Lett. Ed.* **1981**, *19*, 415–418.

9. Feng, D.; Wilkes, G. L.; Leir, C. M.; Stark, J. E. *J. Macromol. Sci. Chem.* **1989**, *A26*, 1151–1181.

10. Hashimoto, T.; Sakurai, S.; Morimoto, M.; Nomura, S.; Kohjiya, S.; Kodaira, T. *Polymer* **1994**, *35*, 2672–2678.

11. Aldebert, P.; Dreyfus, B.; Pineri, M. *Macromolecules* **1986**, *19*, 2651–2653.

12. Aldebert, P.; Dreyfus, B.; Gebel, G.; Nakamura, N.; Pineri, M.; Volino, F. *J. Phys. France* **1988**, *49*, 2101–2109.

13. Eisenberg, A.; Hird, B.; Moore, R. B. *Macromolecules* **1990**, *23*, 4098–4107.

14. Tomita, H.; Register, R. A. *Macromolecules* **1993**, *26*, 2791–2795.

15. Eisenberg, A.; Farb, H.; Cool, L. G. *J. Polym. Sci. A-2* **1966**, *4*, 855–868.

16. Liu, K. J.; Anderson, J. E. *Macromolecules* **1969**, *2*, 235–237.

17. Moacanin, J.; Cuddihy, E. F. *J. Polym. Sci. C* **1966**, *14*, 313–322.

18. Yarusso, D. J.; Ding, Y. S.; Pan, H. K.; Cooper, S. L. *J. Polym. Sci. Polym. Phys. Ed.* **1984**, *22*, 2073–2093.

19. Galland, D.; Belakhovsky, M.; Medrignac, F.; Pineri, M.; Vlaic, G.; Jérome, R. *Polymer* **1986**, *27*, 883–888.

20. Register, R. A.; Foucart, M.; Jérôme, R.; Ding, Y. S.; Cooper, S. L. *Macromolecules* **1988**, *21*, 1009–1015.

21. Visser, S. A.; Cooper, S. L. *Polymer* **1992**, *33*, 930–937.

22. Ding, Y. S.; Yarusso, D. J.; Pan, H. K. D.; Cooper, S. L. *J. Appl. Phys.* **1984**, *56*, 2396–2403.

23. Ding, Y. S.; Register, R. A.; Yang, C.-Z.; Cooper, S. L. *Polymer* **1989**, *30*, 1221–1226.

24. Meagher, A.; Coey, J. M. D.; Belakhovsky, M.; Pinéri, M.; Jérome, R.; Vlaic, G.; Williams, C.; Dang, N. V. *Polymer* **1986**, *27*, 979–985.

25. Pan, H. K.; Meagher, A.; Pineri, M.; Knapp, G. S.; Cooper, S. L. *J. Chem. Phys.* **1985**, *82*, 1529–1538.

26. Pan, H. K.; Yarusso, D. J.; Knapp, G. S.; Cooper, S. L. *J. Polym. Sci. Polym. Phys. Ed.* **1983**, *21*, 1389–1401.

27. Vlaic, G.; Williams, C. E.; Jérome, R.; Tant, M. R.; Wilkes, G. L. *Polymer* **1988**, *29*, 173–176.

28. Grady, B. P.; Cooper, S. L. *Macromolecules* **1994**, *27*, 6627–6634.

29. Grady, B. P.; Cooper, S. L. *Macromolecules* **1994**, *27*, 6635–6641.

30. Tadano, K.; Hirasawa, E.; Yamamoto, H.; Yano, S. *Macromolecules* **1989**, *22*, 226–233.

31. Goddard, R. J.; Grady, B. P.; Cooper, S. L. *Macromolecules* **1994**, *27*, 1710–1719.

32. Wilson, F. C.; Longworth, R.; Vaughan, D. J. *Polym. Prepr. Am. Chem. Soc., Div. Polym. Chem.* **1968**, *9*, 505–514.

33. Longworth, R.; Vaughan, D. J. *Nature* **1968**, *218*, 85–87.

34. Bonotto, S.; Bonner, E. F. *Macromolecules* **1968**, *1*, 510–515.

35. Forsman, W. C. *Macromolecules* **1982**, *15*, 1032–1040.

36. Forsman, W. C. In *Developments in Ionic Polymers, 1*; Wilson, A. D.; Prosser, H. J., Eds.; Applied Science: New York, 1983; Chapter 4.

37. Dreyfus, B. *Macromolecules* **1985,** *18,* 284–292.

38. Datye, V. K.; Taylor, P. L. *Macromolecules* **1985,** *18,* 1479–1482.

39. Delf, B. W.; MacKnight, W. J. *Macromolecules* **1969,** *2,* 309–310.

40. Marx, C. L.; Caulfield, D. F.; Cooper, S. L. *Macromolecules* **1973,** *6,* 344–353.

41. Binsbergen, F. L.; Kroon, G. F. *Macromolecules* **1973,** *6,* 145.

42. Yarusso, D. J.; Cooper, S. L. *Macromolecules* **1983,** *16,* 1871–1880.

43. MacKnight, W. J.; Taggart, W. P.; Stein, R. S. *J. Polym. Sci. Symp.* **1974,** *45,* 113–128.

44. Gouin, J. P.; Williams, C. E.; Eisenberg, A. *Macromolecules* **1989,** *22,* 4573–4578.

45. Gouin, J. P.; Bossé, F.; Nguyen, D.; Williams, C. E.; Eisenberg, A. *Macromolecules* **1993,** *26,* 7250–7255.

46. Nguyen, D.; Varshney, S. K.; Williams, C. E.; Eisenberg, A. *Macromolecules* **1994,** *27,* 5086–5089.

47. Nguyen, D.; Williams, C. E.; Eisenberg, A. *Macromolecules* **1994,** *27,* 5090–5093.

48. Nguyen, D.; Zhong, X.-F.; Williams, C. E.; Eisenberg, A. *Macromolecules* **1994,** *27,* 5173–5181.

49. Lee, D.-C.; Register, R. A.; Yang, C.-Z.; Cooper, S. L. *Macromolecules* **1988,** *21,* 998–1004.

50. Lowry, S. R.; Mauritz, K. A. *J. Am. Chem. Soc.* **1980,** *102,* 4665–4667.

51. Mauritz, K. A.; Hora, C. J.; Hopfinger, A. J. In *Ions in Polymers*; Eisenberg, A., Ed.; Advances in Chemistry 187; American Chemical Society: Washington, DC, 1980; Chapter 8.

52. Gierke, T. D.; Munn, G. E.; Wilson, F. C. *J. Polym. Sci. Polym. Phys. Ed.* **1981,** *19,* 1687–1704.

53. Fujimura, M.; Hashimoto, T.; Kawai, H. *Macromolecules* **1981,** *14,* 1309–1315.

54. Fujimura, M.; Hashimoto, T.; Kawai, H. *Macromolecules* **1982,** *15,* 136–144.

55. Hsu, W. Y.; Gierke, T. D. *Macromolecules* **1982,** *15,* 101–105.

56. Datye, V. K.; Taylor, P. L.; Hopfinger, A. J. *Macromolecules* **1984,** *17,* 1704–1708.

57. Treloar, L. R. G. *The Physics of Rubber Elasticity,* 3rd ed.; Clarendon: Oxford, UK, 1975.

58. Hird, B.; Eisenberg, A. *J. Polym. Sci. Polym. Phys.* **1990,** *28,* 1665–1675.

59. Nyrkova, I. A.; Khokhlov, A. R.; Doi, M. *Macromolecules* **1993,** *26,* 3601–3610.

60. Muthukumar, M.; Edwards, S. F. In *Comprehensive Polymer Science*, Vol. 2; Allen, G.; Bevington, J. C., Eds., Pergamon: Oxford, UK, 1989; Chapter 1.

61. Kim, J.-S.; Jackman, R. J.; Eisenberg, A. *Macromolecules* **1994,** *27,* 2789–2803.

62. Hara, M.; Eisenberg, A.; Storey, R. F.; Kennedy, J. P. In *Coulombic Interactions in Macromolecular Systems*; Eisenberg, A.; Bailey, F. E., Eds.; ACS Symposium Series 302; American Chemical Society: Washington, DC, 1986; Chapter 14.

63. Hird, B.; Eisenberg, A. *Macromolecules* **1992,** *25,* 6466–6474.

64. Jiang, M.; Gronowski, A. A.; Yeager H. L.; Wu, G., Kim, J.-S.; Eisenberg, A. *Macromolecules* **1994,** *27,* 6541–6550.

65. Wignall, G. D.; Ballard, D. G. H.; Schelten, J. *Eur. Polym. J.* **1974,** *10,* 861–865.

66. Peiffer, D. G.; Weiss, R. A.; Lundberg, R. D. *J. Polym. Sci. Polym. Phys. Ed.* **1982,** *20,* 1503–1509.

67. Weiss, R. A.; Lefelar, J. A. *Polymer* **1986,** *27,* 3–10.

68. Yarusso, D. J.; Cooper, S. L. *Polymer* **1985,** *26,* 371–378.

69. Galambos, A. F.; Stockton, W. B.; Koberstein, J. T.; Sen, A.; Weiss, R. A.; Russell, T. P. *Macromolecules* **1987,** *20,* 3091–3094.

70. Fitzgerald, J. J.; Weiss, R. A. *J. Polym. Sci. B: Polym. Phys.* **1990,** *28,* 1719–1736.

71. Wang, J.; Li, Y.; Peiffer, D. G.; Chu, B. *Macromolecules* **1993,** *p.* 2633–2635.

72. Li, Y.; Peiffer, D. G.; Chu, B. *Macromolecules* **1993,** *26,* 4006–4012.

73. Eisenberg, A.; Navratil, M. *Macromolecules* **1973,** *3,* 604–612.

74. Ma, X.; Sauer, J. A.; Hara, M. *Macromolecules* **1995,** *28,* 3953–3962.

75. Wollmann, D.; Williams, C. E.; Eisenberg, A. *Macromolecules* **1992,** *25,* 6775–6783.

76. Yano, S.; Tadano, K.; Jérôme, R. *Macromolecules* **1991,** *24,* 6439–6442.

77. Gao, Z.; Zhong, X.-F.; Eisenberg, A. *Macromolecules* **1994,** *27,* 794–802.

78. Vanhoorne, P.; Jérôme, R.; Teyssié, P.; Lauprêtre, F. *Macromolecules* **1994,** *27,* 2548–2552.

79. Tsagaropoulos, G.; Kim, J.-S.; Eisenberg, A. *Macromolecules* **1996,** *29,* 2222–2228.

80. Tsagaropoulos, G.; Eisenberg, A. *Macromolecules* **1995,** *28,* 396–398.

81. Tsagaropoulos, G.; Eisenberg, A. *Macromolecules* **1995,** *28,* 6067–6077.

82. Chu, B.; Wang, J.; Li, Y.; Peiffer, D. G. *Macromolecules* **1992,** *25,* 4229–4231.

83. Pineri, M.; Duplessix, R.; Gauthier, S.; Eisenberg, A. In *Ions in Polymers*; Eisenberg, A., Ed.; Advances in Chemistry 187; American Chemical Society: Washington, DC, 1980; Chapter 18.

84. Earnest, T. R. Jr.; Higgins, J. S.; Handlin, D. L.; MacKnight, W. J. *Macromolecules* **1981,** *14,* 192–196.

85. Forsman, W. C.; MacKnight, W. J.; Higgins, J. S. *Macromolecules* **1984,** *17,* 490–494.

86. Squires, E., Painter, P., Howe, S. *Macromolecules* **1987,** *20,* 1740–1744.

87. Register, R. A.; Cooper, S. L.; Thiyagarajan, P.; Chakapani, S.; Jérôme, R. *Macromolecules* **1990,** *23,* 2978–2983.

88. Register, R. A.; Pruckmayr, G.; Cooper, S. L. *Macromolecules* **1990,** *23,* 3023–3026.

CHAPTER 4

GLASS TRANSITIONS IN RANDOM IONOMERS

4.1 EFFECT OF IONIC FORCES

Given the morphological picture presented in Chapter 3, it is not surprising that the glass transition behavior of ionomers can be quite complex. Before discussing this topic, it is useful to review the effect of ionic forces on the T_g by describing the behavior of one ionizable homopolymer (not subject to clustering) as a function of the degree of ionization and the type of counterion.

Figure 4.1 shows the T_g for various linear metalphosphate polymers using sodium-poly(metaphosphate) $[(NaPO_3)_x]$ as the starting point. The T_g of that ionic homopolymer is $\sim 280°C$ (1). As one reduces the degree of neutralization from 100% for $(NaPO_3)_x$ down to 0% for $(HPO_3)_x$, the T_g drops progressively to $-10°C$ (Fig. 4.1). Similarly, the behavior of the T_g as a function of the mole percent of metal ion other than Na^+ is shown for $NaPO_3$ and KPO_3, $NaPO_3$ and $Ca(PO_3)_2$, and $NaPO_3$ and $La(PO_3)_3$. Thus the $NaPO_3$–$Ca(PO_3)_2$ line represents the effect of various counterion compositions on the T_g in this counterion copolymer system (in which the backbone is identical, but the counterions are mixed). The abscissa is now the percentage of repeat units neutralized by the nonsodium counterion. The $Ca(PO_3)_2$ homopolymer has a T_g of $520°C$. It is clear that the T_g for the identical $(PO_3^-)_x$ backbone can vary from -10 to $520°C$, depending on whether the polymer is ionized and which counterions are used.

This range of $530°C$ is much larger than the T_g range of the commonly used nonionic organic polymers and illustrates the strong effect of ionic forces on the T_g. It is also noteworthy that copolymers of $NaPO_3$–$NaAsO_3$ in which the occasional phosphorus atom is replaced by an arsenic atom, show no change in the T_g with composition. This result shows that the ionic interactions determine the T_g, because the ionic interactions between the sodium ions and phosphate or arsenate polyions

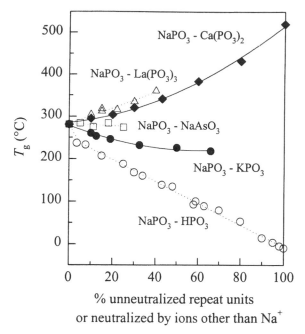

Figure 4.1. Glass transition temperatures of copolymers of NaPO$_3$ with other materials, as indicated. Modified from Eisenberg et al. (1).

are similar; therefore, no effect on the T_g is expected when one replaces phosphorus with arsenic.

A range of homopolymers was also investigated in this context, and the results are given in Table 4.1, which also lists the Pauling radii and the ratio of charge q to the distance a between centers of charge in the materials (1). Empirically, it was found that a plot of the T_g against the q/a ratio was linear for the vast majority of the polyphosphate points. Subsequently, it was found that the same relationship could be used for the silicates (with a higher intercept) (2) and for the acrylates

TABLE 4.1. Characteristics of Polyphosphate Homopolymers

Material	T_g, °C	Pauling Radius r, Å	Charge q	q/a
HPO$_3$	− 10	—	—	0.00
LiPO$_3$	335	0.60	1	0.50
NaPO$_3$	280	0.95	1	0.42
Ca(PO$_3$)$_2$	520	0.99	2	0.84
Sr(PO$_3$)$_2$	485	1.13	2	0.79
Ba(PO$_3$)$_2$	470	1.35	2	0.73
Zn(PO$_3$)$_2$	520	0.74	2	0.93
Cd(PO$_3$)$_2$	450	0.97	2	0.84

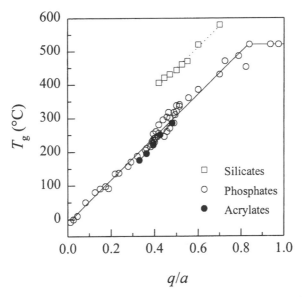

Figure 4.2. Glass transition temperatures versus the charge to distance ratio for silicates ($\square$), phosphates ($\bigcirc$), and acrylates ($\bullet$). Data from Eisenberg et al. (1,3) and Eisenberg and Takahashi (2).

(with a slightly lower intercept) (3). The plots of T_g versus the q/a ratio are shown in Figure 4.2.

4.2. GLASS TRANSITION VERSUS ELECTROSTATIC WORK

The glass transition results can be rationalized if one considers that the T_g, as observed experimentally, is a nearly isokinetic phenomenon, because it occurs at the point at which the volume shrinks (in response to a drop in temperature) with a relaxation time of ~1 min. In normal organic polymers, this temperature is determined by the barrier to bond rotation around the backbone, in addition to intermolecular constraints. In the polyphosphates, where (in the absence of ionic interactions) the energy barrier to rotation around a phosphorus–oxygen bond is low, the rotational barriers are determined by the strengths of the ionic interactions. Thus the T_g should be proportional to the electrostatic work W_{el} of removing an anion from the coordination sphere of the cation or vice versa (4):

$$T_g \propto W_{el} \qquad (4.1)$$

The electrostatic work, in turn, is equal to the integral of the force F_{el} over distance from the distance closest approach a to as far as the counterion can move. While this distance is obviously not large, the nature of electrostatic interactions is such that the removal of an ion from the distance at closest approach to ~ 10 Å expends almost as much work as the removal of the ion to infinity. It is thus simpler to perform the integration from the distance of closest approach to infinity. Thus the equation becomes

$$T_g \propto \int_a^\infty F_{el} \, dx \qquad (4.2)$$

The electrostatic force is proportional to the anion charge q_a times the cation charge q_c divided by the square of the distance between them x:

$$T_g \propto \int_a^\infty \frac{q_a q_c}{x^2} \, dx \qquad (4.3)$$

Performing this integration gives

$$T_g \propto (q_a q_c / a) + \text{constant} \qquad (4.4)$$

Since the anion charge is not well defined here, it was incorporated in the original presentation into the proportionality constant. Thus the T_g is nearly proportional to the cation charge divided by the distance between the centers of charge:

$$T_g \propto (q_c / a) + \text{constant} \qquad (4.5)$$

As can be seen in Figure 4.2, this equation is valid not only for the phosphates but also for the silicates and acrylates.

It is worth noting that there are two deviations from this linear behavior in the phosphates. One is seen with small ions of high charge, such as magnesium. Note the horizontal section of the phosphate curve in Figure 4.2 at 520°C, which represents the limiting case in which the phosphorus–oxygen bonds become the labile moieties. In these materials, the strength of the ionic interaction becomes higher than the phosphorus–oxygen bond strength, and the glass transition mechanism changes from ion hopping to bond interchange.

The other deviation is seen with the transition metal ions, such as cadmium. Different types of bonding (e.g., coordination) may be encountered with transition metal ions, which make the simple approach inapplicable.

It should be pointed out that the three curves in Figure 4.2 are approximately parallel. The curve for the silicates lies at a considerably higher temperature than do the others, which almost coincide, at least over the narrow q/a range for which data for the acrylates are available. The relative positions of the curves are not unexpected, because the silicate repeat unit has two negative charges, whereas the phosphate and acrylate repeat units have only a single negative charge.

4.3. TWO GLASS TRANSITIONS

The best way to demonstrate the existence of two T_g values in many ionomers is to show typical examples of properties as a function of temperature. A wide range of techniques is available for determining T_g values, including differential scanning calorimetry (DSC), differential thermal analysis (DTA), expansion coefficient measurements, and dynamic mechanical measurements.

4.3.1. Two Loss Tangent Peaks

In modulus or loss tangent versus temperature plots, the glass transition is accompanied by a steep descent of the modulus over a narrow temperature region and the appearance of a pronounced loss tangent maximum. Although a detailed discussion of the viscoelastic properties of ionomers will begin in Chapter 5, it is useful here to consider the temperature dependence of the modulus and the loss tangent for a series of poly(styrene-co-sodium methacrylate) [P(S-co-MANa)] ionomers of varying ion contents.

The plots of the modulus and loss tangent against temperature are shown in Figure 1.2 (5). In the modulus and loss tangent plots, the curves for polystyrene (PS) are typical of a normal thermoplastic. The modulus drops by three orders of magnitude over a temperature range of 30°C, and the loss tangent goes through a maximum, reaching a value of 3, and then drops rapidly as the temperature is increased. The plots for the 21.6 mol % sample are similar, except that the drop in the modulus is not quite as steep and the loss tangent curve, while not quite as high, is somewhat broader, giving a similar value for its integrated area. These plots are characteristic of the glass transition and show activation energies (as determined from an Arrhenius plot of peak positions obtained at different frequencies) of approximately 540 kJ/mol for PS and 590 kJ/mol for the 21.6 mol % ionomer (5).

The behavior at intermediate compositions is much more interesting. For example, the 4.5 and 10.7 mol % samples show clearly that there is a two-step descent in the modulus and that there are two peaks in the loss tangent. The activation energies of the peaks for the 4.5 mol % sample are 550 and 220 kJ/mol, in order of increasing temperature; those for 10.7 mol % sample are 630 and 370 kJ/mol. These values are typical of activation energies for glass transitions, which suggest that two glass transitions occur in these materials.

4.3.2. Two Glass Transitions by DSC

In view of the preceding discussion, one would also expect to see two glass transitions by DSC. In the ionomer field, DSC has not, in general, proven to be a preferred method of determining both glass transitions, because phase dimension problems (i.e., sizes of the separated phase) tend to interfere. As discussed in Chapter 3, the sizes of the phase-separated regions may be quite small and highly irregular for some compositions. While two glass transitions are seen in ionomers when determined mechanically, DSC has not been particularly useful in this connection, primarily because the size requirement for the detection of the glass transitions appears to be different for these two methods.

Nonetheless, it has been possible to detect the presence of two T_g values by DSC in a number of cases, including a 40% neutralized poly(styrene-co-methacrylic acid)

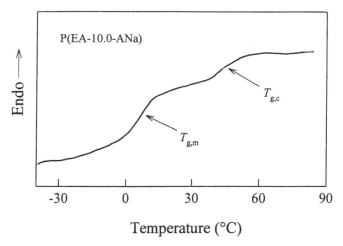

Figure 4.3. A DSC thermogram for P(EA-10.0-ANa). Modified from Tong and Bazuim (8).

sample with a acid comonomer content of 18 mol % (6), the sulfonated-ethylene-propylene-ethylidene norbonene (S-EPDM) system (7), the poly(ethyl acrylate-*co*-sodium acrylate) ionomer system (8), and the *n*-butyl acrylate-zwitterionomer system (9). The results of a DSC study are shown in Fig 4.3 (8). Recently, two glass transitions were detected in the P(S-*co*-MANa) system for which dynamic mechani-

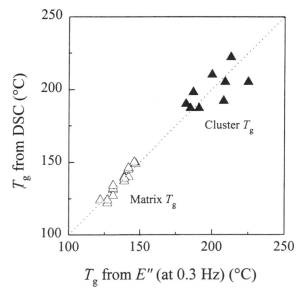

Figure 4.4. Comparison of the glass transition temperatures obtained from E'' (at 0.3 Hz) and DSC for matrix and cluster regions. Modified from Kim et al. (5).

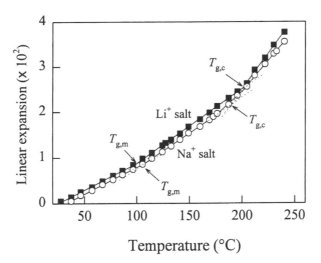

Figure 4.5. Change in relative length of sample versus temperature for lithium (■) and sodium (○) salts of perfluorosulfonate sheets dried under vacuum at 160°C. Modified from Takamatsu and Eisenberg (10).

cal data are also available (5). Thus a direct comparison can be made of the peak positions obtained by dynamic mechanical measurements with the characteristic features of DSC thermograms indicating the glass transitions. This comparison is shown for the P(S-*co*-MANa) system in Figure 4.4, a one-to-one relationship is clear. It should be pointed out that for zwitterionomers based on *n*-butyl acrylate, it is easy to detect the cluster T_g by DSC (9).

4.3.3. Expansion Coefficients

Changes in the values of expansion coefficients have been used frequently to determine the position of the T_g, especially before the advent of DSC or DTA. One such determination was made for the ionomers, specifically the perfluorosulfonates (10) (Fig. 4.5). It is clear that two kinks in the length–temperature plots occur, indicating two glass transitions.

A number of other arguments can be advanced to show that both features are indeed glass transitions. Many of these arguments are based on details of the dynamic mechanical property studies, plasticization studies, and other topics that will be discussed later in the text.

4.4. MATRIX GLASS TRANSITION

4.4.1. The Change/Distance Effect

An extensive study of T_g as a function of ion concentration was performed by DSC for the ethyl acrylate-*co*-acrylic acid ionomers using several alkali and alkaline earth counterions (11). For any counterion, it was found that a sigmoidal curve was obtained in the T_g versus ion concentration plot, which consists of an initially linear segment that is followed by a relatively more rapid increase in T_g at intermediate ion concentrations and, finally, at the highest concentrations, another region with a relatively slow but linear increase in T_g with concentration. Two such plots are shown in Figure 4.6 for the Cs^+ ion and for the Ca^{2+} ion. When the data for the samples are plotted not as q/a, but as cq/a, it is found that all the curves are superimposable (Fig. 4.7) Sigmoidal behavior was also found for sodium-sulfonated polyaryletheretherketone (PEEK) (12).

The sigmoidal behavior arises in heterogeneous polymer systems because DSC experiments integrate over a region of space that may be larger than a single region of the matrix or clustered material (discussed later in this chapter). Thus it is the dominant phase, which is reflected in the graph, if the sizes of the phase-separated regions are below the critical dimension for independent detection by DSC. Because of the possible ambiguity, this section focuses on the early part of the curve, i.e., the initial rise in T_g at low ion contents.

Unlike ethyl acrylate ionomers, some materials do not exhibit the q/a dependence of the T_g, even at very low ion content. Prime examples are the styrene-based ionomers, in which the T_g has been found to rise with ion concentration at approximately the same rate, independent of the nature of the pendent ion or the counterion (13,14). Figure 4.8 shows T_g as a function of ion concentration for styrene-*co*-styrenesulfonate copolymers (13), and Table 4.2, lists dT_g/dc values for a wide range of styrene-based ionomers containing different pendent ions or counterions. It is noteworthy that the values of the dT_g/dc are identical within experimental error for

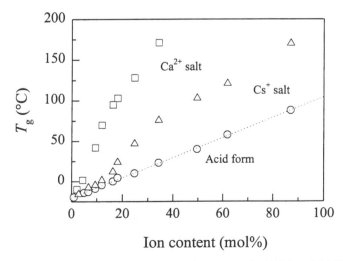

Figure 4.6. Glass transition temperatures versus ion content for P(EA-*co*-MAA) systems of various MAA contents and its completely neutralized Cs^+ and Ca^{2+} form. Modified from Matsuura and Eisenberg (11).

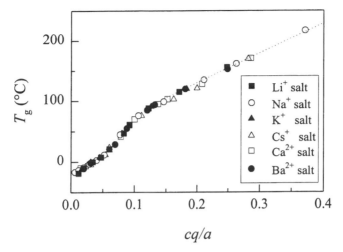

Figure 4.7. Glass transition temperatures versus ion content plotted against *cq/a* using the same data as Figure 4.6 plus data from other cations. ■ Li⁺ salt; ○ Na⁺ salt; ▲, K⁺ salt; △, Cs⁺ salt; □, Ca²⁺ salt; ●, Ba²⁺ salt. Adapted from Matsuura and Eisenberg (11).

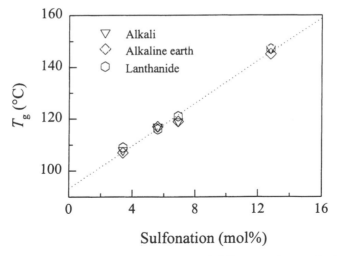

Figure 4.8. Matrix glass transition temperatures (by DSC) versus degree of sulfonation for styrene ionomers containing counterions of various families. ▽, alkali; ◇, alkaline earth; ○, lanthanide. Modified from Yang et al. (13).

TABLE 4.2. Change in Glass Transition Temperature as a Function of Ion Concentration for Styrene-Based Ionomers

Type of Comonomer Unit	dT_g/dc	Reference
4-Styrenesulfonate	2.8	14
	3.0	15
	3.5	13
	3.8	16
4-Styrenecarboxylate	3.3	14
	3.9	17
Methacrylate	3.5	14
	3.3	5
	3.8	16
Acrylate	3.2	18
	3.6	19
4-Vinylpyridinium	3.5	20
4-Phenoxystyrene	2.9	21
4-Benzyloxystyrene	2.4	21
4-C_2 alkyl styrene	3.6	22
4-C_6 alkyl styrene	4.1	22
4-C_{11} alkyl styrene	4.9	22
4-C_2 alkyl ether styrene	4.7	22
4-C_5 alkyl ether styrene	3.2	22
4-C_{11} alkyl ether styrene	4.8	22

the styrenesulfonate anion, the styrenecarboxylate anion, (meth)acrylate anion, and vinylpyridinium cation. On the other hand, values for the alkyl styrenes and alkyl ether styrenes (the combs) range, within that one family, from 3.2 to 4.9. Thus we conclude that, while there is a considerable amount of scatter for the various series, the slope is essentially independent of the ion for this broad range of materials, although it appears that the hydroxystyrene and methoxystyrene values are in fact lower than the others.

It is of interest to ask why only some polymers show a q/a effect; unfortunately, no precise answer can be given. One possibility, however, involves ion hopping. As will be seen in Chapter 5, the temperature of the cluster glass transition is related to ion hopping. The q/a effect is seen, generally, in those situations in which ion hopping participates in the glass transition. In the phosphates, the T_g is determined by the strength of the interactions between anions and cations. For the glass transition to take place, anions must be removed from the coordination spheres of cations at some constant rate, which is a function of the strength of the interaction. This idea immediately implies that a q/a dependence should be observed. By contrast, in the styrene ionomers this behavior is not seen for the matrix T_g, possibly because of the low dielectric constant of the material. The T_g is determined only by the presence of ionic interactions, which act as cross-links (or a filler) that persist far above the matrix T_g. Thus, in systems in which the q/a effect is not observed, the multiplets probably remain intact throughout the matrix glass transition range. Data for the cluster glass transition have not yet been accumulated for a wide enough range of counterions to see whether the q/a relation is obeyed.

In the ethyl acrylate system (and possibly other systems in which a q/a effect is operative for the matrix glass transition) ion hopping may be a contributing factor to the low temperature glass transition even at relatively low ion concentrations. It should be recalled that ethyl acrylate has a dielectric constant that is considerably higher than that of styrene, which weakens the interactions between ion pairs. This explanation is rather speculative. In addition, phase inversion in the ethyl acrylate ionomers does not occur below an ion content of ~14 $\pm$ 2 mol %, which means that, at comparable ion contents, the matrix has a much higher concentration of isolated multiplets than is found in the styrene ionomers, which may make the contribution of ion hopping more important and thus more easily detectable.

4.4.2. Glass Transition Versus Percent Neutralization

Only a few detailed studies have been performed on the effect of the degree of neutralization on the matrix glass transition. Ogura et al. (23) investigated the glass transition of styrene-based ionomers containing 20 mol % of methacrylic acid neutralized to varying extents with NaOH. An interesting aspect of their paper is that they employed infrared (IR) spectroscopy to determine the T_g, observing the temperature dependence of the peak at 1700 cm^{-1} (the dimeric carbonyl stretching vibration) and 1745 cm^{-1} (the monomeric carbonyl stretching vibration). This method found the same T_g as was determined by DSC. The increase in T_g is linear, ranging from 117 (by DSC) and 120°C (by IR) for the unneutralized material to 158 (by DSC) and 162°C (by IR) for the 90% neutralized material. The behavior is quite different for low ion contents. For example, we (24) found that the T_g barely changed as a function of the degree of neutralization for a styrene sample containing 5 mol % of methacrylic acid partly neutralized with Na$^+$. A similar result was obtained by Connolly (25) for ethylene-based ionomers (also of low ion content). It is not surprising that T_g does not change dramatically with percent neutralization at low ion concentration; because even for complete neutralization, the rate of change of the glass transition with ion content is only 3°C/mol % for P(S-co/-MANa) copolymers. Thus the completely neutralized 5 mol % sample shows a T_g increase of only ~ 15°C. Because the entire range of T_g values expected for 0–100% neutralization is only 15°C, this partial neutralization does not change the T_g drastically, especially for low degrees of neutralization.

4.4.3. Copolymerization, Cross-Linking, and Filler Effects

In the vast majority of cases, the T_g of ionomers increases with ion content. Ionic interactions are usually strong in media of low dielectric constant, and the T_g of ionic homopolymers is usually high (26). Therefore, it is of interest to ask whether the increase in T_g with increasing ion content is a copolymerization effect, a cross-linking effect, or a filler effect. All these effects are interrelated. Formally, the question of copolymerization versus cross-linking, which will be tackled first, has a parallel in nonionic polymers. If one copolymerizes a monomer such as styrene with a cross-linking agent such as divinylbenzene, it is known that the T_g increases. Does it increase because of a copolymerization effect or because of a cross-linking effect? The question arises because the divinylbenzene homopolymer has a higher T_g than does polystyrene. It is known that ion pairs like those in sodium methacrylate associate strongly to give multiplets. These multiplets, because of the limitation that they impose on segmental mobility of the polymer backbone, increase the T_g. Therefore, formally, an interpretation of the behavior either as a cross-linking or as a copolymerization effect is not unreasonable; however, it is much more helpful to think of it as a cross-linking effect.

An illustration of this cross-linking effect can be seen for styrene ionomers, in which dT_g/dc is essentially independent of the nature of the ion pair. It is unlikely that the T_g of all the ionic homopolymers should be equal. However, cross-linking would explain the relationship of dT_g/dc. Duchesne (27) compared the relative magnitudes of the copolymerization effect to the cross-linking effect on T_g for the ethyl acrylate-*co*-vinylpyridinium copolymers. He found that the cross-linking effect accounts better for the observed increase in the T_g. Duchesne also pointed out that copolymers of ethyl acrylate with nonionic 4-vinylpyridine (which is not associated and has approximately the same structure as N-methyl-4-vinylpyridinium iodide) give a different variation of T_g with 4-vinylpyridine content (Fig. 4.9a). Specifically, the increase in the T_g with increasing 4-vinylpyridine content was much smaller for a nonquaternized sample than for the ionomer. Figure 4.9b shows the ΔT_g versus 4-vinylpyridine content plots for ionomers and for the same ionomers after subtraction of the copolymerization effect. The value for ΔT_g was obtained by subtracting the contribution of the copolymerization effect (1.6°C/mol %) from the experimental data. In addition, in the calculation of cross-linking density for the ionomers, it was assumed that ionic cross-links consist of two ion pairs. The ΔT_g for an ethyl acrylate polymer cross-linked with ethylene glycol dimethacrylate (EGDMA) is shown in Figure 4.9b for comparison. It was found that the calculated points for the ionic cross-links lie on the same line as do those of the chemically cross-linked system, suggesting that the ionic cross-links are as efficient as the covalent cross-links in raising T_g.

Duchesne (27) also studied the effect of the copolymerization of 2-methyl-5-vinylpyridine with ethyl acrylate on T_g. He showed that the T_g of the nonionic material increases at a rate of approximately 1.3°C/mol %. For the ionized systems, the increase in T_g as a function of ion content was the same as for the nonionic system up to ~6 mol %. Above that, however, the increase was much more rapid (5°C/mol %), which is consistent with that normally observed for other ionic comonomers. This behavior was not observed in the 4-vinylpyridinium case. The difference is instructive. The ionic segment of the 2-methyl-5-vinylpyridinium ion is much less exposed than that of the 4-vinylpyridinium system. Therefore, a higher entropic penalty must be paid for such ion pairs to align properly for maximum interaction. For this reason, at low ion contents the interaction is hindered, and the T_g increases at a rate similar to that of the nonionic copolymer system. By contrast, in the

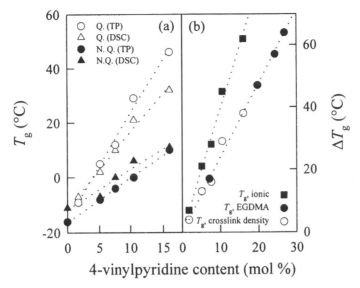

Figure 4.9. **a**, Glass transition temperatures by DSC or torsion pendulum versus functional group content for P(EA-*co*-4VP) copolymers. ○, quaternized by methyl iodide (torsion pendulum); △, quaternized by methyl iodide (DSC); ●, unquaternized (torsion pendulum); ▲, unquaternized (DSC). **b**, Glass transition temperature difference between linear and cross-linked polymers for same matrix. ■, ionomers; ○, the same ionomers after subtraction of the copolymer effect; ● ethylene glycol dimethacrylate. Modified from Duchesne (27).

4-vinylpyridinium case, with ionic segments that are much more exposed, the cross-linking effect is apparent immediately. Because the copolymerization effect for these two ionic systems should be identical, we can conclude that the cross-linking effect is the correct explanation.

We now turn to the filler effect. In Chapter 5, we demonstrate that the level of the ionic modulus can be understood in terms of a filler effect. Therefore, it is reasonable to ask if the increase in the T_g might also be the result of a filler effect, with the clusters acting as the filler. Unfortunately, a quantitative answer to this question is not possible, because the effect of fillers on the T_g is not understood quantitatively for small particle sizes (28–31).

Clearly, all three effects—copolymerization, cross-linking, and filler (which are intimately related)—can lead to an increase in T_g. Cross-linking probably gives the most reasonable explanation. However, the filler effect should be borne in mind, at least in some concentration ranges.

4.4.4. Change in Glass Transition as a Function of Change in Ion Concentration

Changing the ion concentration clearly provides an effective way of changing T_g. It is, therefore, of interest to compare the relative effectiveness of different ions.

TABLE 4.3. Rate of Change in T_g as a Function of Ion Concentration for Various Ionomers

Ionomer	dT_g/dc, °C/mol %	Reference
Poly(butyl acrylate-*co*-N-methyl-4-vinylpyridinium iodide)	1.0	32
Poly(sulfone-*co*-sodium sulfonate) (<12 mol %)	2.0	33, 34
Poly(ethyl acrylate-*co*-sodium acrylate)	3.4	11
Zinc-sulfonated EPDM	5.0	35
Poly(butadiene-*co*-lithium methacrylate)	5.4	36
Poly(ethylene-*co*-sodium methacrylate)	5.7	37
Poly(butadiene-*co*-N-methyl-4-vinylpyridinium iodide)	8.9	36
Poly(ethylene-*co*-magnesium methacrylate)	9.7	37

Perhaps the best way to explore this effect is by looking at values of the rate of change of T_g with ion concentration dT_g/dc in the limit of 0% ion content (Table 4.3). In general, the lower the T_g of the parent polymer, the greater the effect of ion incorporation for the same ion pair. This effect must be combined with the q/a effect (when appropriate) for attempting even a qualitative prediction of the effect of both the type of ion and the ion concentration on the glass transition. Figures 4.6 and 4.7 illustrate the combination of the two effects, by showing the results for a range of cations from Cs^+ to Ca^{2+} for several ethyl acrylate-*co*-acrylic acid copolymers.

The polymers listed in Table 4.3 mostly contain anionic pendent groups. However, cationic pendent ions also provide the opportunity for major variations in T_g. One example is given by Wollmann et al. (38) of glass transition relations in ionomeric "combs" of styrene-*co*-4-vinylpyridinium quaternized with an iodoalkane. Glass transition was studied as a function of both ion concentration and the length of the alkyl chain. An empirical expression was found, which takes into account both of these factors:

$$T_g = 106.5 + c(3.5 - 0.25n) \qquad (4.6)$$

where n is the number of carbon atoms in the iodoalkane chain. This equation was found to work for ionomers quaternized with alkyl chains of 4–10 carbon atoms and for c values up to 10 mol %.

4.4.5. Ammonium Counterions

Ammonium ions can act as counterions for pendent anionic groups. Weiss et al. (15) investigated sulfonated polystyrenes with ammonium counterions of a wide range of structures. They showed that, although the T_g values of the zinc and ammonium salts demonstrate the same behavior as a function of concentration, the alkyl amines are different in that, on occasion, the T_g even drops appreciably. For example, it was found that the monosubstituted, disubstituted, and trisubstituted amines of a relatively short alkyl chain length show trends of T_g versus ion content that are independent of the type of alkyl chain. However, for the longer alkyl chain lengths, T_g is definitely a function of the chain length at constant ion concentration. Thus, although the T_g of the butylamine-containing material at ~22 mol % is ~155°C, that of material containing the 20 carbon salt is ~60°C.

Behavior of the disubstituted and trisubstituted materials is generally similar, except that the deviations from "universal" behavior are seen for shorter and shorter chain lengths with increasing numbers of alkyl substituents. For example, the trisubstituted ammonium salt already shows dramatically different behavior for the butyl and methyl substituents. For the butyl substituent, the T_g does not change at all with ion concentration if the counterion of the sulfonated polystyrene is a tributylammonium counterion. For longer alkyl chains, T_g drops dramatically, as before. For example, for the 22 mol % sulfonated sample with a trisubstituted 12 carbon chain ammonium counterion, T_g is around 20°C. Finally, the dT_g/dc values for the primary, secondary, and tertiary ammonium ions are 2.9, 2.8, and 2.7°C/mol % of ions, respectively.

In a follow-up study, Smith and Eisenberg (39) showed that the T_g of the substituted amine salts could be raised considerably if the amine consisted of a large rigid ring system. For example, while the unneutralized polymer of 8.1 mol % poly (styrene-co-styrenesulfonic acid) has a T_g of 122°C, the 1-adamantanamine neutralized material has a T_g of 137°C. While flexible side chains act as an internal plasticizer, rigid substituents behave in the opposite way.

Fan and Bazuin (40) studied the T_g values by DSC of poly(styrene-co-styrenesulfonic acid) (5.0 mol % of acid groups) neutralized with various bifunctional or multifunctional amines or pyridines. They found that there is no significant effect of the basicity of the organic molecule on the matrix T_g values, which were all found to be 112 ± 2°C. They concluded that the ion content is too low to observe a major effect of the basicity on the matrix glass transition. In summary, ammonium counterions provide a convenient way of varying the T_g of ionomers with pendent anionic groups.

4.4.6. Internal Plasticization Effect

Glass transition temperature can also be varied at a constant ion concentration using the internal plasticizer effect. For such a study, two series of styrene ionomers were prepared that contained phenyl rings with attached alkyl side chains or alkyl ether side chains in the *para* position, the ions being found at the ends of the side chains (22). In this family of materials, T_g increases with increasing ion concentration at approximately the same rate as it does in other styrene ionomers, i.e., at ~4.2°C/mol %. A resemblance to the comb system studied by Wollmann et al. (38) is apparent; and indeed, a similar equation can be developed for the alkyl side chain ionomers:

$$T_g = 124 - 4n + c(3.3 + 0.13n) \tag{4.7}$$

where n is the number of carbon atoms in the alkyl side chain. "Classical" internal plasticization is also possible, i.e., the attachment of nonionic alkyl side chains to the phenyl ring of styrene. An example of that type of plasticization has also been reported (41)(Scheme **4.1**).

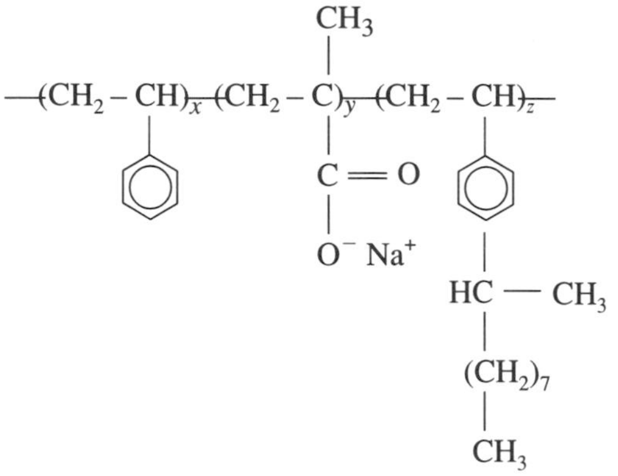

$$
\begin{array}{c}
\text{CH}_3 \\
|\\
-\!\!(\text{CH}_2-\text{CH})_{\overline{x}}\!\!-\!\!(\text{CH}_2-\text{C})_{\overline{y}}\!\!-\!\!(\text{CH}_2-\text{CH})_{\overline{z}}\!- \\
\end{array}
$$

$$
\begin{array}{ccc}
\bigcirc && \text{C}=\text{O} \quad\quad \bigcirc\\
&& | \\
&& \text{O}^- \text{Na}^+ \quad | \\
&& \qquad\qquad \text{HC}-\text{CH}_3 \\
&& \qquad\qquad | \\
&& \qquad\qquad (\text{CH}_2)_7 \\
&& \qquad\qquad | \\
&& \qquad\qquad \text{CH}_3 \\
\end{array}
$$

Scheme 4.1

4.4.7. Semicrystalline Materials

The presence of crystallinity in partly crystalline ionomers, especially at high levels, complicates the determination of the glass transition. This consideration is especially true in materials based on ethylene, in which the degree of crystallinity strongly affects the dynamic mechanical peaks associated with the glass transitions. In the ethylene-based ionomers, it was found that upon neutralization a new peak (labeled β), appears at $\sim -10°C$ in the loss tangent or G'' plots. MacKnight et al. (42) observed a slight increase in peak height but no change in peak position with increasing degree of neutralization. They assigned the β peak to a relaxation in the amorphous branched polyethylene phase, from which most of the ionic components have been excluded. However, Longworth and Vaughan (43) observed that the peak shifts to lower temperatures and that its height increases with increasing ion content. Furthermore, Otocka and Kwei (37) found that the glass transition in ethylene-based ionomers is normal, because the height of the loss tangent peak (related to the glass transition) moves to higher temperatures with increasing ion concentration and decreases in intensity, which is typical of a wide range of ionomers. The increase in the T_g with concentration of ionic materials is 5.7°C/mol % (Table 4.3). Earnest and MacKnight (44) found that as the degree of crystallinity increases, the temperature of the β relaxation increases. A review of the glass transition in the ethylene- and polypentenamer-based ionomers is available (45). The T_g for other specific families of materials will be discussed more extensively in connection with the mechanical properties of ionomers based on materials other than styrene (Chapters 6 and 7).

4.4.8. Theories

A number of attempts have been made to develop theories or models of the glass transition in ion-containing polymers. While a detailed discussion of these is beyond the scope of this book, references are made briefly to three theories that have appeared since the mid 1970s. While the cross-linking effect has also received attention from theorists, it is not included here because it does not refer specifically to ion-containing polymers. Becker (46) developed an additivity rule for the calculation of T_g values of homopolymers or copolymers containing ionic groups:

$$T_g = \frac{Y + \sum Z_i q_i K_i}{Z} \tag{4.8}$$

where Y is the molar T_g function, Z is the sum of the atomic distances of the parent polymer, Z_i is the number of counterions i, q_i is their charge, and K_i is the characteristic increment of the counterion derived from another relation. The equation was applied to phosphates, silicates, acrylates, methacrylates, and some copolymers, with generally good agreement.

Subsequently, Tsutsui and Tanaka (47) and Peiffer (48) used models based on cohesive energy density (CED) approaches. Tsutsui and Tanaka (47) calculated the CEDs of ionic polymers with the assumption that they can be approximated by the electrostatic energy of the system. In this model T_g is given by:

$$T_g = K_1(\text{CED})^{1/2} \tag{4.9}$$

K_1 is a constant that depends on the polymer system. Reasonably good agreement was obtained for phosphates, acrylates, and ionenes. Peiffer (48) proposed

$$T_g = \frac{2N_A e^2}{3CR}(q/a) + \frac{300}{R} \tag{4.10}$$

where N_A is Avogadro's number, e is the electronic charge, C is a constant, and R is the gas constant. It should be stressed that these equations are applied either to completely ionic homopolymers or to those copolymers in which a q/a effect is operative (i.e., to systems in which ion hopping may be an important component of the glass transition).

4.5. CLUSTER GLASS TRANSITION

4.5.1 Effect of Immobilization

Far fewer investigations of the cluster T_g have been undertaken than of the matrix T_g. The most thoroughly studied system is P(S-co-MANa). The T_g values are shown as a function of ion content (determined from tan δ peak positions at 1 Hz) in Figure 4.10 (5). The difference between two T_g values is $>60°C$, and it increases with ion concentration. Also, the cluster T_g shows some curvature with ion content, the best fit being

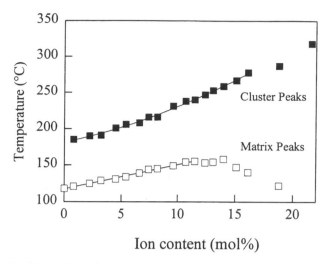

Figure 4.10. Positions of tan δ peaks for cluster (■) and matrix (□) glass transitions for P(S-*co*-MANa) versus ion content. Modified from Kim et al. (5).

$$T_g = 181 + 3.3c + 0.16c^2 \qquad (r^2 = 0.998) \qquad (4.11)$$

This result contrasts with the matrix T_g, which depends linearly on c. At 21 mol %, the cluster T_g is ~320°C. The T_g of pure poly(sodium methacrylate) is 310°C (26). This value was estimated by extrapolation of the T_g of P(S-*co*-MANa) to pure poly(sodium methacrylate) (37). Another estimate of T_g of poly(sodium methacrylate) can be obtained by taking the T_g of poly(sodium acrylate) as 250°C (3) and adding 100°C, the difference of the T_g values of methyl methacrylate (105°C) and methyl acrylate (5°C). By this approach, one obtains a value of 350°C. Although neither of these estimates is exact, we can be reasonably sure that the T_g of pure poly(sodium methacrylate) (either ~310 or ~350°C) is close to that of the 21 mol % ionomer (~320°C), which is surprising. If one views these phenomena from the point of view of the morphology, as discussed in Chapter 3, it is clear that the immobilization effect of the ionic aggregates is considerable.

Because the cluster T_g of the 21 mol % ionomer is close to that of pure poly(sodium methacrylate), it appears that the T_g versus ion content plot must be quite level between 21 and 100 mol % of sodium methacrylate. A similar leveling effect has been observed in the sulfonated styrene–ethylene–butylene terpolymers (49). The cluster glass transition levels at ~8 mol % at a value of ~230°C. The fact that this leveling occurs at a relatively low ion concentration may be related to a possible segregation phenomenon. All sulfonate groups are attached to phenyl rings. Therefore, because multiplets consist of sodium sulfonate ion pairs, the multiplet must be surrounded by phenyl rings. This region immediately surrounding a multiplet would, therefore, be depleted of backbone material, i.e., the ethylene–butylene chains. By contrast, the next shell beyond the phenyl rings would probably be enriched with ethylene–butylene segments. Thus styrene segments, which make up the high T_g component, are effectively segregated from the low T_g component, which contains the aliphatic chain segments. This segregation effect possibly results in an enlargement of the cluster regions without accompanying changes in the detailed arrangement of the ions in the cluster.

4.5.2. Differences in the Two Glass Transition Temperature Values

A comparison was made between the positions of the two T_g values in several styrene-based ionomers and in the poly(vinylcyclohexane-*co*-sodium acrylate) [P(VCH-*co*-ANa)] based system by Kim et al. (19). Figure 4.11 shows that at relatively low ion contents, the T_g values vary with the ion content in a similar manner. For the P(S-*co*-MANa) case, this trend extends from an ion content of 2–10 mol % (15–85% cluster volume fraction). The relative volume fractions of clusters and multiplets will be discussed in Chapter 5. When the two T_g values are plotted against each other, a slope of 1 is obtained for all the materials mentioned above (see Fig. 4.12). The T_g values for some of the materials were taken from both E'' peaks and tan δ peaks. A linear relation between two T_g values was also observed by Tomita and Register (17) for the zinc-neutralized *p*-carboxylated polystyrene system and

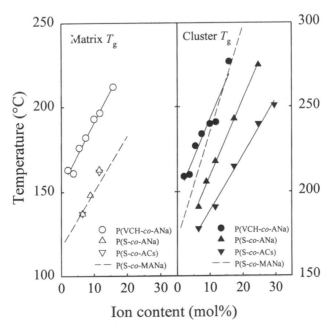

Figure 4.11. Tan δ peak positions (1 Hz); matrix and cluster glass transitions for random ionomers versus ion content. ○ and ●, P(VCH-*co*-ANa); △ and ▲, P(S-*co*-ANa); ▽ and ▼, P(S-*co*-ACs); − −, P(S-*co*-MANa). Modified from Kim et al. (19).

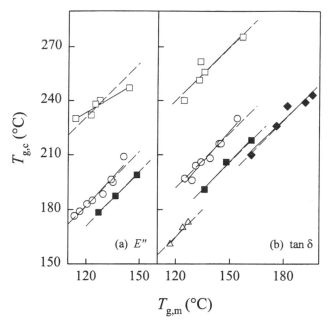

Figure 4.12. Cluster glass transition temperatures versus matrix glass transition temperatures from (**a**) E'' and (**b**) tan δ peak position for various ionomers. ■, P(S-*co*-ANa); ◆, P(VCH-*co*-ANa); □, P(S-*co*-SSNa); △, P(S-*co*-SCNa); and ○, P(S-*co*-MANa). Modified from Kim et al. (19).

by Dulac and Bazuin (50) for the sulfonated polystyrene ionomers neutralized with bifunctional organic cations.

The slope of 1 means that the T_g values of the two phases increase at a similar rate as a function of ion content. This result is not surprising when one recalls that the two materials that give rise to glass transitions are essentially the same polymer, differing only in the degree of immobilization induced by the multiplets. The parallelism breaks down, however, at high ion contents, i.e., for a cluster volume fraction >0.85.

The differences in the T_g values are instructive. The difference between cluster and matrix glass transitions is ~120°C for the poly(styrene-*co*-sodium styrenesulfonate) [P(S-*co*-SSNa)] system. This value is much higher than in carboxylate systems; for the styrene-*co*-p-carboxystyrene [P(S-*co*-SCNa)] ionomers, the difference is ~ 45°C, reflecting much weaker interactions. For other styrene-based ionomers, P(S-*co*-MANa) has a difference of ~70°C; styrene-*co*-sodium acrylate [P(S-*co*-ANa), ~55°C; and P(VCH-*co*-ANa), ~50°C. The differences between P(S-*co*-MANa), P(S-*co*-ANa), and P(VCH-*co*-ANa) have been ascribed to differences in multiplet size (19).

The absolute value of the cluster T_g depends not only on the nature of the ion pairs but also, to some extent, on the glass transition of the matrix. In one study of the dynamic mechanical properties of partly sulfonated styrene–ethylene–butylene copolymers, it was found that both matrix and cluster glass transitions drop appreciably below those of the pure polystyrene ionomer (49). For example, at 5% sulfonation of a 45 wt % styrene–ethylene–butylene copolymer (45 is the wt % of styrene in the styrene–butadiene copolymer), it was found that the matrix T_g was ~0°C, while that of the cluster was 180°C; but at 9% sulfonation the matrix T_g was ~10°C, while that of the cluster was 230°C. Thus it is clear that the strength of interaction of the ion pairs within the multiplets determines T_g and that the T_g of the matrix polymer, the chains of which exert an elastic force on the multiplets, is also important. This aspect will be discussed more extensively in connection with plasticization (Chapter 8).

4.5.3. Homoblends

A significant development was Douglas et al.'s (51) finding of a cluster T_g in homoblends consisting of mixtures of sulfonated polystyrene and poly(styrene-*co*-4-vinylpyridine) [P(S-*co*-4VP)]. A cluster peak was also found in blends of poly(styrene-*co*-styrenesulfonic acid) and poly(ethyl acrylate-*co*-4-vinylpyridine) [P(EA-*co*-4VP)]. Ionomer blends will be discussed in Chapter 9; however, a brief discussion of the cluster glass transition in homoblends is in order at this point. In blends of this type, proton transfer from styrenesulfonic acid to 4-vinylpyridine renders one of the chains polycationic and the other chain polyanionic (52), which provides ionic cross-linking. The same behavior is found when zinc-sulfonated polystyrene is mixed with P(S-*co*-4VP) random copolymers. The cluster T_g values found by Douglas et al. (51) for the various types of blends are given in Figure 4.13.

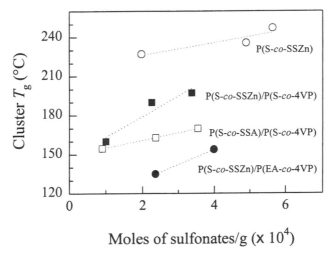

Figure 4.13. Cluster glass transition versus degree of sulfonation for various polystyrene ionomers and blends with PEA-based ionomers. ○, P(S-*co*-SSZn); ■, P(S-*co*-SSZn)/P(S-*co*-4VP); □, P(S-*co*-SSA)/P(S-*co*-4VP); ●, P(S-*co*-SSZn)/P(EA-*co*-4VP). Modified from Douglas et al. (51).

The existence of the cluster tan δ peak in these styrene homoblends shows that the materials are clustered in exactly the same way as the simple ionomers. However, Douglas et al. (51) showed that while the zinc ionomer shows a small-angle x-ray scattering (SAXS) peak, the homoblend containing the zinc ionomer does not. The aggregation behavior of the Zn^{2+} ions in the blend must be different from that in the unblended ionomers. The authors suggested that in the blend the functionality of the cross-links is approximately 8, both for the Zn^{2+}-containing blends and for some of the blends that do not contain zinc ions (i.e., the protonated blends). Application of the kinetic theory of rubber elasticity to partly clustered ionomer systems is probably valid as long as the system is not too far above the percolation threshold (see Chapter 5). Douglas et al. (51) suggest that the effect can be considered as plasticization, since the T_g of the clusters of the zinc ionomer decreases. However, because the amount of clustered material does not change appreciably (Chapter 9), it is a special type of plasticization that is involved here. It must be akin to the effect of nonpolar plasticization but differing from the latter in that the morphology of the aggregates changes drastically but without destroying the existence of clustered material.

The existence of a cluster tan δ peak, combined with the functionality of 8, and the absence of the SAXS peak can be understood within the Eisenberg–Hird–Moore (EHM) model (53). Structures of the type presented in Figure 4.14 exhibit a functionality of 8, because each phenyl group in the multiplet acts as the origin of two chains. Within the EHM model, it seems reasonable to suggest that such a site would act as a multiplet and would reduce the mobility of the styrene chains immediately surrounding it. At an ion concentration of ~7 mol %, the distance between Zn^{2+} ions or between the four ion aggregates formed via proton transfer in the protonated blends (Fig. 4.14b) is on the order of 20 Å. This distance is well within two persistence lengths (2 2 10 Å) between multiplets needed for substantial overlap of regions of restricted mobility. Single Zn^{2+} ions distributed more or less randomly in space would not be expected to give rise to a SAXS peak (which is seen in the

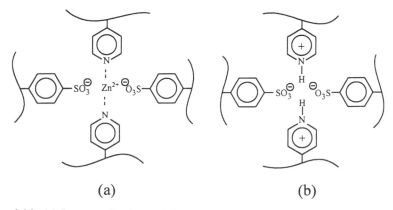

(a)   (b)

Figure 4.14. (a) Ion–coordination and (b) H-bond-assisted ion–ion interactions in blends.

unblended sulfonated styrene ionomers, for which there are much larger aggregates of ionic groups). Thus the absence of a SAXS peak is also consistent with the structure of Figure 4.14.

4.5.4. Unclustered Materials

Some ionomers do not show any clustering behavior, while exhibiting a strong effect of ionic forces on T_g. The poly(sodium metaphosphate) polymers are an excellent example of such behavior, because a drastic increase in T_g is seen if the polymer is ionized. $(HPO_3)_x$ has a T_g of $-10°C$, and the T_g of $(NaPO_3)_x$ is 285°C. However, no evidence of clustering is seen (1); we know that the polyphosphates are unclustered above the glass transition, because the viscoelastic properties of $(NaPO_3)_x$ do not show any indication of an usual ionic modulus but merely a shift of the transition region to higher temperatures. It should be pointed out that this effect may also be a consequence of the fact that $(NaPO_3)_x$ is a homopolymer with a relatively small backbone repeat unit, in which the charge is delocalized.

The styrene-co-N-methyl-4-vinylpyridinium iodide [P(S-co-4VPMeI)] copolymers show similar behavior, in that the log E versus temperature curve shifts to higher temperatures as a function of ion content, without an appreciable change in shape. Here the material has a relatively low dielectric constant and is known to exhibit clustering if the ionic interaction is strong (e.g., sodium carboxylate ion pairs). The behavior of the vinylpyridinium system is consistent with the idea that the multiplets are too unstable to survive above the glass transition. As was pointed earlier, the materials do show an increase in T_g with increasing ion content but no clustering. The large increase in the glass transition suggests that ionic cross-linking exists. However, the viscoelastic properties above T_g show no evidence of aggregation, which suggests that the multiplets fall apart in the region of the glass transition.

4.5.5. Differences in Clustering Behavior of Styrene Ionomers

We now consider the clustering behavior at the glass transition of styrene ionomers with four different ion pairs: sodium sulfonate, sodium carboxylate (methacrylate), pyridinium sulfonate ion pairs produced by proton-transfer in blends, and N-methylvinylpyridinium iodide. The ionic interactions in the multiplets are strongest in the sulfonates. Typical styrene ionomers are sulfonated in the *para* position of the benzene ring. The combination of the strength of the ionic interaction and the position of ionic groups results in the formation of stable and relatively large multiplets (~10–20 ion pairs). The stability of the multiplets is seen in the high value of the cluster T_g. However, the large sizes (and the consequent small number) of multiplets result in a relatively small volume fraction of material located in the regions of restricted mobility. Therefore, the volume fraction of clusters in the sulfonate ionomers at a particular ion content is relatively low compared to the methacrylate system.

The ionic interactions are weaker in P(S-co-MANa) ionomers than in P(S-co-SSNa) ionomers. For this reason, the cluster glass transition occurs at much lower temperatures. The proximity of the sodium carboxylate ion pair to the backbone means that the multiplets are smaller in the sodium methacrylate case than in the sulfonate case (Chapter 3). The extent of clustering (the amount of material of reduced mobility) is, therefore, higher than in the sulfonates at a comparable ion content, i.e., the area under the loss tangent peak connected with the cluster glass transition is larger in methacrylates than in sulfonates (14).

Lefelar and Weiss (54) explored the difference in interaction strengths and ascribed it to the details of the packing of the anion and the cation. Bazuin et al. (55) explored this packing of ions and the details of the differences between Na^+ and Zn^{2+} as related to degree of miscibility between polystyrene ionomers and poly(phenylene oxide).

The styrene homoblends (51) discussed above are subject to similar considerations. The smaller effective multiplet size in the homoblends means that a larger amount of material is involved in cluster formation than in the sulfonates; the volume of the clusters is probably comparable to that in the carboxylates, as are the cluster peak areas. Also, because the ionic interactions are weaker than in the sulfonates, the cluster T_g values are considerably lower.

Finally, in P(S-co-4VPMeI), the ionic interactions are so weak at T_g that the multiplets pull apart and no clustering is observed. One interesting feature of the 4VP ionomers (which will be discussed in detail in Chapter 8) is that the addition of either an internal or an external plasticizer induces clustering. The reason for the appearance of clustering on plasticization seems to be the strength of the ionic interactions relative to kT at T_g. At T_g of P(S-co-4VPMeI) (~ 110–140°C at ion concentrations of 2–10 mol %), there is enough thermal energy kT_g to disrupt the multiplets (20). However, when plasticizers are introduced that depress T_g to the range of 30–70°C, kT_g becomes smaller relative to the strength of the interaction in the multiplets, and clustering is induced (56).

4.5.6. Mechanism of the Cluster Glass Transition

Three possible mechanisms for the cluster glass transition have been suggested (14). In the first, the glass transition occurs under conditions in which both the rigidity and the stability of the multiplets are preserved. As the temperature increases, chain mobility increases and eventually reaches a level consistent with a glass transition. The multiplets remain intact and rigid throughout.

In the second mechanism, the multiplets are stable, but their rigidity decreases with temperature. No ion hopping between multiplets would be observed at the glass transition. The rigidity of the multiplets holding the chains decreases, and this decrease is accompanied by an increase in mobility, which induces the glass transition.

In the third mechanism, ion hopping can occur near the glass transition, leading to a loss of stability of the multiplets and a reduction in rigidity. These features are observed in all the experiments to date. As will be seen in the next chapter, both styrenesulfonates and styrenecarboxylates (or styrenemethacrylates) have been investigated in this context, and ion hopping has, indeed, been found in the temperature range in which the cluster glass transition are observed.

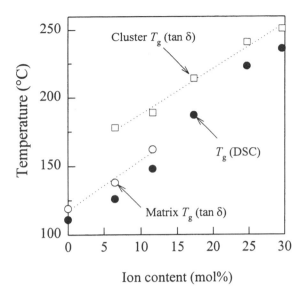

Figure 4.15. Glass transition temperatures versus ion content for P(S-*co*-ANa) ionomers obtained from DSC (●), cluster tan δ peak position (□), and matrix tan δ peak position (○).

As mentioned earlier, the cluster T_g has been successfully observed by DSC in a number of ionomer systems. However, DSC detection is by no means universal for ionomers that exhibit clustering as determined by dynamic mechanical techniques. One possible reason for the different detection efficiencies is the difference in the sensitivity of the various techniques to different domain sizes. Figure 4.15 illustrates the T_g values from dynamic mechanical properties of the matrix regions and the cluster regions of P(S-*co*-ANa) in the range of 2–16 mol % of ions. In that range, DSC shows only one glass transition, which can be taken as either that of the dominant phase or as an average of the two phases. This result suggests that dynamic mechanical tests are sensitive to smaller domain sizes than is DSC. This condition should be borne in mind when interpreting DSC and dynamic mechanical results if discrepancies in the number of transitions are observed.

4.6. REFERENCES

1. Eisenberg, A.; Farb, H.; Cool, L. G. *J. Polym. Sci. A-2* **1966,** *4,* 855–868.

2. Eisenberg, A.; Takahashi, K. *J. Noncryst. Solids* **1970,** *3,* 279–293.

3. Eisenberg, A.; Matsura, H.; Yokoyama, T. *J. Polym. Sci. A-2* **1971,** *9,* 2131–2135.

4. Eisenberg, A. In *Physical Properties of Polymers;* Mark, J. E.; Eisenberg, A.; Graessley, W. W.; Mandelkern, L.; Samulski, E. T.; Koenig, J. L.; Wignall, G. D., Eds.; American Chemical Society: Washington, DC, 1993; Chapter 2.

5. Kim, J.-S.; Jackman, R. J.; Eisenberg, A. *Macromolecules* **1994,** *27,* 2789–2803.

6. Kanamoto, T.; Hatsuya, I.; Ohoi, M.; Tamaka, K. *Makromol. Chem.* **1975**, *176*, 3497–3500.

7. Maurer, J. J. In *Thermal Analysis: Proceedings of the Seventh International Conference on Thermal Analysis*, Vol. 2; Miller, B., Ed.; Wiley: New York, 1982; pp. 1040–1049.

8. Tong, X.; Bazuin, C. G. *Chem. Mater.* **1992**, *4*, 370–377.

9. Ehrmann, M.; Muller, R.; Galin, J.-C.; Bazuin, C. G. *Macromolecules* **1993**, *26*, 4910–4918.

10. Takamatsu, T.; Eisenberg, A. *J. Appl. Polym. Sci.* **1979**, *24*, 2221–2235.

11. Matsuura, H.; Eisenberg, A. *J. Polym. Sci., Polym. Phys. Ed.* **1976**, *14*, 1201–1209.

12. Bailly, C. H.; Leung, L. M.; O'Gara, J.; Williams, D. J.; Karasz, F. E.; MacKnight, W. J. In *Structure and Properties of Ionomers;* Pineri, M.; Eisenberg, A., Eds.; NATO ASI Series, Series C: Mathematical and Physical Sciences 198; Reidel: Dordrecht, 1987; pp. 511–516.

13. Yang, S.; Sun, K.; Risen, W. M. Jr. *J. Polym. Sci. B Polym. Phys.* **1990**, *28*, 1685–1697.

14. Hird, B.; Eisenberg, A. *Macromolecules* **1992**, *25*, 6466–6474.

15. Weiss, R. A.; Agarwal, P. K.; Lundberg, R. D. *J. Appl. Polym. Sci.* **1984**, *29*, 2719–2734.

16. Fan, X. -D.; Bazuin, C. G. *Macromolecules* **1993**, *26*, 2508–2513.

17. Tomita, H.; Register, R. A. *Macromolecules* **1993**, *26*, 2791–2795.

18. Suchocka-Gałaś, K. *Eur. Polym. J.* **1990**, *26*, 1203–1206.

19. Kim, J. -S.; Wu, G.; Eisenberg, A. *Macromolecules* **1994**, *27*, 814–824.

20. Gauthier, S.; Duchesne, D.; Eisenberg, A. *Macromolecules* **1987**, *20*, 753–759.

21. Clas, S. -D.; Eisenberg, A. *J. Polym. Sci. B Polym. Phys.* **1986**, *24*, 2767–2777.

22. Gautheir, M.; Eisenberg, A. *Macromolecules* **1990**, *23*, 2066–2074.

23. Ogura, K.; Sobue, H.; Nakamura, S. *J. Polym. Sci. Polym. Phys. Ed.* **1973**, *11*, 2079–2088.

24. Kim, J.-S.; Eisenberg, A. *J. Polym. Sci. P B Polym. Phys.* **1995**, *33*, 1967–209.

25. Connolly, J. M. Ph.D. Dissertation, University of Massachusetts at Amherst, 1990.

26. Peyser, P. In *Polymer Handbook*; Brandrup, J.; Immergut, E. H., Eds.; 1988 J Wiley: New York, 1988.

27. Duchesne, D. Ph.D. Dissertation, McGill University, 1985.

28. Ecker, R. *Kautsh. Gummi Kunstst.* **1968**, *21*, 304.

29. Yim, A.; Chahal, R. S.; St. Pierre, L. E. *J. Colloid Interface Sci.* **1973**, *43*, 583–590.

30. Landry, C. J. T.; Coltrain, B. K.; Landry, M. R.; Fitzgerald, J. J.; Long, V. K. *Macromolecules* **1993**, *26*, 3702–3712.

31. Tsagaropoulos, G.; Eisenberg, A. *Macromolecules* **1995**, *28*, 6067–6077.

32. Duchesne, D.; Eisenberg, A. *Can. J. Chem.* **1990**, *68*, 1228–1232.

33. Noshay, A.; Robeson, L. M. *J. Appl. Polym. Sci.* **1976**, *20*, 1885–1903.

34. Orzecuimski, M.; MacKnight, W. J. *J. Appl. Polym. Sci.* **1989**, *30*, 4753–4770.

35. Agarwal, P. K.; Makowski, H. S.; Lundberg, R. D. *Macromolecules* **1980**, *13*, 1679–1687.

36. Otocka, E. P.; Eirich, F. R. *J. Polym. Sci. A-2* **1968**, *6*, 921–932.

37. Otocka, E. P.; Kwei, T. K. *Macromolecules* **1968**, *1*, 401–405.

38. Wollmann, D.; Gauthier, S.; Eisenberg, A. *Polym. Eng. Sci.* **1986**, *26*, 1451–1456.

39. Smith, P.; Eisenberg, A. *J. Polym. Sci. B Polym. Phys.* **1988**, *26*, 569–580.

40. Fan, X.-D.; Bazuin, C. G. *Macromolecules* **1995**, *28*, 8216–8223.
41. Gauthier, M.; Eisenberg, A. *Macromolecules* **1989,** *22,* 3751–3755.
42. MacKnight, W. J.; McKenna, L. W.; Read, B. E. *J. Appl. Phys.* **1967,** *38*, 4208–4212.
43. Longworth, R.; Vaughan, D. T. *Polym. Prepr. Am. Chem. Soc. Div. Polym. Chem.* **1968,** *9*, 525–533.
44. Earnest, T. R. Jr.; MacKnight, W. J. *Macromolecules* **1977,** *10*, 206–210.
45. MacKnight, W. J. In *Structure and Properties of Ionomers;* Pineri, M.; Eisenberg, A., Eds., NATO ASI Series C: Mathematical and Physical Sciences 198; Reidel: Dordrecht, 1987; pp 267–277.
46. Becker, R. Z. *Phys. Chemie, Liepzig* **1976,** *4*, S667–S677.
47. Tsutsui, T.; Tanaka, T. *Polymer* **1977,** *18*, 817–821.
48. Peiffer, D. G. *Polymer* **1980,** *21*, 1135–1138.
49. Nishida, M.; Eisenberg, A. *Macromolecules* **1996,** *29*, 1507–1515.
50. Dulac, L.; Bazuin, C. G. *Acta Polym.* **1997,** *48*, 25–29.
51. Douglas, E. P.; Waddon, A. J.; MacKnight, W. J. *Macromolecules* **1994,** *27*, 4344–4352.
52. Smith, P.; Eisenberg, A. *J. Polym. Sci. Polym. Lett. Ed.* **1983,** *21*, 223–230.
53. Eisenberg, A.; Hird, B.; Moore, R. B. *Macromolecules* **1990,** *23*, 4098–4107.
54. Lefelar, J. A.; Weiss, R. A. *Macromolecules* **1984,** *17*, 1145–1148.
55. Bazuin, C. G.; Rancourt, L.; Villeneuve, S.; Soldera, A. *J. Polym. Sci. B Polym. Phys.* **1993,** *31*, 1431–1440.
56. Wollman, D.; Williams, C. E.; Eisenberg, A. *Macromolecules* **1992,** *25*, 6775–6783.

CHAPTER 5

STYRENE IONOMERS

The introduction of ionic groups into a polymer such as styrene changes its mechanical properties, as we have already seen. This effect is one of the reasons why ionomers have become the subject of extensive industrial and academic research. This chapter is devoted to the properties of the styrene ionomers. The starting point is a description of stress–relaxation, dynamic mechanical properties, and melt rheology as functions of ion content in the styrene-sodium methacrylate [P(S-*co*-MANa)] system. This is followed by a description of the effect of changing the position of the ionic group, specifically its distance from the backbone. Next, the effect of changing the ionic group at approximately the same distance from the backbone is considered, along with a discussion of homoblends. The effects of other parameters are then described, e.g., nitration of the backbone, the type of counterion, and molecular weight. Ion hopping is discussed, followed by an examination of physical properties, such as density, expansion coefficients, dielectric properties, various spectroscopic properties, chain orientation, and small molecule transport. Engineering properties—specifically fatigue, fracture, and crazing—are covered, and then time dependence is discussed. Finally, chemical aspects are considered, specifically the stability of styrene ionomers and the reactions that occur in these materials.

Some of the topics discussed in this chapter have been investigated much more fully for nonstyrene-based ionomers, e.g., chemical reactions have been studied much more extensively in the perfluorosulfonates, as have electron spin resonance (ESR), nuclear magnetic resonance (NMR) spectroscopies, and small molecule transport. However, a presentation of similar, though less extensive studies on the styrene ionomers, for which extensive mechanical property data are also available, will give the reader an understanding that can be applied to other ionomers. Even if the study of a particular topic is not the most extensive, the correlation of the results, with the picture of the morphology and mechanical properties presented in the preceding chapters, may improve the overall understanding of structure–property relations. In Chapters 6 and 7, when appropriate, the more detailed studies will either be discussed or referred to.

Even though studies of viscoelastic properties of P(S-*co*-MANa) polymers started in the 1960s (1–3), wherever possible, we present data taken from more recent studies, because the data tend to be more detailed. However, not all of the measurements have been performed recently, and whenever necessary or advisable from a historical point of view, the older literature will also be cited.

5.1. STRESS–RELAXATION OF SODIUM METHACRYLATE SYSTEMS

Eisenberg and Navratil (4,5) performed a detailed study of the stress–relaxation behavior of P(S-*co*-MANa) copolymers in bulk. The results for two ion concentrations are shown in Figures 5.1 and 5.2. Time–temperature superposition is obeyed for the 3.8 mol % sample, but fails for the 7.7 mol % material; therefore, no unambiguous master curve can be constructed for the latter. The failure of time–temperature superposition can be seen from the appearance of tailing segments on the pseudo-master curve constructed by superposition of short time segments. The difficulty of constructing a real master curve is also seen when different sections of the original stress–relaxation curve are shifted to obtain superposition for those specific ranges.

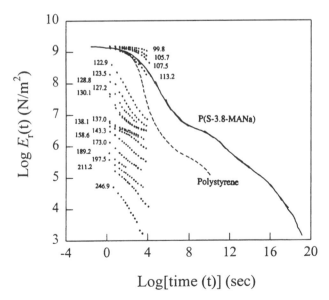

Figure 5.1. Stress relaxation curves (*dotted lines*), the master curve for P(S-3.8-MANa) (*solid line*), and the master curve for pure polystyrene (*dashed line*) T_{ref} equals T_g for each sample; M_n equals 700,000 g/mol for P(S-3.8-MANa) and 230,000 g/mol for polystyrene. Modified from Eisenberg and Navratil (4).

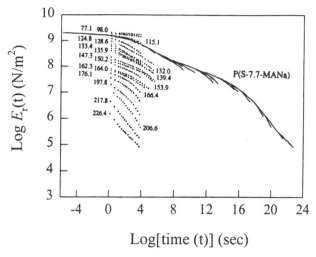

Figure 5.2. Stress relaxation curves (*dotted lines*) and attempted master curve (*solid lines*) for P(S-7.7-MANa). M_n equals 250,000 g/mol; T_{ref} equals T_g for each sample. Modified from Eisenberg and Navratil (4).

Different master curves are obtained, depending on which time segments are picked. Williams–Landel–Ferry (WLF) constants (C_1 and C_2) seem reasonably normal up to about 6 mol %, as shown in Table 5.1 (5). Above that concentration, the values of C_1 and C_2 increase dramatically. We will show later in this chapter that the breakdown of time–temperature superposition coincides with the onset of percolation; therefore, no theoretical significance should be attached to the values of the WLF constants for ion concentrations above the percolation threshold. Because fail-

TABLE 5.1. WLF Constants

Sample	C_1	C_2	Time–Temperature Superposition
PS	14	46	Applicable
P(S-3.8-MAA)	14	46	Applicable
P(S-0.6-MANa)	13	40	Applicable
P(S-1.9-MANa)	19	71	Applicable
P(S-2.5-MANa)	19	71	Applicable
P(S-3.7-MANa)	23	73	Applicable
P(S-3.8-MANa)	23	73	Applicable
P(S-4.6-MANa)	23	73	Applicable
P(S-6.2-MANa)	47	227	Not applicable
P(S-7.7-MANa)	60	240	Not applicable
P(S-7.9-MANa)	47	227	Not applicable
P(S-9.7-MANa)	66	300	Not applicable

ure of time–temperature superposition is seen in a wide range of ionomers, at least in some concentration and temperature ranges, it seems that the phenomena observed here may be quite general.

5.2. DYNAMIC MECHANICAL STUDIES OF SODIUM METHACRYLATE SYSTEMS

Dynamic mechanical studies of P(S-*co*-MANa) copolymers date back 30 years (1–3,5). A recent reinvestigation (6) has produced a detailed picture, which will be summarized here.

5.2.1. Glassy Modulus

Results of a recent dynamic, mechanical study of P(S-*co*-MANa) were shown in Figure 1.2 (6). One finding concerned the behavior of the glassy modulus as a function of ion content: In the range of 0–21 mol % of ionic groups, the glassy modulus on a log scale was constant at 8.90 ± 0.05 if measured at 100°C and at 8.88 ± 0.06 if measured at $(T_g - 40)$°C. The constancy of the modulus in the glassy range can be understood when one realizes that the volume fraction of ionic material ($COO^- Na^+$ groups only), even at 21 mol %, is only 6 vol % (6). For 6 vol % of filler, taking any of the well-known filler equations (e.g., the Guth equation) (7), one finds that the change in the modulus should be no greater than 20%. This value is within the experimental error and would not show up for data plotted on a log scale over five orders of magnitude (as the modulus is usually presented). It should be noted, however, that the constancy of the glassy modulus with ion content is not universal. In some other systems like the ionomers based on methyl methacrylate, an increase is observed (8) (Chapter 7).

5.2.2. Loss Tangent Peaks

The change of the peak positions in the loss tangent curves was discussed earlier (Fig. 4.10). The areas under the curves, however, provide extensive additional information. The results of a deconvolution study, in which the approximate peak areas were determined as a function of the ion concentration for the two loss tangent peaks, are shown in Figure 5.3 (6). The two areas associated with the low and high temperature peaks are identified with glass transitions in the matrix and cluster phases. It is seen that the two areas are equal at ~4 mol % of ions, and that at ~12 mol % only 5 vol % of the low T_g phase remains. The sum of the two areas is constant, which suggests that the relative areas under the two loss tangent peaks are proportional to the relative volume fractions of the materials involved in the two glass transitions. This proportionality will be of crucial importance when filler and percolation effects are discussed.

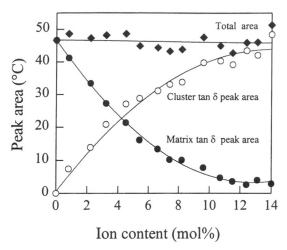

Figure 5.3. Areas under loss tangent peaks (1 Hz) and the sum of the areas versus ion content. ◆, total area; ○, cluster tan δ peak; v, matrix tan δ peak. Modified from Kim et al. (6).

5.2.3. Ionic Modulus

A characteristic feature in the modulus plots is the appearance of a nearly horizontal (or plateau) region in the intermediate range between the glassy and classical rubbery regions. This plateau is related to the presence of ionic groups, and its modulus increases with increasing ion content. The modulus in the plateau region is referred to as the ionic modulus. No one explanation can relate unambiguously the value of the ionic modulus to the ion content. As the ion content changes, the relative number of multiplets in the matrix phase and in the cluster phase changes, and the size(s) of the clusters also changes. The effects of these two different ionic environments on the properties, specifically the ionic modulus, are different, i.e., cross-linking for multiplets and filler for clusters. Therefore, we should not expect that any one simple equation will fit the curve in the plot of ionic modulus versus ion content. At low ion contents, there are isolated multiplets that act as cross-links. At ~15 mol %, most of the material (.95%) is present as clusters. The material has undergone a phase inversion. Several regions of behavior are, therefore, expected. Three possible relationships between the plateau modulus and the ion content can be suggested: (a) the ionic modulus reflects cross-link formation resulting from ion aggregation, (b) the modulus reflects the presence of a filler that consists of the cluster regions, and (c) the modulus can be related to percolative behavior.

5.2.4. Cross-Linking

It is tempting to relate the ionic modulus to cross-linking by ionic groups. Such an attempt was made for the styrene ionomers (5), using the equation

$$G = \rho RT/M_c \qquad (5.1)$$

where G is the shear storage modulus at the inflection point of the ionic plateau, R is the gas constant, T the absolute temperature, and M_c the average molecular weight between cross-links. It was shown that up to ~6 mol % of ions, equation 5.1 was valid within a factor of two. The data of Kim et al. (6) are also described by this equation within the same margin of error. A plot of the experimental ionic modulus versus ion content is given in Figure 5.4, which also shows the rubber elasticity, filler effect, and percolative behavior curves.

In the next approximation, an attempt was made to calculate the functionality of cross-links (Table 5.2). Values of functionalities <4 are not physically meaningful, because a functionality of 2 is essentially a chain middle, and a functionality of 3 implies that three chains emanate from one cross-link. A functionality of 4 represents a minimum multiplet of two ion pairs, with each pendent anion having two chains emanating from it. Calculated functionalities of <4 can possibly be interpreted as incomplete association of ionic groups, because the existence of isolated ion pairs would increase the apparent M_c. Therefore, the simplest way to understand the low modulus values within the framework of cross-linking is to assume that not all ion

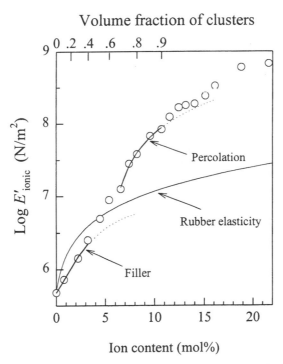

Figure 5.4. Log of the ionic modulus E'_{ionic} (1 Hz) versus ion content for P(S-*co*-MANa) of various ion contents; Points: experimental; lines calculated from filler, rubber elasticity, and percolation approaches. Volume fractions of clusters at various ion contents are indicated.

TABLE 5.2. Apparent Functionality and Degree of Incorporation of Ionic Groups in Multiplets of P(S-*co*-MANa) Ionomers as a Function of Ion Content

Ion Content, mol %	Functionality	Fraction of Ionic Groups in Multiplets, %
0.8	2	30
2.2	2	40
3.4	2	60
4.5	2	100
5.4	3	—
6.6	3	—
7.4	6	—
8.2	11	—
9.6	∞	—

pairs are in multiplets. Using this assumption, we derived the numbers in the third column of Table 5.2. If the assumption is valid, these numbers show that the degree of incorporation of ionic groups in multiplets increases with increasing ion content. It should be stressed, however, that this calculation is approximate, and that other interpretations of the discrepancy may be invoked. We give the results of this calculation to demonstrate one possible explanation of why the ionic modulus is lower than that predicted by equation 5.1.

Functionalities >4 can explain some of the deviations from a plot of equation 5.1. For the 8.2 mol % ionomer, a functionality of 11 was obtained. Interpretations other than straightforward multiple functionality are available, such as the formation of sizable cluster regions (Section 3.2.2). For ion contents ≥9.6 mol %, the concept of multiple functionality breaks down, because the calculated functionality is infinite, so it is more appropriate to speak of phase inversion or percolation.

5.2.5. Filler

It was recognized early in the studies of ionomers that these materials show many properties of filled systems, especially at intermediate ion contents. One publication treats the filler behavior of ionomers in considerable detail (6). Both the multiplets and the clusters can be regarded as filler. To ascertain which of the entities is acting as the filler, it is necessary to obtain the volume fraction of both species. The volume fraction of clusters can be obtained from the tan δ peak area. From a knowledge of the volumes of ion pairs (Chapter 1), one can calculate the total volume of the multiplets and, in turn, their volume fraction as a function of ion concentration, assuming that either all, or some reasonable fraction, of the ion pairs are in the multiplets.

A number of relations exist between the volume fraction of filler and the modulus of the filled system. One of the earliest is Guth's equation (7):

$$E^* = E(1 + 2.5V_f + 14.1V_f^2) \qquad (5.2)$$

where E^* is Young's modulus of the filled material, E is Young's modulus of the unfilled material, and V_f is the volume fraction of filler. Equation 5.2 assumes spherical particles (not rods or lamellae) up to ~ 30 vol % of filler. Here, the rubbery modulus of polystyrene is taken as E, while E^* is the ionic modulus of the materials. A graph of this equation as a function of the volume fraction of filler for both clusters and multiplets is shown in Figure 5.5a. It is clear that the equation fits the cluster data directly (without any adjustable parameters) but does not fit the data for the multiplets. Thus up to 50 vol % of clusters, which is found at 3.4 mol % of ionic groups, the equation with clusters as filler gives a good fit (6).

A number of other equations have been developed that relate the modulus to the volume fraction of the filler over a wide range of filler content. The Halpin–Tsai equation (9,10) is one example:

$$\frac{M}{M_1} = \frac{1 + AB\varphi_2}{1 - B\varphi_2} \qquad (5.3)$$

where

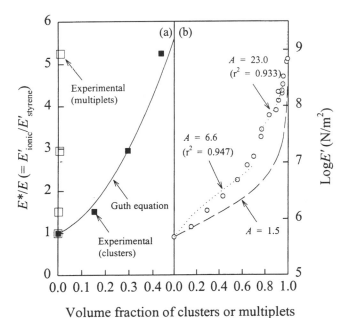

Figure 5.5. **a,** The ratio of the moduli versus volume fraction of filler. **b,** Ionic moduli versus volume fraction of clusters. *Dashed line,* Halpin–Tsai regular system; *dotted lines,* calculated values for regions below and above the volume fraction of 0.7; ◯, observed experimental data. Modified from Kim et al. (6).

$$B = \frac{M_2 / M_2 - 1}{M_2 / M_1 + A} \tag{5.4}$$

M is the modulus of the material; subscripts 1 and 2 refer to the continuous phase and the dispersed phase, respectively; φ_2 is the volume fraction of component 2; and A is a constant that depends on the morphology of the material:

$$A = \frac{7 - 5v_1}{8 - 10v_1} \tag{5.5}$$

and v_1 is Poisson's ratio for the matrix.

Figure 5.5b shows the application of the Halpin–Tsai equation over the entire range of volume fractions of the clustered material (the filler) (6). It is seen that the best fit is obtained with two different A values, one operative at a volume fraction <0.70 ($A = 6.6$), and the other >0.80 ($A = 23.0$). If the filler consists of isolated multiplets surrounded by regions of restricted mobility, A should be 1.0–1.5 for regularly shaped (e.g., spherical) particles. Figure 5.5b also shows the result using $A = 1.5$. The A values obtained from the best fit are clearly much higher, suggesting that in this system the shapes of the filler particles are highly irregular, as suggested by the Eisenberg–Hind–Moore (EHM) model (11). The large change in the value of A also suggests that geometrical and size changes occur with increasing volume fraction of clusters.

5.2.6. Percolation

The concept of percolation was developed to determine the point of attainment of a continuous path if a percolating species (e.g., conducting spheres) is placed randomly in space with a probability p (12). It was found that just above the percolation threshold (p_c), the conductivity increased as a power law $(p - p_c)^t$, where p is the concentration, p_c the critical concentration at the percolation threshold, and t is the critical exponent (13). Several theoretical studies have been performed on this and related areas (14–16). It has been found that plots of the logarithm of the property in question versus log $(f - f_c)$ give straight lines with a slope of n (the critical exponent), where f is the volume fraction of the percolating species and f_c is the critical concentration or volume fraction. The critical concentrations are in the range of 0.12–0.43, depending on the lattice type (14); and the critical exponents are in the range 0.4–4, depending on the system. The most reasonable value for the critical exponent was suggested to be 1.5, and this value is often used in judging whether ideal percolative behavior is involved (13). Bond percolation has also been considered. In that context, it is the connection between the lattice sites that are now filled with conducting bonds. Few papers apply percolation concepts to the conductivity of ionomers and ionomer blends (17–21).

The EHM model suggests that percolation concepts can be applied to the modulus of clustered ionomers. This idea is particularly relevant if one considers that phase inversion most likely occurs at ~ 6 mol % of ionic groups, as suggested from other

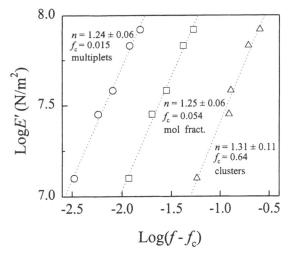

Figure 5.6. Log E' versus log $(f\text{-}f_c)$ for multiplets ($\bigcirc$), mole fraction of ions ($\square$) and volume fraction of clusters ($\triangle$). The critical exponent n indicates the slope of the line. Modified from Kim et al. (6).

considerations. A study of the modulus in terms of site percolation was performed by Kim et al. (16), who found that a log–log plot of the ionic modulus versus $(f-f_c)$, shows a linear relationship for samples of 5–10 mol % of ions (Fig. 5.6). This suggests that percolation concepts are, indeed, applicable in this system. The percolation threshold f_c was found to occur at 64 vol % of clusters, which corresponds to 5.4 mol % of ions; the slope was determined to be 1.3, which is close to the universal critical exponent for conductivity percolation of 1.5.

The value of 5.4 mol % coincides with discontinuities in various properties: onset of failure of time–temperature superposition in stress–relaxation data and water uptake data (5) and various parameters obtained from the dynamic mechanical property data. Discontinuities at 5.4 mol % of ions were found in the slope of the log E' curve in the matrix T_g range, in the cluster tan δ peak width and height in plots of tan δ as a function of temperature (6), and in the crossover of the areas under the tan δ peaks as a result of the matrix and clusters (6,22). The 64 vol % of clusters at the percolation threshold is higher than values observed for other properties. However, in the ionomers one should consider the coordinated growth of clusters as a function of ion content and the difference in mobility (or modulus) in different regions of the clustered material. Therefore, the difference in the critical volume fraction between that seen in this study and those found from other investigations is not unreasonable.

5.3. MELT RHEOLOGY OF SODIUM METHACRYLATE SYSTEMS

The melt rheology of P(S-*co*-MANa) ionomers was first investigated by Shohamy and Eisenberg (23), who showed that for relatively low ion concentrations and nar-

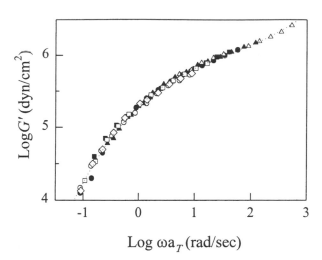

Figure 5.7. Master curve of G' versus ωa_T for P(S-1.5-MAA), its Na$^+$ salts, and its methyl ester. $T_{ref} = 164°C$ for the ester (150 $\triangle$ and 160 $\blacktriangle$), 172°C for the acid (170 $\bigcirc$ and 180 $\bullet$), and 202°C for the salt (195 $\blacksquare$, 205 $\square$ and 210 $\diamondsuit$). Modified from Shohamy and Eisenberg (23).

row temperature ranges, time–temperature superposition could be applied to plots of G' versus ω, as shown in Figure 5.7. Furthermore, they showed that over the narrow range of frequency and temperature employed in the study, the same master curve is obtained not only for the ionomer but also for the acid and the methyl ester, the only difference being the temperature shift factor. The plot of the temperature shift factor between the ester and the salt of identical composition and molecular weight is shown in Figure 5.8.

Our laboratory performed a much more thorough study recently. G' and G'' were determined as functions of ion concentration and frequency (Fig. 5.9). Table 5.3 shows G^o_{ionic} from the ionic modulus and G^o_{ent} from the entanglement modulus. In general, time–temperature superposition was found to be applicable < 5.2 mol %, but clearly failed for the 9.0 mol % sample. The shift factors were found to be of the WLF type; failure of time–temperature superposition is suggested by the values of the WLF constants for the 5.2 and 9.0 mol % samples; the 5.2 mol % sample seems to be on the verge of failure because the value of C_2 is considerably higher than that for the samples of lower ion content. The values of C_1 and C_2 obtained in this study are not the same as those obtained in Eisenberg and Navratil's (5) study, because the reference temperatures were different. The melt rheology of a number of other ionomers has been investigated (24–34); some of the results will be discussed in connection with the specific ionomers in Chapter 6.

5.4. VISCOELASTICITY OF OTHER CARBOXYLATED IONOMERS

The preceding sections dealt with mechanical properties of P(S-*co*-MANa) ionomers. Clearly, however, the carboxylate anion can be placed in many different positions

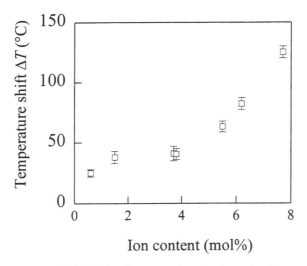

Figure 5.8. Temperature shift ΔT, for the ester–salt pairs determined at $\omega = 1$ rad/s and at $G' = 2 \times 10^4$ N/m^2, as a function of ion content. Modified from Shohamy and Eisenberg (23).

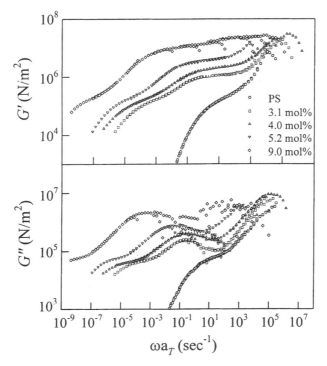

Figure 5.9. LogG' and logG'' at 160°C versus ωa_T for PS and various P(S-*co*-MANa) ionomers. ○, PS; □, 3.1 mol %; △, 4.0 mol %; ▽, 5.2 mol %; ◇, 9.0 mol %.

TABLE 5.3. Rheological Characteristics and WLF Parameters of the P(S-*co*-MANa) Samples

Ionomer	$G^\circ_{ionic} \times 10^{-6}$, N/m^2	$G^\circ_{ent} \times 10^{-5}$, N/m^2	T_g by DSC, °C	C_1	C_2
PS	—	2.0	103	13	63
P(S-3.1-MANa)	1.3	2.1	106	17	72
P(S-4.0-MANa)	2.3	2.3	111	19	81
P(S-5.2-MANa)	4.4	2.4	114	22	130
P(S-9.0-MANa)	12.1	2.8	122	54	520

relative to the backbone. This section describes three such polymers: styrene-*co*-acrylic acid [P(S-*co*-AA)] copolymers, in which the ion placement is the same as in the methacrylic acid copolymers: *p*-carboxystyrene [P(S-*co*-SC)]; and styrene-based combs, in which the ionic groups are placed at varying distances from the backbone using spacers attached to the *para* position of the benzene ring. The results for these materials are compared with those for the sodium methacrylate-based system.

5.4.1. P(S-*co*-ANa) Versus P(S-*co*-MANa)

Kim et al. (35) studied [P(S-*co*-ANa)] copolymers. On the basis of the multiplet picture, it is clear that in the acrylates the contact surface area is smaller than it is in the methacrylates because of the absence of the methyl group in the acrylates (Section 3.1.1.3). Therefore, the multiplet is expected to be somewhat larger in the acrylates than in the methacrylates. As a consequence, the volume fraction of material of reduced mobility per ionic group is expected to be somewhat lower in the acrylates. Therefore, the ionic modulus of the clustered systems would also be expected to be lower at comparable ion contents. The volume fractions of clustered material at such two ion contents have been compared quantitatively by deconvolution of the areas under the loss tangent peaks. For the 6.5 mol % samples, it was found that the volume fraction of clustered material was 0.60 for the acrylates and 0.7 for the methacrylates. For the 11 mol % ionomers, the volume fractions of clusters were 0.85 and 0.93, respectively. One indication that the difference between the cluster and matrix T_g values is larger in the methacrylates than in the acrylates was shown in Figure 4.12; this phenomenon is illustrated further in Figure 5.10 for 11 mol % samples (35). Similar trends were also found for 6.5 mol % samples.

5.4.2. P(S-*co*-SCNa) Versus P(S-*co*-MANa)

P(S-*co*-SC) is another styrene ionomer in which the ionic group is based on the carboxylate anion, but the ionic group is farther removed from the chain than in the methacrylate or acrylate cases. Increasing the distance from the chain backbone results in an increase in multiplet size. A direct consequence is that the volume fraction of clustered material is expected to be lower in the *p*-carboxylate case than in the methacrylate (or acrylate) case. The mechanical properties of P(S-*co*-SC) were investigated by Hird and Eisenberg (36) and by Tomita and Register (37). Hird and Eisenberg found that the cluster T_g values of the P(S-*co*-SCNa) ionomers were consistently lower than those of the P(S-*co*-MANa), but not by much. The ionic modulus of the P(S-*co*-SCNa) was lower than that of the P(S-*co*-MANa) which is consistent with the lower volume fraction of clustered material. The Bragg spacings—related to intermultiplet distances and determined from the maxima in the SAXS peak for these ionomers—are on the order of 30 Å and are relatively insensitive to changes in type of counterion or ion content (36). This result is in general agreement with those of Tomita and Register (37), who investigated both sodium and zinc salts of P(S-*co*-SC). The Bragg spacing for the methacrylates is on the order of 23 Å, which clearly illustrates that the multiplets are significantly larger in P(S-*co*-SC) than in P(S-*co*-MA). At temperatures above the cluster glass transition, the viscoelastic properties of the *p*-carboxystyrene and the methacrylate ionomers are similar, which is not surprising, given that the backbone is identical and that the multiplet consists of sodium carboxylate pairs in both cases.

5.4.3. Combs

Perhaps the most effective way of varying the distance of the carboxylate group from the main chain is by attaching the group to a side chain at, e.g., the *para*

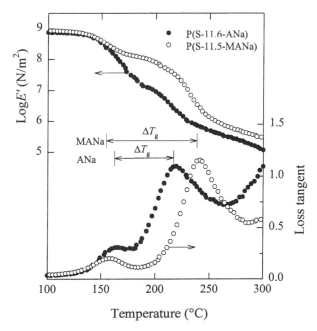

Figure 5.10. Log E' and loss tangent data measured as a function of temperature at 1 Hz for P(S-11.6-ANa) (●) and P(S-11.5-MANa) (○). Modified from Kim et al. (35).

position of the benzene ring. Gauthier and Eisenberg (38) studied a series of such materials (Scheme **3.1**). For the C_{11} ether series, the modulus and loss tangent are shown as a function of temperature in Figure 5.11. All the materials had the same side chain content, and the ion content was determined by the degree of hydrolysis of the methyl ester, which ranged from 0% for the nonionic sample, to 100% for the 14 mol % sample. Several features emerge from these two plots. The first is the appreciable rise in the glass transition temperature with increasing ion content. The T_g is ~ 70°C for the nonionic (ester) sample and increases with ion content at a rate of 5°C/mol % (Fig. 5.11). Another interesting feature is the value of the ionic modulus. In the C_{11} ether ionomers, the ionic moduli are similar to those of P(S-*co*-MANa) series, suggesting that the material is highly clustered, which is borne out by the areas under the loss tangent peaks (Fig. 5.11).

The extent of clustering, however, is not uniform as a function of side chain length in the comb series. The extent of clustering can be estimated from the ratio of the apparent molecular weight between cross-links calculated from the simple rubber elasticity equation (eq. 5.1) and from the values of the ionic inflection point. A plot of the ratio of these two calculated quantities as a function of chain length of the comb is given in Figure 5.12 for 7.5 mol % of ions. In the methacrylate series this ratio is ~3; it initially decreases with side chain length, reaching a minimum at 7 carbon atoms, but increases again as the alkyl chain length increases.

Clearly, the degree of clustering is not a linear function of the side chain length. Clustering is extensive for materials with short side chains (e.g., the methacrylates) but is also quite strong for those with the C_{11} side chain. Clustering is marginal in the C_5 side chain case. The reason for this behavior is not completely apparent, but a combination of two different factors is suggested. As the side chain length increases,

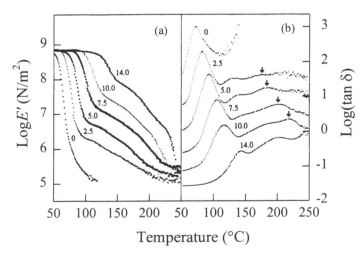

Figure 5.11. (a) Log E' and (b) log tan δ curves (1 Hz) versus temperature for the C_{11} ether series. Numbers indicate ion content in mole percent. Curves in (b) displaced vertically for clarity. Modified from Gauthier and Eisenberg (38).

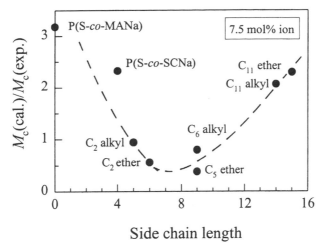

Figure 5.12. Deviation from rubber elasticity expressed as M_c(calculated) (38)/ M_c(experimental) versus side chain length for the ionic modulus of the alkyl and ether series (7.5 mol %) (shown in scheme **3.1**).

starting with the methacrylate system, clustering gets weaker because of the increased backbone mobility and the increase in the multiplet size. With a further increase in the side chain length, however, the packing of ion pairs in the multiplets becomes more efficient. For C_{11}, the multiplets are so large that crowding reduces the mobility of the backbone to the level of the methacrylates. The C_2 to C_5 ethers represent a minimum, in the sense that the reduction of clustering (correlated with an increase in multiplet size) is observed but crowding has not become severe. The C_{11} system, therefore, probably shows a situation in which crowding is the predominant mechanism of reduction of mobility in clustering. By contrast, in the P(S-*co*-MANa) ionomers, in which crowding may not be severe because of the small size of the multiplet, the direct attachment of the chain to the multiplets is probably the factor that limits mobility, at least, within the persistence length of the polystyrene chain.

5.5. PENDENT ION EFFECT ON VISCOELASTICITY

The first part of this chapter dealt with carboxylated ionomers. This section deals with the effect of changing the pendent ion, while keeping polystyrene as the polymer matrix.

5.5.1. Sulfonated Ionomers

It was recognized early that substitution of a sulfonate for a carboxylate group changed the properties of the ionomers drastically. Lundberg and Makowski (39) showed that the melt viscosity at 220°C, is more than two orders of magnitude greater for P(S-*co*-SSNa) than for P(S-*co*-SCNa) for ion contents between 2 and 4 mol % and identical molecular weight (Fig. 5.13). They also showed that the position of the major drop in the modulus–temperature curves increases by ~100°C for the sulfonates over carboxylates (at 5 mol %). In a subsequent study, Rigdahl and Eisenberg (40) showed that the cluster glass transition in the sulfonated ionomers occurred at higher temperatures than in the P(S-*co*-MANa) system and ascribed the differences in bulk viscosities found by Lundberg and Makowski (39) to this increase in the cluster glass transition temperature. The stress–relaxation behavior was found to be generally similar to that of P(S-*co*-MANa) in terms of both the applicability of time–temperature superposition at low ion concentrations and its failure at high ion contents.

More recently, a number of investigators addressed the dynamic mechanical properties of sulfonated polystyrene (36,41,42). Figure 5.14 shows plots of G' and G'' as a function of reduced frequency for various ion contents (41). The modulus at 1 Hz for different ion contents is shown as a function of temperature in Figure 5.15 (36). There are two features of interest: the much more extended ionic plateau for the sulfonates and that this plateau lies at a lower modulus than it does in the carboxylates for corresponding ion contents. For example, the inflection point in the carboxylate (9.9 mol %) is found at a modulus of $10^{7.8}$ N/m^2, whereas in the sulfonate (11.0 mol %) it is at a modulus of $10^{7.45}$ N/m^2.

In the sulfonates, ion hopping in the multiplets starts at higher temperatures, which induces the cluster glass transition at higher temperatures. This topic was

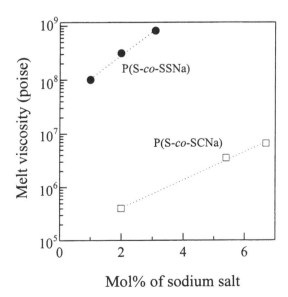

Figure 5.13. Melt viscosity of lightly sulfonated polystyrene compared with lightly carboxylated polystyrene sodium salts at 220°C. Modified from Lundberg and Makowski (39).

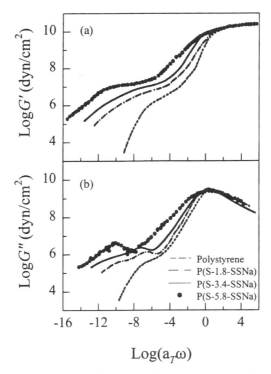

Figure 5.14. Master curves of log G' and log G'' versus ωa_T for sulfonated polystyrene of various ion contents. $T_{\text{ref}} = T_g$; *dashed line*, polystyrene; *dashed and dotted line*, P(S-1.8-SSNa); *solid line*, P(S-3.4-SSNa); *dotted line*, experimental points for P(S-5.8-SSNa). Modified from Weiss et al. (41).

discussed earlier in connection with the glass transition of various polystyrene iono-mers (Chapter 4). At 2–10 mol % of ions, the cluster glass transitions in the sulfonates are 230–280°C, whereas in the carboxylates they lie between 180 and 220°C; the matrix glass transitions occur at essentially the same temperatures in both systems. These features can be understood in terms the sizes and stabilities of the multiplets in the systems, as proposed by Hird and Eisenberg (36).

In the sulfonate case, the ion is at the *para* position of the benzene ring, and in the methacrylate case, it is right on the polymer backbone; thus one expects the multiplets to be larger for sulfonates. An increase in the size of the multiplet, while keeping the persistence length of the polymer chain constant, reduces the total amount of clustered material. Therefore, a higher ion content is needed in the sulfo-nates to achieve a comparable volume fraction of clustered material; conversely, at the same ion content, the amount of clustered material and hence the ionic modulus are lower. Morphological studies suggest that in the *p*-sulfonates the multiplets are somewhat larger than in the *p*-carboxylates, even though the distances of the pendent ions from the backbone in these two systems are identical (36). This difference in

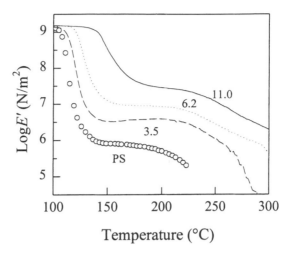

Figure 5.15. Log E' versus temperature curves (1 Hz) for P(S-*co*-SSNa) ionomers with a range of ion contents, indicated for each curve. Modified from Hird and Eisenberg (36).

multiplet size can, again, be ascribed to the stronger interactions in the sulfonate case, which probably results in an increase in the surface energies and, therefore, an increase in the multiplet size.

The underlying reason for the difference in the strength of interaction between carboxylates and sulfonates was studied by Lefelar and Weiss (43), as noted in Section 4.5.5. These two factors—the strength of interaction and the size of the multiplets—are the essential features needed to understand the differences between the methacrylates, the *p*-carboxylates, and the *p*-sulfonates.

5.5.2. Benzyloxy or Phenoxy Ionomers

Few studies have been performed with alkoxide anions as the ionogenic group. Two such studies involved copolymers of styrene with 4-phenoxystyrene and styrene with 4-benzyloxystyrene (44–46). The modulus as a function of temperature for these two ionomers at the 10 mol % level is shown in Figure 5.16, along with the plot for P(S-*co*-MANa) of comparable ion content. Phenoxystyrene shows clear evidence of an ionic inflection point, which lies almost an order of magnitude below that of the P(S-*co*-MANa) copolymer. The strength of the ionic interaction for the carboxy and alkoxy ionomers can also be gauged from the position of the cluster peak at low ion contents. It was observed that the cluster glass transitions do not vary appreciably with structure for these materials, suggesting that the onset of ion hopping (which reflects the strengths of the interactions) is not significantly different. By contrast, the hydroxymethylstyrene shows no such inflection point, which probably indicates that clusters are not present.

In the light of the preceding discussion, these observations can be understood in terms of the relative sizes of the multiplets. In the P(S-*co*-MANa) case, the multiplets are the smallest of the three copolymers because of the proximity of the ion to the backbone, and the total volume of the clustered material at a given ion content is the largest. In the *p*-hydroxystyrene case, the greater distance of the ionic group from the backbone increases the size of the multiplet and decreases the volume fraction of clustered material. The value of the modulus at the ionic inflection point, therefore, decreases considerably. Finally, the hydroxymethylstyrenes are analogous to the carboxylated systems in which the ions are positioned at the end of a spacer. In this particular case, no evidence of clustering is seen, which is quite similar to the situation for the comb ionomer with a carboxylate at the end of a two or five carbon spacer. It is conceivable that the multiplets fall apart right above the glass transition. It should be also pointed out that the parent compounds of the series, i.e., the alcohols, show no unusual features relative to those of unmodified styrene, except for a slight increase in the glass transition temperature as a result of the presence of hydrogen bonds.

5.5.3. Vinylpyridinium Methyl Iodide Ionomers

The viscoelastic properties of *N*-methyl-4-vinylpyridinium iodide [P(S-*co*-4VPMel)] ionomers in the range of 0–10 mol % ion content were first investigated by Gauthier ct al. (47), who found that the plots of the moduli as a function of temperature have

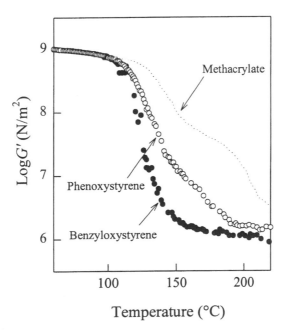

Figure 5.16. Log G' versus temperature for styrene ionomers based on 9.6 mol % sodium methacrylate (•), 10.4 mol % sodium phenoxystyrene (○), and 10.2 mol % sodium benzyloxystyrene (●). Modified from Clas and Eisenberg (46).

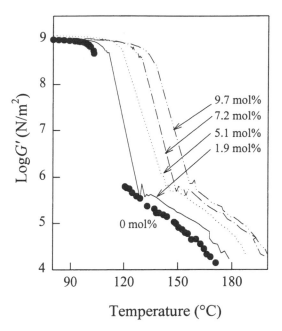

Figure 5.17. Log G' versus temperature for P(S-*co*-4VP)MeI of various ion contents. Experimental points are shown only for PS. Modified from Gauthier et al. (47).

identical shapes and merely shift to higher temperatures with increasing ion content (Fig. 5.17). They concluded that the aggregates do not survive above the glass transition temperature, although the ionic interactions manifest themselves as an increase in the glass transition temperature, with dT_g/dc being approximately the same as in a wide range of other ionomers (3.5°C/mol %). From the shapes of the modulus curves, notably the absence of an inflection point which is related to the ion concentration (i.e., the ionic modulus), it is clear that clustering or even ion aggregation is absent in these materials above T_g. The shapes of the loss tangent curves as a function of temperature also reflect this absence.

A subsequent study of the small-angle x-ray scattering (SAXS) pattern of these materials revealed that no ionic peak was present (48). This behavior, in combination with viscoelastic results, suggests that below T_g the multiplets must be small, not large enough to show a SAXS peak (which requires the aggregation of several iodide ions per multiplet) but large enough to have an effect on the T_g.

5.5.4. Homoblends

The homoblends to be discussed here consist of a mixture of the two polystyrene-based materials, one of which has pendent cations and the other, pendent anions (Chapter 2). A typical example involves styrenesulfonic acid [P(S-*co*-SSA)] and 4-vinylpyridine [P(S-*co*-4VP)]. On mixing, proton transfer occurs from the sulfonic acid group to the pyridine nitrogen atom, resulting in the formation of a pyridinium sulfonate ion pair or possibly a higher aggregate. The glass transition temperatures of such polymers were discussed in Section 4.5.3, and a possible multiplet was shown in Figure 4.14.

The viscoelastic properties of these materials have been investigated by dynamic mechanical thermal analysis (DMTA), melt rheology, and stress—relaxation techniques. Figure 5.18 shows the results obtained by the latter two techniques (49), spanning a temperature range of 105–252°C for polystyrene samples made of a blend of P(S-5.1-SSA) and P(S-5.2-4VP) for stress–relaxation experiments and one of P(S-5.1-SSA) and P(S-4.8-4VP) for melt rheology experiments. The curves at higher modulus (lower temperature) were obtained by stress–relaxation, while those at lower modulus (higher temperature) were obtained in a Rheometrics mechanical spectrometer. The shapes of the curves suggest that time–temperature superposition might well be applicable. However, it should be noted that these data have not been obtained over a wide enough range of times or of frequencies for an unambiguous judgment to be made in this regard. The WLF constants are not significantly different from those of normal polystyrene: For the sample shown in Figure 5.18, $C_1 \approx 10$

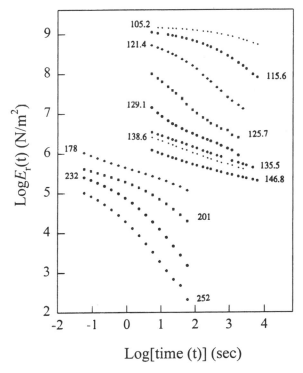

Figure 5.18. Log E_r versus log time for a homoblend of P(S-5.1-SSA) with P(S-5.1-4VP) for stress relaxation (105.2–146.8°C) data and for a homoblend of P(S-5.1-SSA) with P(S-4.8-4VP) to melt rheology (178–252°C) data. Modified from Smith (49).

and $C_2 = 180$; by contrast, in polystyrene $C_1 = 8$ and $C_2 = 150$ ($T_{\text{ref}} = T_g + 65°C$).

Dynamic mechanical studies in the glass transition region for these systems were performed by Douglas et al. (50) who showed that the materials were clustered, because two glass transitions were observed. A subsequent study by our laboratory was undertaken to obtain more precise data at low temperatures; we found that the relative volume fractions of clustered and unclustered materials are similar to those seen in the P(S-co-MANa) ionomers. The primary difference between P(S-co-MANa) ionomers and the homoblends lies in the position of the tan δ peak from the clustered material. In the homoblends, the difference in the glass transition temperatures at the 10 mol % level is ~ 20°C, whereas for P(S-co-MANa) it is 70°C and for P(S-co-SSNa) it is 100°C. An illustration of the tan δ peak deconvolution for the 10 mol % sample is shown in Figure 5.19, along with that for the P(S-co-MANa) system for comparison. Although the general shapes of deconvoluted peaks are similar, the cluster peak in the homoblend is somewhat broader than that of P(S-co-MANa) and the matrix peak is somewhat narrower. The relative areas are also similar.

The most important effect observed here is the difference in the glass transition temperatures in these two systems. The fact that the homoblend shows only a small difference between the matrix and cluster glass transition temperatures may explain the apparent applicability of time–temperature superposition in stress–relaxation in these systems. The homoblends are noteworthy because two materials that either

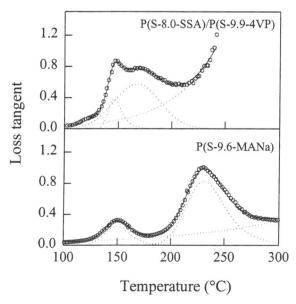

Figure 5.19. Tan δ vs temperature (1 Hz) for a homoblend of P(S-8.0-SSA) with P(S-9.9-4VP) and for P(S-9.6-MANa).

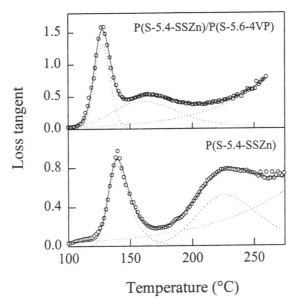

Figure 5.20. Tan δ versus temperature (1 Hz) for a homoblend of P(S-5.4-SSZn) with P(S-5.6-4VP) and for P(S-5.4-SSZn).

have a low degree of clustering—the P(S-*co*-SSA) case—or are completely non-ionic—the P(S-*co*-4VP) case—give a highly clustered system upon mixing as a result of the formation of ion pairs.

Another homoblend that has been studied is the blend of P(S-*co*-4VP) with zinc-neutralized sulfonated polystyrene [P(S-*co*-SSZn)]. P(S-*co*-SSZn) is highly clustered, though not as highly as the sodium methacrylate analog (judging from the area of the cluster peak). Douglas et al. (50), also investigated these materials; the glass transitions obtained from that study were discussed in Chapter 4. This system was reinvestigated by our laboratory. Figure 5.20 shows the deconvoluted loss tangent curve for P(S-*co*-SSZn) and for the homoblend. It is clear that the glass transition temperature of the cluster regions drops dramatically on blending, as Douglas et al. (50) suggested. However, the extent of clustering (as measured by the area under the tan δ peak for the cluster T_g) increases considerably. These phenomena can be understood within the framework of the EHM model of ionomer morphology.

5.6. EFFECTS OF VARIOUS OTHER PARAMETERS

5.6.1. Dielectric Constant

Gauthier and Eisenberg (51) investigated the effect of the matrix dielectric constant on the viscoelastic behavior of styrene ionomers. The matrix polarity was changed by partial nitration, and the cluster and matrix T_g values were determined as a function of the degree of nitration for a 7 mol % sodium methacrylate ionomer. It was shown that the cluster T_g values remain essentially constant, whereas the matrix T_g values increase from ~140 to 170°C when the degree of nitration increases from 0 to 32%. This result is surprising because the nitro group is extremely polar; and at 32% nitration, it raises the dielectric constant of styrene from 3.2 to 7.3, which is higher than that of the acrylates. At such a high dielectric constant, one might expect an effect on the cluster T_g.

The increase in the matrix T_g on nitration results from the interaction of the dipoles of the nitro groups in the *para* position of the benzene ring. Size alone is not enough to explain the increase because, e.g., *p*-methylstyrene has a glass transition temperature identical to that of unfunctionalized polystyrene (52).

To understand why the cluster T_g does not go up in parallel with the matrix T_g, recall that the multiplet holds the chain at a fixed point. Therefore, the chain is less flexible, and the formation of dipole–dipole pairs is strongly hindered in the vicinity of the multiplets. The effect of dipole–dipole interactions in the immediate vicinity of the multiplet is not as effective as in the bulk (only the size effect is operative).

In a parallel study (51), nitrobenzene was used as a plasticizer for the styrene system (Chapter 8). No cluster T_g decrease was seen in that study, which indicates that a change in matrix polarity alone is not sufficient to accomplish that; instead, specific interactions are required, at least in the polarity range investigated here.

5.6.2. Counterions

The investigation of the effect of a change of the counterions on the viscoelastic properties of the styrene ionomers has been only fragmentary. In general, the effects observed are as expected. In one study, Hara et al. (42) looked at the modulus as a function of temperature for Cs$^+$, K$^+$, and Ca^{2+} neutralized sulfonated polystyrene ionomers of 4.1 mol % sulfonate content (Fig. 5.21). It is clear that the calcium

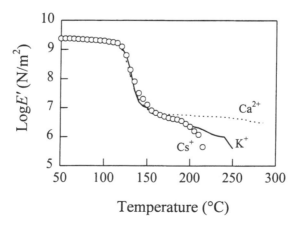

Figure 5.21. Log E' versus temperature for P(S-4.1-SSX), when X equals Ca^{2+}, K$^+$, and Cs$^+$. Modified from Hara et al. (42).

ionomer, which contains a relatively small divalent cation, is subject to the strongest interactions, leading to a constant ionic modulus over the temperature range explored. The cesium ion, which is large and monovalent, has the weakest interaction, as shown by the drop in the modulus accompanying the cluster T_g.

Lefelar and Weiss (43) compared the Zn^{2+} and Na^+ salts of the sulfonated polystyrene ionomers and also found that the monovalent cation shows a drop in the modulus at a lower temperature than the divalent cation. Unlike the Ca^{2+} multiplet, which retains in its integrity to about 300°C, the zinc salt shows a drop at lower temperatures (~230°C for the 5.8 mol % ionomer), indicating the earlier onset of ion hopping. However, the modulus of the Zn^{2+} salt is retained to higher temperatures than in the sodium salt. Zn^{2+} as a cation is particularly interesting in the sulfonated ethylene–propylene–diene torpolymer (EPDM) system (27), which will be discussed in Chapter 7. The ammonium counterion (30,53) was discussed in Chapter 4 along with the aliphatic amines in connection with the glass transition temperatures. A more extensive discussion of counterion effects will be given in connection with the properties of the ethylene and urethane ionomers in the next chapters.

5.6.3. Molecular Weight

The effect of a change in molecular weight on the viscoelastic properties of the styrene ionomers was studied by Kim et al. (54), who investigated both the methacrylate and the sulfonate systems. In the methacrylate ionomers, the ion content was ~ 4 mol %, whereas in the sulfonate system it was ~5 mol %. Figure 5.22 and 5.23 show the data for G' and E' and for tan δ as a function of temperature for several different molecular weights. The loss tangent curves for the P(S-*co*-MANa) system show a fairly strong molecular weight dependence, over essentially the entire temperature range. A comparison of the curves for the highest and the lowest molecular weights is particularly instructive, specifically with regard to the cluster glass transition. As the deconvoluted curves in Figure 5.24 show, cluster glass transitions are present in both samples (clearly visible only in the high molecular weight material and barely so in the lowest molecular weight sample). The relative magnitudes of the areas are approximately the same. The cluster glass transition in the lowest molecular weight sample occurs approximately 15°C lower than in the highest molecular weight sample. This observation is surprising, because the change is considerably larger than for the matrix T_g; it could be an artifact arising from deconvolution. The primary difference in the shapes of the curves for the sulfonates and the methacrylates arises because, in the sulfonates, all the molecular weight–dependent phenomena occur at higher temperature as a result of the increased stability of the multiplets. This fact means that the window for the observation of the temperature-dependent phenomena before the onset of decomposition is considerably smaller in the sulfonate system.

5.6.4. Degree of Neutralization

Navratil and Eisenberg (55) studied the effect of the degree of neutralization using stress–relaxation for a relatively high molecular weight P(S-4.6-MANa) sample.

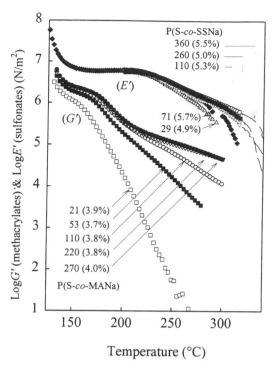

Figure 5.22. Log G' for P(S-*co*-MANa) and log E' for P(S-*co*-SSNa) versus temperature (1 Hz) for styrene ionomers. The molar ion contents and molecular weights ($\times 10^3$) are indicated for each sample. Modified from Kim et al. (54).

They showed that the ionic inflection point does not appear until ~60% neutralization, and that it becomes more pronounced as the degree of neutralization increases up to 100%. The observation that the ions do not cause clustering until a degree of neutralization of 50% is reached probably means that either the multiplets are too transient to form long lived cross-links or that no multiplets exist below that degree of neutralization.

We recently investigated the dynamic mechanical properties of a P(S-5.1-MANa) ionomer (56). The results of that study are shown in Figure 5.25. The loss tangent peak owing to the cluster transition begins to appear only at 50% neutralization, which is reflected also in the modulus curves, consistent with the previous observation. The tan δ peak height attributed to the matrix glass transition decreases from ~3 for the acid form (not shown) to ~1 for the ionomer, whereas the cluster peak moves to higher temperatures and increases in intensity between 50 and 100% neutralization. A similar result was also observed in the P(S-9.6-SSA) system neutralized with 1,6-hexanediamine, a bifunctional organic counterion (57).

Hara et al. (58) studied the effect of overneutralization as part of an investigation of the fatigue properties of styrene ionomers. They found that a large excess of neutralizing agent formed microsized phase-separated crystallites (discussed below).

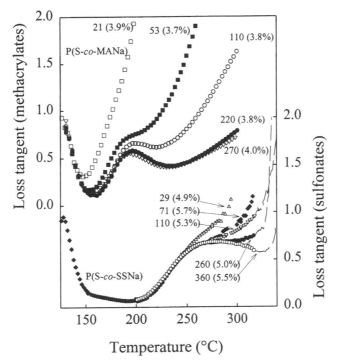

Figure 5.23. Tan δ versus temperature (1 Hz) for P(S-*co*-MANa) (*top*) and P(S-*co*-SSNa) (*bottom*) ionomers. The molar ion contents and molecular weights ($\times 10^3$) are indicated for each sample. Modified from Kim et al. (54).

Register and Cooper (59) investigated the effect of the presence of excess neutralizing agent on the anomalous small-x-ray scattering from P(S-*co*-SSNi) ionomers. They found that, while the positions of the small-angle x-ray peaks were similar, the intensity of the small-angle upturn decreased as the amount of excess neutralizing agent increased. The authors concluded that the small-angle upturn was not owing to the presence of excess neutralizing agent or of crystallites resulting from its presence, but that other explanations for that peak must be found. In an NMR study, O'Connell et al. (60) also suggested that excess NaOH aggregates form separate crystallites (discussed below).

We found that the presence of 50% excess neutralizing agent had only a small effect on the dynamic mechanical properties of P(S-*co*-MANa) ionomers, whereas the addition of >100% excess neutralizing agent had a major effect (56). In the plot of tan δ versus temperature, the high temperature side of the cluster peak showed an upswing. A SAXS study (56) also found that when the degree of neutralization increased from 100 to 200%, the peak position remained constant, but its intensity increased slightly. However, the intensity of the small-angle upturn for the 200% sample increased appreciably.

From these results, we concluded that the intermultiplet distance and the size of

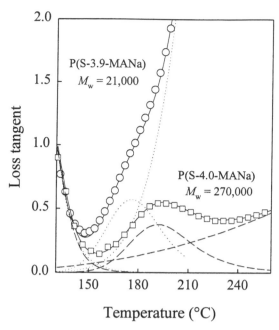

Figure 5.24. Deconvoluted tan δ curves versus temperature (1 Hz) for P(S-3.9-MANa) (M_w = 21,000) and P(S-4.0-MANa) (M_w = 270,000) samples. Modified from Kim et al. (54).

multiplets did not change significantly but that the ionic groups were more inhomogeneously distributed. We speculated that small amounts of excess neutralizing agent reside in or near the ionic cores, up to ~ 200% neutralization, resulting in a slightly higher intensity of the SAXS peak. Above 200%, however, the excess neutralizing agent can accumulate in the form of separate crystallites in the hydrocarbon phase, which could increase the heterogeneity of the distribution of ionic groups, resulting in a shift of the small-angle upturn to a higher angle. In addition, we suggested that the microcrystallites (which adhere poorly to the matrix) act as sites at which ionic groups can reside temporarily; the binding of the ionic groups to the NaOH crystallites, however, probably weakens at higher temperatures, and thus the material flows more readily above the cluster T_g than it does without the excess neutralizing agent.

The influence of excess neutralizing agent on the properties of styrene ionomers differs from study to study, in part owing to differences in sample treatment, which could influence the size of the crystallites. The area clearly needs further investigation, preceded by a consensus on uniform sample preparation conditions.

5.7. ION HOPPING

The first semiquantitative investigation of ion hopping in sulfonated styrene ionomers was performed by Hird and Eisenberg (36), who studied the modulus–tempera-

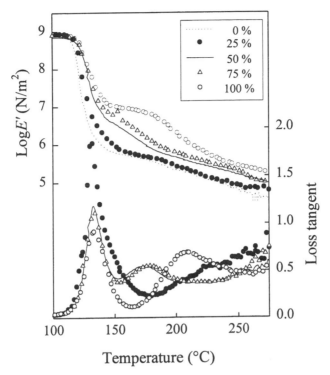

Figure 5.25. Log E' and tan δ versus temperature (1 Hz) for underneutralized P(S-5.1-MAA) ionomers. *Dotted line*, 0% neutralized; ●, 25%; *solid line*, 50%; △; 75%; ○, 100%. Modified from Kim and Eisenberg (56).

ture behavior in sulfonated styrenes, such as the 6.2 mol % ionomer, shown in Figure 5.26. In the ionic modulus region, the modulus does not vary much with temperature. As the temperature is increased beyond a certain point, the modulus decreases. The point of intersection between the horizontal segment of the ionic modulus and the linear decrease beyond the horizontal region gives a well-defined temperature, which depends on the frequency of the measurement. The plots in Figure 5.26 are constant frequency plots. The network (with multiplets as crosslinks) remains intact over the temperature range for which the plot is horizontal. The decrease in the modulus is the result of ion hopping. Thus the intersection of the two straight line segments, which is illustrated for the 30 Hz curve, can be taken as the onset of ion hopping (at a given frequency). From an Arrhenius plot of the onsets (log frequency) versus reciprocal temperature, an activation energy E_a of 190 kJ/mol is obtained. If the first deviation from the horizontal line (Fig. 5.26, arrow) is taken as the onset of ion hopping rather than the intersection of straight line segments, the activation energies obtained are similar. The ion hopping activation energies for various ion contents obtained by both methods are given in Table 5.4. The activation energies

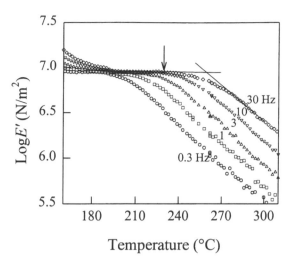

Figure 5.26. Log E versus temperature curves measured at several frequencies for a P(S-6.2-SSNa) sample. ◇ 30 Hz; ▽, 10 Hz; △, 3 Hz; □ 1 Hz; ○, 0.3 Hz. Modified from Hird and Eisenberg (36).

are independent of ion concentration and the method of determination, with a typical value of 190 kJ/mol.

Kim et al. (6) performed a similar analysis for the P(S-*co*-MANa) system. In this case, the activation energies for ion hopping start at approximately 200 kJ/mol for the 2 mol % sample and increase linearly to ~400 kJ/mol for the 10 mol % sample. This difference in behavior between the sulfonates and the methacrylates shows that, although the position of the loss tangent peaks is a function of the strength of the interaction, the peak for the sulfonates is located at a higher temperature than that for the methacrylates, the activation energies for ion hopping are not related to the strength of the interaction. The activation energies are probably more dependent on the degree of cooperativity of the process, because they are ~190 kJ/mol for the sulfonates (which interact strongly), but 200–400 kJ/mol for the methacrylates (which interact less strongly); the activation energies are probably also related to the presence of other mechanisms that may be contributing, e.g., flow.

TABLE 5.4. Ion Hopping Activation Energies for P(S-*co*-SSNa) Ionomers

Ion Content, mol %	E_a from Intersection, kJ/mol	E_a from First Deviation, kJ/mol
3.5	170	175
6.2	190	170
7.0		220
8.0	220	195
11.0	185	180

In connection with ion hopping, the reader's attention should be called to a theoretical paper that discusses the dynamics of reversible networks (61). While none
of the experimental ionomer systems studied to date is suitable for direct comparison
with that theory at present, it is anticipated that future theoretical and experimental
work will follow along these lines.

Another type of ion hopping—alkali counterion transfer from carboxylic ionomer
to a polymer containing sulfonic acid in solution—has been explored by Mikeš et
al. (62). Because solution properties of ionomers are not covered in this book, the
reader is referred to the original literature.

5.8. OTHER PHYSICAL PROPERTIES AND SPECTROSCOPY

5.8.1. Density

Only a few determinations of the density of styrene ionomers have been performed.
Yano et al. (63) measured the density as a function of degree of neutralization styrene
ionomers containing 4.4 mol % methacrylic acid neutralized by copper. They showed
that the density of the ionomers increases with increasing degree of neutralization,
from 1.046 to 1.074 g/mL over the range of 7–79% neutralization. Above this degree
of neutralization, the density levels off up to 139% neutralization, beyond which no
data were given. In another study published in the same year, Arai and Eisenberg
(64) looked at the densities of completely neutralized P(S-*co*-MANa) copolymers
of 0–9 mol % ion content (Fig. 5.27). The best-fit second-order polynomial to the
data was

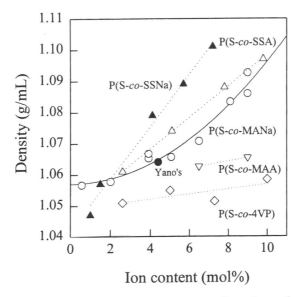

Figure 5.27. Density versus ion content at room temperature for various polystyrene ionomers
(63–66).

$$\text{Density (g/mL)} = 1.057 + 6.53 \times 10^{-5} x + 3.85 \times 10^{-4} x^2 \quad (5.6)$$

where x is the mol % of ions. The line is shown in Figure 5.27, along with one datum point from Yano et al.'s work (65). The densities of two samples of styrene-co-methacrylic acid copolymers, of sulfonated polystyrene in the acid form, of un-quarternized P(S-co-4VP) copolymers and of P(S-co-SSNa) ionomers (66) are also shown.

5.8.2. Dielectric Properties

Hodge and Eisenberg (67) conducted a study of the dielectric properties of P(S-co-MANa) ionomers >100°C over a frequency range of 10^1–10^4 Hz and an ion concentration range of 2–9 mol %. Not unexpectedly, the dielectric constant increased from ~2.7 to 3.2 as the concentration of ionic groups increased from 2 to 9 mol %. In the glass transition region, the conductivity losses were found to be high and dominated the data at the lowest frequencies. However, it was also found that the limiting low frequency conductivity could be evaluated directly from the high temperature data and from the low frequency invariant and could thus be subtracted. The observed conductivity activation energies were found to be independent of composition at 26 ± 2 kcal/mol. The conductivity was found to increase uniformly with increasing salt concentration, with a ratio of conductivities for the 9.0 mol % and the 3.9 mol % samples of ~3.

A plot of the dielectric loss tangent as a function of temperature for the 6.5 mol % sodium methacrylate sample at four different frequencies is given in Figure 5.28.

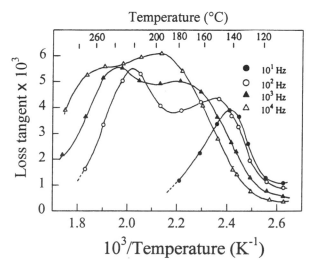

Figure 5.28. Dielectric tan δ versus reciprocal temperature for a P(S-6.5-MANa) sample at various frequencies, after subtraction of the conductivity contribution. ●, 10 Hz; ○, 10^2 Hz; ▲ 10^3 Hz; △ 10^4 Hz. Modified from Hodge and Eisenberg (67).

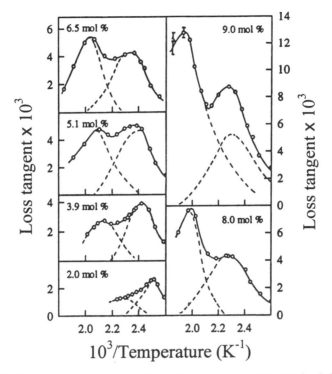

Figure 5.29. Dielectric tan δ versus reciprocal temperature (100 Hz) for P(S-*co*-MANa) ionomers of various ion contents. The *dotted lines* represent the deconvolution into two symmetric peaks. Modified from Hodge and Eisenberg (67).

It is seen clearly that after subtracting the dc conductivity components, two peaks are present that, as might be expected, shift with frequency by different amounts. The 100 Hz loss tangent data for various ion contents are shown in Figure 5.29. The attempts at deconvolution in the figure are primitive, and the curves should be taken only as indications of trends rather than of absolute values. The loss tangent peak height as functions of ion content is shown in Figure 5.30. Because the half widths are approximately the same, the heights are a close representation of the areas under the curves. The sum of the areas of the two peaks increases linearly with increasing ion content. This result is in marked contrast to dynamic mechanical properties, for which the integrated areas are independent of ion content. This difference is not surprising when one considers that in dielectric measurements it is the polar groups that respond, whereas in mechanical studies it is the polymer as a whole. Therefore, because the number of polar groups increases with increasing ion content, the absolute value of the dielectric loss should also increase. The peak positions were also investigated as a function of ion content. The trends observed are essentially the same as those seen for the mechanical property measurements.

Arai and Eisenberg (64) investigated the dielectric loss ϵ'' as a function of tempera-

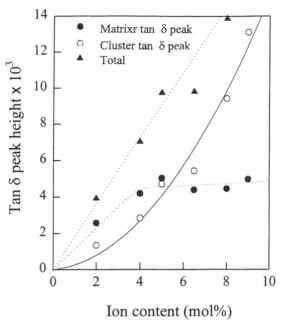

Figure 5.30. Dielcctric tan δ peak heights versus ion content for P(S-*co*-MANa) ionomers. ●, matrix tan δ peak; ○, cluster tan δ peak; ▲ total. Modified from Hodge and Eisenberg (67).

ture for different degrees of neutralization. From Figure 5.31, it is clear that the unneutralized P(S-9.0-MAA) sample shows only a low temperature peak, which increases in both intensity and temperature with increasing degree of neutralization. At 50% neutralization, one sees the beginning of a second peak at higher temperature, which manifests itself as a broadening of the single peak. The second peak becomes evident for the 75% neutralized sample, and dominant in the 90 and 100% samples. Again, both the intensity and position of the high temperature peak increase with increasing degree of neutralization, in parallel to the observations in dynamic mechanical studies.

Figure 5.32 shows conductivity as a function of temperature for samples of various degrees of neutralization. It is clear that the ionomers, in the absence of water, possess low conductivity, not very different from that of normal dielectric materials, such as nonionic polystyrene. In this particular case, a mixture of the 100% neutralized ionomer and the acid copolymer was also investigated; the curve labeled "50% Blend" in Figure 5.32. One significant result of the study of the dielectric strength in the sodium-neutralized ionomer of 9 mol % suggests that only ~2% of—COO⁻ Na⁺ groups are present as simple ion pairs that are dielectrically active; all the others are incorporated in the multiplets. This concept should be borne in mind when one considers the degrees of incorporation of ionic groups into multiplets; however, estimates from results of different techniques of the number of isolated ion pairs differ widely for the various ionomers.

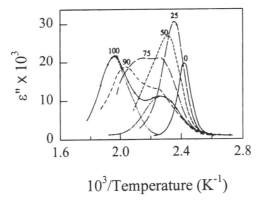

Figure 5.31. Dielectric loss versus reciprocal temperature (100 Hz) for the partly neutralized P(S-9.0-MANa) copolymers. Numbers indicate degree of neutralization. Modified from Arai and Eisenberg (64).

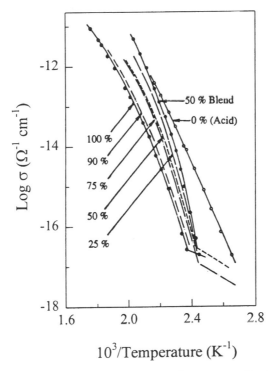

Figure 5.32. Log σ (dc conductivity) versus reciprocal temperature for the partly neutralized P(S-9.0-MANa) ionomers and a 50% blend. Numbers indicate the degree of neutralization. Modified from Arai and Eisenberg (64).

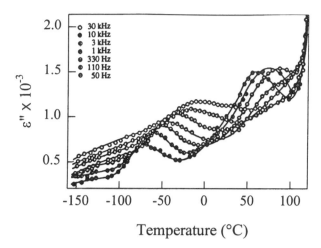

Figure 5.33. Dielectric loss versus temperature for Cu²⁺ 47% neutralized P(S-4.4-MAA) copolymer at various frequencies. ○, 30 kHz; ◑, 10 kHz; ◐, 3 kHz; ◓, 1 kHz; ◒, 330 kHz; ◔, 110 Hz; ⊖, 50 Hz. Modified from Yano et al. (63).

Partly neutralized styrene-*co*-methacrylic acid ionomers were investigated by Yano et al. (63,65). The results for the 47% neutralized Cu²⁺ sample for an ionomer containing 4.4 mol % of acid groups are shown in Figure 5.33. The low temperature peak, called the γ relaxation, has been ascribed to a local motion below T_g. By contrast, the higher temperature β relaxation has been ascribed to local motion of short segments, which involve —O—Cu—O— bonds. Remember, however, that in a subsequent study (65), the β peak was also found, to a small extent, in the acid form of the copolymer. A review of dielectric behavior and related molecular processes in ion-containing polymers is available (68); it includes some of the studies on the styrene and other ionomers.

5.8.3. Spectroscopy

5.8.3.1. Infrared. In an early study of the far-infrared spectra of ionic groups in ionomers, Tsatsas and Risen (69) showed that bands in the 100–500 cm⁻¹ range were sensitive to the type of cation. They found that the Cs⁺ salt of ethylene methacrylic acid exhibited a broad band in the 100–250 cm⁻¹ range, centered around 135 cm⁻¹, which for Na⁺ shifts to 230 cm⁻¹ and for Li⁺, to 450 cm⁻¹. They suggested that the breadth of the cation motion band indicates a multiplicity of environments for the cations, which differ in the forces exerted on the cation. In a subsequent study (70), the styrene ionomers were investigated, especially the styrene-*co*-methacrylate system. Again, a cation-sensitive band was found in the 100–300 cm⁻¹ range. For the Na⁺ salt, the band centers around 250 cm⁻¹; for the Ba²⁺ salt, around 185 cm⁻¹; and for the Cs⁺ salt, around 115 cm⁻¹. When the materials were studied as a function of Na⁺ ion content, the 250 cm⁻¹ band was found to exhibit a strong shoulder at

~170 cm^{-1} for ion contents >3.8 mol %. The 250 cm^{-1} band was assigned to a low-order multiplet mode and the shoulder at 170 cm^{-1}, to a cluster mode, involving vibrations of aggregates with a number of cationic and anionic sites close together. Two additional bands are seen in the carboxylate system. One is located at 405 cm^{-1}, with an intensity that is independent of ion content and is assigned to vibrations of the polymer backbone. The other, at ~385 cm^{-1}, increases in intensity with ion content and is assigned to vibrations of the carboxylate anion (Fig. 5.34).

In a study of sulfonated systems, Mattera and Risen (71) found that the positions of bands in the 100–300 cm^{-1} range were very much a function of the type of cation present, with the Cs$^+$ band centered around 100 cm^{-1}, rising progressively with decreasing mass of the cation to ~210 cm^{-1} for Na$^+$. Plots of the peak positions against the reciprocal square root of the mass of the counterion are shown in Figure 5.35 for both divalent and monovalent cations. It was suggested that the vibration is primarily the result of motion of the cation under the influence of the electrostatic field of the anionic groups. The force constants obviously differ for the monovalent and divalent cations. The bands for the lithium- and magnesium-neutralized sulfo-

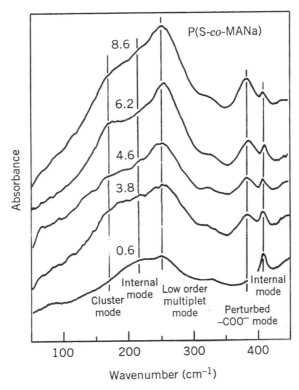

Figure 5.34. Far-infrared spectra of a series of high molecular weight P(S-*co*-MANa) iono-mers with varying ion contents. Numbers refer to the ion concentration (mol %). Modified from Rouse et al. (70).

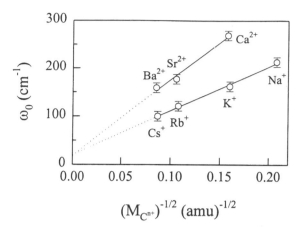

Figure 5.35. Cation-motion frequency versus $(M_c^{n+})^{-1/2}$ for films of the dehydrated P(S-6.9-SSA) copolymers neutralized with various cations. Modified from Mattera and Risen (71).

nated polystyrene ionomers (at 470 and 440 cm^{-1}, respectively) appear at much higher vibrational frequencies, which suggests that the cation–sulfonate forces are covalent, unlike the other cations. It was also, found that the band moves to a slightly lower wave number as the ion content increases. This behavior was explained by assuming that with increasing ion content, more of the cations are in large ionic aggregates.

The earliest study of the near-infrared spectroscopy of ionomers was MacKnight et al.'s (72), which dealt with ethylene-co-methacrylic acid copolymers and their sodium salts. They observed dimerization of the carboxylic acids (via hydrogen bonding) and measured the heat of dissociation of the dimers ΔH from the slope of the van't Hoff plot (73) as 48.5 kJ/mol for the dimer, or 24.3 kJ/mol per hydrogen bond at room temperature. MacKnight et al. (72) also used the 1700 cm^{-1} band (the dimerized carbonyl stretching vibration) to estimate the degree of ionization and found that the degree of neutralization was inversely related to the integrated absorbance of the 1700 cm^{-1} band. The 1750 cm^{-1} band was assigned to the monomeric carbonyl stretching vibration. They showed that the peak absorbance at 3540 cm^{-1}, the free hydroxyl stretching band, was also inversely proportional to the degree of ionization.

In a subsequent investigation, devoted to glass transitions in partially neutralized styrene-based ionomers, Ogura et al. (74) found kinks in the plot of the intensities of 1700 and 1745 cm^{-1} bands against temperature, which they ascribed to the glass transition (Chapter 4). The intensities of these two bands move in opposite directions as a function of temperature, and so do the discontinuities. A plot of the ratio of the intensities of the 1745 and the 1700 cm^{-1} bands shows a pronounced discontinuity, from which the glass transition temperature was estimated as a function of salt content or degree of neutralization. They correctly determined that the glass transition that they observed (which turned out to be the matrix glass transition) was not in any way related to the onset of mobility in the ionic domains.

Following the work of Zundel (75), Fitzgerald and Weiss (76) provided detailed peak assignments for the infrared spectra (in the $500-1400$ cm^{-1} range) of sulfonated styrene ionomers neutralized with Zn^{2+}, Mn^{2+}, Cu^{2+}, and Na^+ and the acid form. In another study, Fan and Bazuin (77) correlated (after Zundel) some of the IR bands for sulfonated polystyrene neutralized with 10 different organic cations involving bifunctional, trifunctional, and quadrifunctional amine or pyridine groups.

5.8.3.2. Raman.

In early Raman spectroscopic studies of styrene ionomers, Neppel et al. (78–80) investigated both P(S-*co*-MANa) and P(S-*co*-SCNa) with a range of ion contents. Two bands (at 170 and 250 cm^{-1}) were present in the neutralized materials for both copolymer families that are not present in the acid. The band at ~ 250 cm^{-1} was found to be relatively concentration independent, whereas that at ~ 170 cm^{-1} increases in intensity with increasing ion content. The lower wave number band was assigned to ionic motions in the clusters and the higher, to motions of ions within unclustered multiplets. From the relative intensities, the percentage of ions in the various environments was estimated as a function of ion content and as a function of temperature. For P(S-*co*-MANa), it was shown that the relative intensities in the Raman spectrum of the cluster band correspond with those of the integrated intensity of the dielectric cluster band. These bands were also seen in the ethyl acrylate system and were interpreted in the same manner as those in styrene (81). The results of all these Raman studies suggest that in some systems, ions are present in two different environments. In the original paper (78), this was taken to indicate ions in clusters or in multiplets. However, in the light of the EHM model, this explanation may need to be revised. One possible explanation of the two bands could be ions in isolated pairs and ions in multiplets. The effect of water might also be involved. A reinvestigation seems warranted. It should be pointed out that no Raman bands associated with ionic species such as multiplets or clusters were seen in polyethylene ionomers (82).

5.8.3.3. Nuclear Magnetic Resonance.

Only a few solid-state NMR studies have been performed on styrene ionomers. In the first of these, Dickinson et al. (83) found a broad peak at -40 ppm in the magic angle spinning ^{23}Na NMR spectrum of 100% Na^+-neutralized sulfonated polystyrene. They also found that, upon exposure of the sample to air of 45% humidity for 12 h, the intensity of the broad peak significantly decreased and new narrow peak appeared at 0 ppm. A similar peak was also found in the spectrum of the dried 250% neutralized sample.

In a more recent study, Park et al. (84) showed by using the spin echo technique that the sodium salts of sulfonated styrene ionomers exhibited microphase separation, because the magnetization decays could be resolved into two components. Not surprisingly, it was found that the volume fraction of the ionic phase was significantly influenced by the ion concentration and the temperature. It was suggested that the ionic regions contain a substantial amount of hydrocarbon material.

O'Connell et al. (60) used ^{23}Na NMR to study the morphology of lightly sulfo-

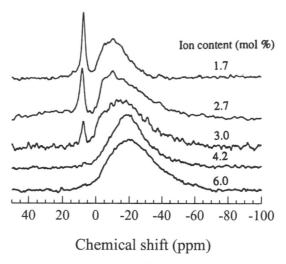

Figure 5.36. Effect of ionization level on the ^{23}Na NMR spectrum of P(S-*co*-SSNa) copolymers. Numbers indicate ion concentration (mol %). Modified from O'Connell et al. (60).

nated polystyrene. Specifically, magic angle spinning was used to investigate the environments of the sodium ions. In the dry materials, two different environments were detected: single sodium ions and those present in higher aggregates. In Figure 5.36, two peaks are seen: a sharp peak at ~7 ppm, and a broad peak at − 10 to − 20 ppm. It is clear that with increasing ion content, the intensity of the sharp peak decreases and disappears completely at ~4 mol % ions. This peak has been assigned to single ions, and the broad peak has been assigned to ionic aggregates. It is evident that the number of single ions decreases as the ion content increases and that eventually all the ions are present in aggregates. The authors also investigated hydration and found that all ions are subject to hydration, whether single or in aggregates. Finally, they examined the effect of overneutralization and noted that the ionic aggregates can incorporate excess NaOH up to 200%; beyond that the excess NaOH aggregates into separate crystallites. It should be pointed out that NMR has been used extensively to study polymer blends (Chapter 9).

5.8.3.4. Electron Spin Resonance.

ESR has been used only to a limited extent in the styrene ionomers, primarily to probe the environments of the single ions or ions in aggregates. Since ESR requires the presence of an unpaired electron, it has primarily been used for acids neutralized with transition metal ions. The first study on ESR of ionomers was performed by Pineri et al. (85), who investigated butadiene-*co*-methacrylic acid copolymers. The first study on styrene ionomers was performed by Yamauchi and Yano (86), who investigated styrene-*co*-methacrylic acid copolymers neutralized with copper acetate. They found that Cu^{2+} was present in two different environments. In one, the Cu^{2+}—Cu^{2+} pair distance corresponded to a Cu^{2+}–acetate-type system, but there were also other Cu^{2+}—Cu^{2+} pairs. Evidence for the existence of isolated Cu^{2+} ions was noted. The authors suggested that the relative proportions of ions in these environments change with the mole fraction of the functional groups in the polymers.

Weiss et al. (87) investigated sulfonated polystyrene neutralized with Mn^{2+}; the sulfonate content was 0.9, 3.2, and 5.5 mol %. For samples that were precipitated into methanol and dried at room temperature under vacuum, the lowest ion content material exhibited six distinct lines in its spectrum, characteristic of isolated ions. As the ion content increased to 5.5 mol %, the spectrum changed to a broad single peak, indicating that the Mn^{2+} ions were no longer present singly but in aggregates. Similar behavior was also found for the lowest ion content sample, upon heating the dried precipitated material to 280°C followed by quenching in liquid nitrogen. The unheated sample showed the six-line spectrum; heating induced a coalescence into a single broad peak, indicating the formation of aggregates. ESR thus allows one to follow the process of a conversion of single ion pairs into ionic aggregates as the precipitated powder coalesces into a bulk sample.

A more quantitative analysis appeared a follow-up paper by Toriumi et al. (88). A much larger number of ion concentrations was investigated, ranging from 0.26 to 5.5 mol %, so the detailed shape of the spectrum could be followed as a function of ion concentration at constant temperature and as a function of temperature for a constant ion concentration. The same general trends as those found in the first study were observed. For the 0.9, 3.2, and 5.5 mol % (as precipitated) samples, it was found that 16, 56, and 75% of Mn^{2+} ions were present in aggregates rather than singly.

In a subsequent publication by Fitzgerald and Weiss (89), the effect of the presence of low molar mass diluents (plasticizers) on the microstructure of Mn^{2+}-neutralized ionomers was explored. Glycerol was used as the plasticizer. The ESR results on plasticized bulk ionomers indicated that the P(S-1.82-SSMn) sample plasticized with 10 wt % glycerol retained ionic aggregates. At this level of plasticization, glycerol completely changes the mechanical properties by eliminating the high-temperature loss tangent peak. These results, therefore, suggest that, although the aggregates remain intact, the mobility of ion pairs into and out of multiplets has increased to the point at which the cluster glass transition has been eliminated. These results are essentially confirmed by some SAXS results, which show that on plasticization, the ionomer peak moves to lower q values, suggesting an increase in the distance between multiplets and probably also in their size.

5.8.3.5. Extended X-Ray Absorption Fine Structure.

In typical extended x-ray absorption fine structure (EXAFS) studies, the intensity of absorption coefficient as a function of ejected photoelectron wave vector (or radial structure function versus radius) is monitored near the absorption edge of the metal cation being investigated. The detailed shape of the plot provides information about the environment of the ion in question. EXAFS has been used extensively to study the local environment around moderately heavy ions; the technique is well developed, and its theory is understood, but the results depend on the quality of the fit of the experimental spectrum for a particular model (90).

A series of papers reporting the EXAFS studies of styrene ionomers originated from the Cooper's group, starting in 1980 (91). In a more recent paper (92), the environment of the Rb^+, Sr^{2+}, Zn^{2+}, and Ni^{2+} cations in sulfonated polystyrene was explored in detail. It was found that the local order around the cations varies appreciably, depending on the type of cation. Thus for Zn^{2+}-sulfonated polystyrene, it was suggested that the first coordination shell around the zinc has a crystal structure of ZnO (i.e., fourfold coordination) with distance between the zinc and the near oxygen atom being 0.197 nm. The second shell was much more difficult to pinpoint, but the best fit was obtained with four sulfur atoms and four oxygen atoms at nearly the same distance from the zinc cation, at about 0.315 nm (92,93). The second shell peak is relatively small, because sulfur and oxygen have phase shifts that nearly cancel each other, even though the structure suggests a well-ordered second coordination shell. Furthermore, it was suggested that the Zn^{2+} cations were bridged by two sulfonate groups. The situation is quite different in Ni^{2+}, which has a larger radius and shows an apparent coordination number of six. The nickel ionomer exhibits a structure similar to nickel benzenesulfonate, unlike its zinc counterpart. In the rubidium salts of sulfonated polystyrene, the oscillations in the EXAFS spectrum are weak, so it is difficult to distinguish peaks in the radial structure function. It was, therefore, suggested that no well-ordered local structure exists in this material. In the strontium ionomer, on the other hand, a reasonable first shell is seen, but no second coordination shell.

It is clear that the local structure of various cations can differ drastically, for the same base polymer, which is not surprising because the electronic structures of the transition metal ions are different. For rubidium (the only alkali metal ion investigated in the work of Cooper et al.) the absence of the local structure arises because the ion is spherical. Finally, as the authors pointed out, the local environment is not an important factor in the ion aggregation, as judged, for example, from the SAXS peak, because all the ions are known to lead to strong ion aggregation.

In a subsequent publication (94) plasticization of zinc-neutralized sulfonated polystyrene ionomer with glycerol, dioctyl phthalate, acetonitrile, and toluene was investigated by EXAFS. It was found that the addition of glycerol changed the coordination shell of Zn^{2+} ions dramatically. At full plasticization the Zn^{2+} ion was coordinated to three glycerol molecules. By contrast, the other plasticizers (i.e., dioctylphthalate, acetonitrile, toluene) did not affect the local structure of the cation, suggesting that the coordination of the sulfonate group to the ion is not altered.

In a more recent study (95) Mn^{2+}-neutralized sulfonated polystyrene ionomers with different levels of sulfonation were investigated. It was found by SAXS that no ionic aggregates were present in the samples cast from tetrahydrofuran (THF) and water solutions (Chapter 3) and that ion aggregation did not manifest itself by the appearance of an ionomer SAXS peak until the sample was annealed at high enough temperature (96). The EXAFS pattern of the cast samples, however, was independent of the thermal treatment of the samples and of the ion concentration.

More recently, the coordination environment of Ni^{2+}, Cd^{2+}, and Zn^{2+} was explored in lightly sulfonated polystyrene (97). It was found that Ni^{2+} is octahedrally coordinated (i.e., has a coordination number of 6) to the oxygen atoms in the first shell, in contrast to Zn^{2+}, which is tetrahedrally coordinated. The second-shell atoms in both cases are sulfur, with a metal–oxygen–sulfur angle of 135–140°. It is possible that Ni^{2+} has two waters of hydration in the coordination structure. In the Cd^{2+} system, the ions are also octahedrally coordinated to oxygen. In a companion paper (98), the effect of uniaxial orientation was investigated. It was found that, in the zinc-neutralized materials, the metal–oxygen first-shell distance increased parallel to the stretch direction. In contrast, in the nickel-neutralized materials, the metal–oxygen distances did not increase in the first shell, although the nickel–oxygen bond vectors aligned in the stretch direction. The sulfur atoms in the second coordination shell did not seem to be affected by uniaxial orientation. No changes were found in the coordination environment of the Cd^{2+}-neutralized material.

5.8.4. Orientation

Infrared dichroism measurements were used to determine the second moment of the orientation distribution function $\langle P_2(\cos\theta)\rangle$, which is defined as $(3\langle\cos^2\theta\rangle-1)/2$, where θ is the angle between the chain axis and the stretch direction. The relationship between the orientation distribution function and the dichroic ratio R is given by

$$\langle P_2(\cos\theta)\rangle = [(R-1)(R_0+2)]/[(R+2)(R_0-1)] \qquad (5.7)$$

where R_0 equals $2\cot^2\alpha$ and α is the angle between the dipole moment vector of the vibration and the chain axis; and R equals $A_\parallel/A_\perp$, the ratio of the absorbances for the electric vector parallel and perpendicular to the stretching direction. The first study of orientation in styrene ionomers was performed by Zhao et al. (99). No orientation effect was observed in the case of the acid, suggesting that hydrogen bonds above T_g are not effective in maintaining orientation. By contrast, in the neutralized systems, a considerable enhancement in the orientation function was observed as the draw ratio λ increased. The rate of increase was highest for the sodium salt, somewhat lower for the cesium salt, still lower for the partly neutralized systems, and even lower for the acid or the unfunctionalized styrene when the experiment was performed at $(T_g+20)°C$. It was also observed that the greater the ion content, the greater the orientation. These enhancements in the orientation function have been attributed to the presence of ionic aggregates (multiplets and clusters), which act as effective cross-links thus modifying the density of the temporary network structure.

Fan and Bazuin (100) compared sulfonated polystyrene and the styrene-co-methacrylate ionomers, and found major differences between the sulfonates and the methacrylates, as expected in view of the relative interaction strengths. In both cases, when stretching was performed at $(T_g+75)°C$ (where T_g is the matrix value determined by DSC) the slope of $\langle P_2(\cos\theta)\rangle$ versus λ plot (i.e., the increase of orientation) increased with ion content; however, at comparable ion contents, the slopes for the methacrylates were much smaller than those for the sulfonates.

A comparison of the $d\langle P_2(\cos\theta)\rangle/d\lambda$ curves for the sulfonates and the methacrylates as a function of ion content at $(T_g + 75)°C$ is shown in Figure 5.37. Not surprisingly, the amount of orientation at $(T_g + 75)°C$ is considerably lower than that observed in the first study at $(T_g + 20)°C$ for comparable ion contents, because the chains can relax more during stretching at $(T_g + 75)°C$ than at $(T_g + 20)°C$. The differences in orientation between the sulfonates and the methacrylates was attenuated when $d\langle P_2(\cos\theta)\rangle/d\lambda$ was plotted against the difference between the stretching temperature and the cluster glass transition temperature $(T - T_g)$, although a major difference is still present (Fig. 5.38).

The considerably higher value for the sulfonate systems indicates that ion hopping, even when normalized against the cluster glass transition, is still slower in the sulfonates than in the methacrylates. Not only the ion-hopping kinetics but also the volume fraction of clustered material must be considered here. In the sulfonate case, the larger aggregates, which result in a smaller volume fraction of the material of reduced mobility, lead to a larger volume fraction of unclustered material at constant ion content than in the methacrylates. Therefore, because orientation takes place primarily in the unclustered regions of the material (because they are softer), the sulfonated polystyrene would be expected to show a higher orientation function at comparable ion contents, quite aside from any considerations of ion hopping kinetics.

Another study was devoted to the orientation behavior of the P(S-5.0-SSA) mixed with bifunctional or multifunctional organic molecules based on amines and pyridines (101). For the range of basic molecules used, the degree of proton transfer varied from 98 ± 1 to $35 \pm 10\%$, depending on basicity. Not surprisingly, it was found that the slope of the orientation function versus draw ratio plot was a direct function of the degree of proton transfer. If stretching was done at $(T_g + 15)°C$,

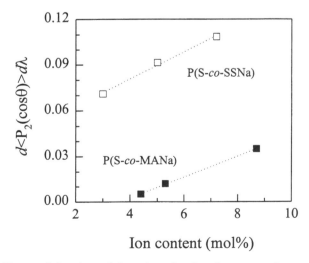

Figure 5.37. Slopes of the plots of the orientation function versus draw ratio for P(S-co-SSNa) (□) and P(S-co-MANa) (■) of various ion contents. Modified from Fan and Bazuin (100).

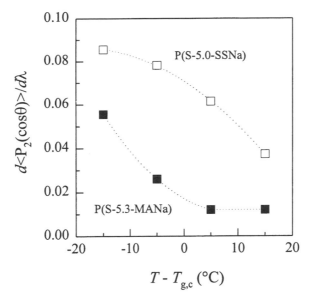

Figure 5.38. Slopes of the plots of orientation function versus draw ratio for P(S-5.0-SSNa) (□) and for P(S-5.3-MANa) (■) for various temperatures. Modified from Fan and Bazuin (100).

the slope ranged from ~11 for the 100%-protonated material down to 4.5 for the 35%-protonated system. This study shows clearly that orientation is primarily a function of the degree of proton transfer, which is directly related to the number of effective intermolecular cross-links. The greater the effective cross-linking, the greater the orientation achieved. A similar linear relationship was found between the Young's storage modulus at $(T_g + 15)°C$ and the degree of proton transfer in the ionomers, despite greater scatter in the datum points.

Blends of P(S-co-SSA) and P(S-co-4VP), discussed earlier, also show orientation behavior that is similar to that of the ionomers (102). In this case, the cross-links are provided by the pyridinium sulfonate ion pairs. As for the ionomers, the rate of increase of the orientation function with draw ratio increases with the degree of functionalization. In contrast, blends of P(S-co-MAA) and P(S-co-4VP) show orientation behavior that, at 5 mol % functionalization, is comparable to that of polystyrene and marginally lower than that of P(S-co-SSA), indicating again that hydrogen bonds are too labile to produce orientation under the conditions of the experiment. However, orientation of P(S-co-SSA) does increases somewhat with sulfonic acid content (although less than for blends).

5.8.5. Gas Transport

A range of polymer types has been under study as candidate materials in gas separation membranes; however, few studies have been performed on gas transport in styrene ionomers (103,104). Chen and Martin (104) explored the diffusion of several gases (N_2, O_2, H_2, CO_2, and CH_4) in sulfonated polystyrene neutralized with either Na^+ or Mg^{2+} ions over a sulfonate concentration range of 0–27.5 mol %. In general, they found that gas transport decreases appreciably with increasing sulfonation; in some cases the decrease in the permeability coefficient was as high as an order of magnitude, although mostly it was somewhat lower. Because, in general, polymers with high gas transport selectivities usually show low permeabilities (and vice versa), it was suggested that the selectivities should increase with sulfonate content because permeabilities decrease. This result was, indeed, found to be the case. For example, the selectivity coefficients for CO_2/CH_4 increased from ~20 to 70 as the ion content increased from 0 to 27.5 mol % for the Mg^{2+} salt. The H_2/CH_4 selectivity coefficients increased from ~20 to 200 for the Mg^{2+} salt, and from 20 to ~120 for the Na^+ salt over the same range of ion contents. The same trend was observed for O_2–N_2, though the variation was considerably smaller. In all cases, the Mg^{2+} salt showed higher selectivities and lower permeabilities than did the equivalent Na^+ salts, because the diffusion coefficient of a gas in a polymer generally decreases with the size of the gas molecule. The solubility coefficients, on the other hand, are governed by the condensability of the gases, being highest for the most condensable (CO_2) and lowest for the least condensable gas (N_2).

It was noted that for the sodium-neutralized polymers, the solubilities of all gases increase with the degree of sulfonation. Sodium neutralization introduces polar groups ($—SO_3^- Na^+$) into the polymer, thus increasing the polarity of the material and hence the dipole-induced dipole interactions. The polarity of the styrene is not increased in a comparable way when magnesium sulfonate groups are introduced, because Mg^{2+}, being divalent, is associated with two sulfonate groups; thus, as was suggested, it lacks the polar character of the sodium sulfonate. The solubility coefficient does not increase with sulfonation in the magnesium sulfonate as it does in the sodium sulfonate case.

5.8.6. Water Absorption

Few water absorption studies have been performed on styrene ionomers. In an early study of the P(S-*co*-MANa) system, Eisenberg and Navratil (5) investigated water uptake as a function of time. The samples were ~1 mm thick and rectangular in shape; water uptake was followed for >4000 h. It was found that in samples containing <6 mol % of ions, the water uptake at steady state was close to one water molecule per ion pair. However, >6 mol % of ions, the water content depended on the ion content. The 7.9 mol % sample adsorbed about three water molecules per ion pair at equilibrium, whereas 9–10 mol % samples had about five water molecules per ion pair.

In a subsequent study, Escoubes et al. (105) investigated water uptake for thinner molded specimens and freeze-dried samples over considerably longer periods. In the molded samples, they found that, at room temperature, the log of the H_2O/Na^+ ratio increased with the log of time, initially at a relatively high rate; the slopes of the initially linear plots were independent of ion content and the samples with the highest ion content showed higher absorption than those of lower ion contents. After 1 day, the slope of the log of water uptake versus the log of time curve changed and remained constant for about 1 year, at which point the work was discontinued. In the long-time range, the slope showed some relation to the ion content; it was highest for the highest ion content at room temperature and at 45°C. The plots, however, were essentially parallel to each other at 70°C.

Sorption isotherms for freeze-dried samples were also determined for different ion contents. It was found that the higher the ion content, the greater the sorption capacity and that the higher the desorption temperature, the smaller the hysteresis, especially for low salt contents. It was also found that the total hydration capacity increased with increasing ion content. Not surprisingly, the amount of nondesorbable water (which was still present under 1.3×10^{-2}Pa at 20°C but could be eliminated by heating the sample under vacuum at 180°C) was found to increase slightly with ion content. In addition, it was shown that at low partial pressures, the hydration energies of ions were of the same order of magnitude as the liquefaction energy of water, which indicated that the interaction between water and the ionic aggregates is of about the same strength as the interaction between water molecules. At high partial pressures, by contrast, the hydration energy drops considerably. It is believed that mechanical rearrangement of chains is needed to accommodate larger amounts of water into the salt.

Water uptake was also studied in the P(S-*co*-SCNa) ionomers by Brockman and Eisenberg (106), who found that, in contrast to P(S-*co*-MANa) ionomers, the water/salt ratio was a linear function of the ion content. The values for the two ionomers, however, were the same at ~7 mol % (at ~2.8 water molecules/ion pair). The reason for the difference in the behavior of two ionomers is not clear but may be related to the fact that the multiplets in the P(S-*co*-SCNa) ionomer are larger and the cluster regions are smaller at all ion contents. The diffusion of water in the P(S-*co*-SCNa) ionomers was also investigated. The diffusion coefficient drops with ion content, but only by ~1.5 orders of magnitude over the ion concentration range of 2–13 mol % (Fig. 5.39).

In a subsequent investigation by Yasuda et al. (107) of P(S-*co*-MANa) ionomers, it was suggested that diffusion could be interpreted by two-domain behavior, i.e., diffusion of water in hydrophilic domains (presumably the cluster) and in hydrophobic domains (presumably the matrix). It was shown that the diffusion coefficients increased upon swelling of the ionic aggregates at a low relative vapor pressure but decreased at a high relative vapor pressure, mainly because of the formation of water clusters. The study of water absorption by styrene ionomers of low ion content is certainly not complete at this time.

5.8.7. Transport of Nongaseous Materials

Perfluorosulfonate ionomers are frequently used as membranes, so extensive studies of water diffusion into and through the perfluorosulfonates have been performed and summarized (108–110). The styrene ionomers, however, were not used as membranes; thus few studies deal with the transport through or into these materials.

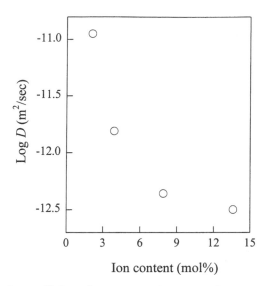

Figure 5.39. Diffusion coefficient of water versus ion content for P(S-*co*-SCNa) copolymer. Modified from Brockman and Eisenberg (106).

In one such study by Jiang et al. (21), the materials used were copolymers of styrene and methacrylic acid, acrylic acid, and styrenesulfonic acid. Film membranes of these materials were prepared, and their transport properties were studied by following membrane equilibration, electrolyte and water absorption, and self-diffusion. The studies were accompanied by measurements of the near-infrared spectra, primarily in the -OH band region. The morphology of the sample, determined by SAXS, was also followed as a function of water and ion contents.

The specific conductivity and the water:ion exchange site ratio for P(S-*co*-MANa) are shown as a function of ion content in Figure 5.40. The material undergoes a transition not unlike an insulator to conductor transition at an ion content of ~17 mol %, where the conductance changes by five orders of magnitude over a narrow range of ion content. The water content does not change nearly as drastically in the same region. The acrylate and sulfonate ionomers behave in a similar way, except that in the sulfonates the transition occurs at ~6 mol % of ionic groups, whereas in the acrylates it occurs at ~14 mol %. The onset of high conductivity was ascribed to percolation.

The transport behavior was interpreted within the percolation framework, using the following:

$$\kappa = \kappa_0 (c - c_0)^n \tag{5.8}$$

where κ and κ_0 are the specific conductivities of the membrane and the conducting phase components, c is the volume fraction of the conducting phase, c_0 is its volume fraction at the percolation threshold, and n is the critical exponent. For the acrylate

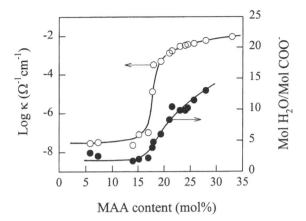

Figure 5.40. Log of specific conductance and equal water uptake versus ion content for P(S-co-MANa) of various ion contents in water-swollen membranes. Modified from Jiang et al. (21).

and methacrylate copolymers, c_0 is approximately 15% and n is close to 1.5. This result shows that the styrene ionomers behave essentially as a classical percolating system.

Unlike dry ionomers, the water-equilibrated ionomers show no SAXS peak at ion concentrations of 14 mol % or lower. In the dry state, it should be recalled that a SAXS ionomer peak is visible even for the 7 mol % sample, with a Bragg spacing of ~20 Å. As the ion content increases in the water-swollen state, a peak appears somewhere between 14 and 18 mol %, the intensity of which then decreases with further increases in the ion (and water) content. The absence of a peak in the water-swollen sample below 14 mol % indicates that no scattering centers are present with dimensions corresponding to the range of q values studied here. This must be related to the hydration of the ions and the absence of a strong enough driving force for aggregation and phase separation. Between 14 and 18 mol %, the equilibrated water content increases from ~6 to 13 wt % (i.e., from two to five water molecules per ion pair). The appearance of a peak has been ascribed to the onset of phase separation between regions containing the aqueous solution of high ion content and the regions of high hydrocarbon component on a size scale of 35 Å. The phase separation is no longer driven by cesium carboxylate ion pair interactions but by the hydrophobicity of the water-poor regions and the hydrophilicity of the regions that contain large amounts of water and hydrated ions. It seems that a minimum quantity of water is necessary for this phase separation to become apparent. It should be mentioned that phase separation of this type is not a prerequisite for high conductivity, because a similar insulator-to-conductor transition is observed in the poly(methyl methacrylate) ionomers at somewhat lower water content. However, in that case, there is no accompanying morphological transition of the type observed here.

5.9. FATIGUE AND FRACTURE PROPERTIES

The fatigue behavior of sulfonated polystyrene was investigated extensively by Hara et al. (58, 111–115). Their first publication in this area dealt with the effect of ion content on the fatigue properties (111). They found that the fatigue resistance initially decreased with ion content up to ~5 mol %; above that value, the fatigue resistance increased dramatically up to 6 mol % and thereafter gradually. The improvement in fatigue resistance was related to the presence of large ionic aggregates or clusters. The decrease in fatigue resistance with increasing ion content at low ion contents parallels the behavior of polymers at low cross-linking densities, showing again that ionic aggregates at low ion contents have some of the properties of cross-links. Above 5 mol % a filler effect takes over, with the filler acting as a reinforcing material.

In a subsequent publication, Hara et al. (112) investigated the effect of the counter-ion on the fatigue behavior. The sulfonate content of the sample was 4.1 mol %. They observed that K^+ and Cs^+ salts exhibited identical fatigue behavior. However, the Ca^{2+} salt showed an increase by a factor of three in the average number of cycles relative to that of the monovalent cations. The behavior of the Ca^{2+} sample at 4.1 mol % was reminiscent of the behavior of the Na^+ salt at considerably higher ion contents. This was ascribed to the stronger ionic interaction in the Ca^{2+} sample.

A third paper was devoted to the study of excess neutralizing agent (58). The authors found that excess neutralizing agent was present in the material in the form of small crystallites, which improved the fatigue behavior appreciably. By contrast, a 200% excess of neutralizing material produced a drop in the fatigue resistance, presumably as a result of the presence of large crystallites. This suggestion also explains the earlier finding that the preparation method had an effect on fatigue resistance (113). In the earlier study, a precipitation method was used to prepare the ionomers, which resulted in the presence of excess neutralizing agent. This excess improved the properties not as a result of the inherent behavior of the ionomers but, presumably, because of the presence of small crystalline particles that were found in samples intentionally neutralized to excess. In an NMR study by O'Connell et al. (60), it was also found that excess NaOH aggregated to form separate crystallites.

The tensile fracture properties of sulfonated styrene ionomers were discussed by Bellinger et al. (114). The tensile strength as a function of ion content for the sodium-neutralized samples is shown in Figure 5.41. It is seen that the tensile strength increases with increasing ion content up to an ion content of ~ 7 mol %. Thereafter it flattens, and even decreases. It was also found that the deformation mechanism of thin films of sulfonated polystyrene ionomers changes from crazing at low ion contents to crazing/shear deformation at higher ion contents (114,115). The plot of toughness versus ion content parallels that of tensile strength. The toughness for pure styrene is approximately 0.4 MJ/m^2, which increases to a plateau at ~ 0.55 MJ/m^2 (in the range of 1 to 4 mol % of ions) and reaches a peak at ~ 0.8 MJ/m^2 (at 7 mol % of ions).

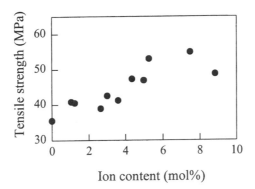

Figure 5.41. Tensile strength versus ion content for P(S-*co*-SSNa) ionomers. Modified from Bellinger et al. (114).

5.10. NONEQUILIBRIUM BEHAVIOR

In this section we discuss nonequilibrium phenomena, i.e., those in which annealing effects are important. An early study of nonequilibrium effects concerned the expansion coefficients of styrene-*co*-sodium methacrylate systems at different ion contents (116). It was found that the expansion coefficient below the matrix glass transition temperature α_g was essentially independent of whether it was measured on heating or cooling. However, in a first heating run from low temperature to above the glass transition temperature, it was observed that the expansion coefficient α'_L was considerably lower than the usual liquid expansion coefficients α_L of polystyrene or the styrene ionomers. If the sample was heated to what is now recognized as the region of the cluster glass transition and subsequently cooled, the expansion coefficient was much higher and typical of the liquid expansion coefficient of nonionic polystyrene. Cycling above the matrix glass transition temperature after cooling from the region of the cluster glass transition produced a reproducible expansion coefficient typical of polystyrene above its T_g.

Cooling from a high temperature to below the matrix glass transition resulted in an expansion coefficient below the matrix T_g (α_g), which is typical of that of polystyrene below its T_g. Thus the only nonequilibrium behavior occurred on the first heating run above the matrix T_g, and not the subsequent cycling above T_g (Fig. 5.42). The value of α_g was $(6.9 \pm 0.9) \times 10^{-5}/°C$; α_L was $(16.0 \pm 2.0) \times 10^{-5}/°C$; and α'_L was $(12.0 \pm 1.9) \times 10^{-5}/°C$, independent of ion content in the range investigated (0–9 mol %). A low first expansion coefficient was observed only for samples >6.4 mol %; this value was much later identified as being above the percolation threshold of clustering.

These phenomena can be explained by the EHM model. Below T_g, one observes the expansion coefficient typical of that of normal polystyrene below its T_g, because the material is completely glassy. If the material is a percolating system, the first expansion coefficient (α'_L) above the matrix T_g is dominated by the percolating hard

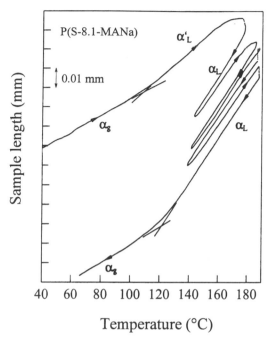

Figure 5.42. Sample length versus temperature for P(S-8.1-MANa) for several succes-
sive heating runs, keeping the sample above T_g. Modified from Eisenberg and Trepman
(116).

component (i.e., the cluster regions). However, once the cluster glass transition is
exceeded, the clusters have diminished or even disappeared, so that on cooling the
behavior is governed not by the clustered regions, which are now no longer percolative, but by the liquid matrix regions. Only storage at or near the glass transition
temperature of the matrix over an extended period of time reestablishes percolation
and hence the reemergence of the α'_L value on first heating from below T_g.

In another example of nonequilibrium behavior (65), the dc conductivity of a 50%
neutralized sample of styrene-*co*-sodium methacrylate copolymer was compared to
that of a blend of 100 and 0% neutralized materials. The results are shown in Figure
5.32. The blend sample was prepared in solution: The two powder samples (100
and 0% neutralized materials) were dissolved in benzene–methanol (90/10 v/v), the
solutions were mixed, stirred vigorously for 20 s, and freeze-dried; subsequently,
the sample was dried further at a temperature of 20°C above the glass transition,
under vacuum, overnight. In spite of the exposure to temperatures above the glass
transition for an extended period, the difference between the dc conductivity versus
temperature curves for the blend and the 50% neutralized sample persisted, suggesting that the exposure to temperatures above T_g was not sufficient to achieve the
equilibrium morphology. This, again, indicates that cation exchange between the
acid and the salt is slow.

Weiss (117) determined the change in heat capacity ΔC_p at the glass transition for sulfonated polystyrene as a function of the aging time of the sample. For pure polystyrene, an excess enthalpy peak was observed owing to the enthalpy relaxation process accompanying aging, similar to that found by Petrie (118). The 2.3 mol % sample showed a slight decrease in ΔC_p in the early stages of aging, but after ~ 5 days, ΔC_p leveled off. For the 5.5 mol % sample, a different picture emerged: ΔC_p dropped steeply and continued to decrease even up to 100 days. A DSC thermogram of the 5.5 mol % sample showed differences between the first and the second heating. The first heating produced a slight sub-glass transition endotherm, whereas the second did not. From this behavior, together with the change in ΔC_p with aging time, one can draw conclusions that are similar to those of the reinterpretation of Eisenberg and Trepman's (116) study.

Both of these studies (116,117) point to a development of the cluster phase with time at low temperatures. This cluster development aspect is important because it shows that large-scale mobility is not required for evolution of the clusters, although there must obviously be small-scale mobility. In the P(S-*co*-SSNa) at 5.5 mol %, the system is below the percolation threshold, whereas the P(S-*co*-MANa) system at 5.4 mol % ion content is above it. Therefore, the effects on heat capacity (sub-T_g endotherm) in sulfonates should not be expected to be as dramatic as those that were observed in the expansion coefficient study of the P(S-*co*-MANa) copolymers.

A different study related to nonequilibrium sample preparation conditions was performed by Lundberg and Phillips (119). Samples of sulfonated ionomers of 1.7 mol % were freeze-dried from a 4 wt % ionomer solution and also from a more dilute solution (0.3 wt %). The viscosities are shown as a function of annealing time in Figure 5.43. The higher concentration sample had a viscosity an order of magnitude greater than that of the more diluted sample; however, the viscosity of the second sample increased with storage time at an appreciable rate. A number of conclusions can be drawn from this experiment. First, the coil dimensions in the sample freeze-dried under high dilution must be smaller than those formed at high concentration, suggesting that the low-concentration sample has, most likely, a higher degree of intramolecular cross-linking. Furthermore, the slow rate of change in the viscosity of the low concentration sample shows that it takes a long time (much longer than 1 h at 220°C) for this sample to reach a viscosity comparable to that of the sample freeze-dried at higher concentration, i.e., for the coil dimensions to relax to their larger "equilibrium" values. The multiplets need not be appreciably different in the two systems, merely the degree of chain loopback into the same or neighboring multiplets.

In another study of time-dependent properties of styrene ionomers, the SAXS profile was measured as a function of temperature for a sulfonated polystyrene ionomer containing 7.6 mol % of Mn^{2+} sulfonate, which had been cast from a THF–water (90/10, v/v) mixture (96). It was found that the sample as-cast at room temperature showed essentially no SAXS ionomer peak, but a relatively high upturn at small angles. When the temperature increased to about 165°C a classical small-angle ionomer peak emerged, which was accompanied by a decrease in the intensity

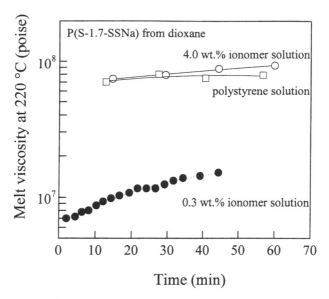

Figure 5.43. Effect of freeze-drying P(S-1.7-SSNa) at different polymer concentrations from dioxane on melt viscosity at 220°C. Melt viscosities of polystyrene are also shown. Modified from Lundberg and Phillips (119).

of the small-angle upturn. The intensity of the peak increased with increasing temperature to the maximum temperature studied (238°C). An accompanying ESR study of Mn^{2+} ions showed that throughout the entire temperature range (25–150°C), the manganese was present as an associated species. It was concluded that while multiplets are present at room temperature under these preparation conditions, they are either small or few in number. As the temperature increases, either the size or the number of the multiplets increases to the point at which a SAXS peak develops. This conclusion is supported by the fact that the relative values of the SAXS invariant remain constant with temperature until a temperature slightly higher than the matrix T_g of the sulfonated styrene ionomer. Above that temperature, they increase.

The degree of ion aggregation is small in the as-made sample because water keeps the ions hydrated, preventing them from aggregating while the THF is still in the sample; thus the sample has a low T_g value. The water is expected to evaporate more slowly than the THF because it interacts more strongly with the ionic groups; the THF interacts only weakly with the polystyrene. Thus, when the THF has evaporated and the styrene has undergone its glass transition, some water still remains, weakening ion aggregation. However, as the THF evaporates, the glass transition temperature of the ionomer increases, making it more and more difficult for the ions to aggregate even after the water has evaporated. Once the sample at room temperature is below its T_g, even if more and more water is driven off, aggregation becomes impossible at room temperature over reasonable time scales and is only possible on heating above T_g.

5.11. CHEMISTRY IN IONOMERS

In this section we discuss two aspects of the chemistry of ionomers: thermal stability and chemical reactions.

5.11.1. Stability

In an early study, Bukin (120) found that an increase in the degree of neutralization from 10 to 90% of the carboxylate groups in a P(S-10.0-MAA) copolymer increased the softening temperature and oxidative thermal stability (in oxygen at 300°C) with little effect on the thermal stability (in vacuum at 350°C). In general, neutralization with group I metal was found to increase the thermal stability, whereas group II metal increased the oxidative stability.

A thorough investigation of the thermal and thermal oxidative stability of P(S-*co*-AA) ionomers appeared in 1982 (121). Weight loss was investigated as a function of time at several different temperatures for P(S-*co*-AA) copolymers (in the range of 2.3–14.6 mol %). Figure 5.44 shows data at 340°C for P(S-*co*-ANa) copolymers of various sodium acrylate contents. The maximum rate of weight loss was found to decrease with increasing ion content both at 280°C and 340°C. For example, at 340°C, the P(S-2.3-ANa) sample was found to lose weight at a maximum rate of 3.9%/min, the P(S-5.3-ANa) sample at 2.2%/min, and the P(S-9.7-ANa) sample at 0.6%/min. Also, a large endotherm was observed in the differential thermograms at ~400°C and was identified with a substantial weight loss, indicating decomposition.

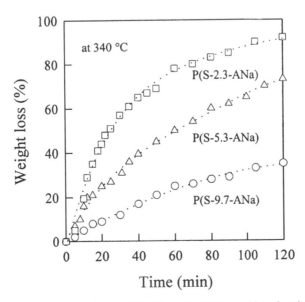

Figure 5.44. Weight loss versus time at 340°C in air for P(S-*co*-ANa) of various ion contents. Modified from Suchocka-Gałas (121).

A more recent paper (122) reported that in the P(S-*co*-ANa) the onset of decomposition and the maximum rate of decomposition were independent of the sodium acrylate content. The temperatures at which the maximum loss rate occurred were somewhat higher in the ionomers than for the acid copolymers. It was suggested that these results also indicate that the thermal stabilities of the sodium salts of the copolymers are almost the same as for the polystyrene.

The kinetics of the decomposition reaction were also investigated. It was found that copolymers containing <6 mol % of sodium acrylate had an order of reaction of 3/2, whereas for those with ion contents >6 mol % the reaction was second order. The higher order of reaction for the sodium salts suggests that there is a difference in the mechanism of degradation of the acid copolymers and their sodium salts. For the P(S-*co*-AA), the order was equal to 1, i.e., independent of the amount of acrylic acid. The increase at 6 mol % in the order of reaction for the sodium salts also suggests a change in the mechanism of degradation at that point. The kinetics of thermal degradation of styrene containing divalent cation acrylates were investigated by Rozwadowska et al. (123), who found that, in contrast to the behavior of the P(S-*co*-AA) ionomers containing monovalent cations (which are more stable than styrene), the introduction of a low concentration of divalent cation salts into polystyrene as a comonomer reduces the thermal stability.

The thermal stability of P(S-*co*-MAA) ionomers was studied by Suchocka-Gałas (124), who found that the thermal stabilities of P(S-*co*-MANa) copolymers were practically independent of the amount of either methacrylic acid or sodium methacrylate introduced into the polymer in the range of 0.8–10.0 mol % of ions. The behavior thus contrasts with that of the styrene-*co*-metal acrylate systems studied earlier.

Kim et al. (54) investigated P(S-*co*-MANa), P(S-*co*-MACs), and P(S-*co*-SSNa) at ~5.5 mol % of the ionic comonomer. At this ion content, the ionomers are no less stable than is styrene. In summary, it seems that at low ion concentrations, sodium salts do not reduce the stability of the styrene ionomers appreciably, whereas divalent salts do.

In the styrene-*co*-vinylpyridinium ionomers, Gauthier et al. (47), in a differential scanning calorimeter (DSC) study found a substantial endotherm in the 160–190°C range. An infrared study suggested that this endotherm was connected with dequaternization of the vinylpyridinium ionomers in vacuum. The reaction could be detected at temperatures as low as 100°C over long time periods. The chemical structure of the pyridinium ion influences the rate of the dequaternization reaction, which proceeds most easily for the 2-vinylpyridinium system, but is much slower for 4-vinylpyridinium and 2-methyl-5-vinylpyridinium moieties. Steric hindrance seems to be an important factor in the stability of the vinylpyridinium salts, as was pointed out by Charlier et al. (125) (in connection with telechelic polystyrenes, which contain quaternary ammonium iodide or ammonium sulfonate end groups). They found that ammonium iodide functional groups are stable <200°C. It was suggested that at higher temperatures a Hofmann-type degradation is most likely to occur, producing unsaturated olefins at both chain ends. It was also found that the α,ω-sulfonic acid polystyrene telechelics neutralized with aliphatic tertiary amines are stable <180°C.

5.11.2. Reactions in Ionomers

It has been known for a considerable time that multiplets in the ionomers could be used as microreactors for a range of reactions after swelling with a suitable solvent, such as water (126–133). Few studies of this type have been performed on styrene ionomers. In one example, Risen's group inserted rhodium ions (Rh^{3+}) or ruthenium ions (Ru^{3+}) into styrene-*co*-styrenesulfonate ionomers by an ion-exchange process (71,134–136). In this series of experiments, the transition metal ions in the ionomer aggregates were used as catalysts in reactions involving carbon monoxide, methanol, or ethanol to give various carbonyls or with H_2 to form hydrides. In another example, partial or complete reduction of Rh^{3+} of Ru^{3+} ions by H_2 was achieved to give metallic particles. It was shown that the metal particles were small <10 Å, and uniformly distributed in all cases. It was suggested that these metal particles could be used as catalysts in chemical reactions.

Another example involves the formation of cadmium sulfide nanoparticles in random ionomers (137). As shown in Chapter 3, ionic aggregates increase in size as the ionogenic unit is placed farther from the backbone. This fact was used as the primary mechanism for controlling the size of the nanoparticles. Cadmium methacrylate ionomers were prepared by using cadmium acetate as the neutralizing agent for P(S-*co*-MAA) copolymers. These ionomers, after deposition on a glass surface as a thin film, were exposed to hydrogen sulfide; and the resulting cadmium sulfide aggregates were analyzed by UV/VIS (ultraviolet/visible) spectroscopy, exploiting the known relationship between aggregate size and the position of the absorption edge (138,139). It was shown that the size of nanoparticles was, indeed, directly related to the size of the primary ionic aggregates and that size control over the range of 18–25 Å could be achieved to within ± 1 Å. It should be remembered that, although the change in the radius of the aggregate is relatively small, the change in aggregation number is quite large, because the aggregation number increases as the cube of the radius. Work of this type in perfluorosulfonate (126–128,131,132) and ethylene carboxylate ionomers (129,130) antedates the styrene work.

As these examples indicate and as is well known for other ionomers, especially the perfluorosulfonates (Chapter 7), a range of reactions can be performed in ionic aggregates. However, as will be shown in Chapter 7, there are a number of advantages to using block copolymers rather than random ionomers for this type of study, especially in connection with size control.

5.12 REFERENCES

1. Erdi, N. Z.; Morawetz, H. *J. Colloid Sci.* **1964,** *19*, 708–721.

2. Fitzgerald, W. E.; Nielsen L. E. *Proc. R. Soc. Ser. A.* **1964,** *282*, 137–146.

3. Ide, F.; Kodama, T.; Hasegawa, A.; Yamamoto, O. *Kobunshi Kagaku* **1969,** *26*, 873–882.

4. Eisenberg, A.; Navratil, M. *J. Polym. Sci. Polym. Lett.* **1972,** *10*, 537–542.

5. Eisenberg, A.; Navratil, M. *Macromolecules* **1973,** *6*, 604–612.

6. Kim, J.-S.; Jackman, R. J.; Eisenberg, A. *Macromolecules* **1994**, *27*, 2789–2803.

7. Guth, E. *J. Appl. Phys.* **1945**, *16*, 20–25.

8. Ma, X.; Sauer, J. A.; Hara, M. *Macromolecules* **1995**, *28*, 3953–3962.

9. Halpin, J. C. *J. Composite Mater.* **1969**, *3*, 732–734.

10. Halpin, J. C.; Kardos, J. L. *Polym. Eng. Sci.* **1976**, *6*, 344–352.

11. Eisenberg, A.; Hird, B.; Moore, R. B. *Macromolecules* **1990**, *23*, 4098–4108.

12. Hammersley, J. M. *Proc. Cambridge Phil. Soc.* **1957**, *53*, 642.

13. Kirkpatrick, S. *Rev. Mod. Phys.* **1973**, *45*, 574–588.

14. Zallen, R. *The Physics of Amorphous Solids*; Wiley: New York, 1983.

15. *Percolation Structures and Processes*; Deutscher, G.; Zallen, R.; Asdler, J., Eds.; Annals of the Israel Physical Society 5; Israel Physical Society: Jerusalem, 1983.

16. Stauffer, D. *Introduction to Percolation Theory;* Taylor and Francis: London, 1985.

17. Hsu, W. Y.; Bakley, J. R.; Meakin, P. *Macromolecules* **1980**, *13*, 198–200.

18. Hsu, W. Y.; Gierke, T. D.; Molnar, C. J. *Macromolecules* **1983**, *16*, 1947–1947.

19. Hsu, W. Y.; Berzins, T. *J. Polym. Sci., Polym. Phys. Ed.* **1985**, *23*, 933–953.

20. Gronowski, A. A.; Jiang, M.; Yeager, H. L.; Wu, G.; Eisenberg, A. *J. Membr. Sci.* **1993**, *82*, 83–97.

21. Jiang, M.; Gronowski, A. A.; Yeager, H. L.; Wu, G.; Kim, J.-S.; Eisenberg, A. *Macromolecules* **1994**, *27*, 6541–6550.

22. Hird, B.; Eisenberg, A. *J. Polym. Sci. B Polym. Phys.* **1990**, *28*, 1665–1675.

23. Shohamy, E.; Eisenberg, A. *J. Polym. Sci., Polym. Phys. Ed.* **1976**, *14*, 1211–1220.

24. Sakamoto, K.; MacKnight, W. J.; Porter, R. S. *J. Polym. Sci., Part A-2* **1970**, *8*, 277–287

25. Erhardt, P. F.; O'Reilly, J. M.; Richards, W. C.; Williams, M. W. *J. Polym. Sci. Symp.* **1974**, *45*, 139–151.

26. Earnest, T. R. Jr.; MacKnight, W. J. *J. Polym. Sci. Polym. Phys. Ed.* **1978**, *16*, 143–157.

27. Agarwal, P. K.; Makowski, H. S.; Lundberg, R. D. *Macromolecules* **1980**, *13*, 1679–1687.

28. Weiss, R. A.; Agarwal, P. K. *J. Appl. Polym. Sci.* **1981**, *26*, 449.

29. Connelly, R. W.; McConkey, R. C.; Noonan, J. M.; Pearson, G. H. *J. Polym. Sci. Polym. Phys. Ed.* **1982**, *20*, 259–268.

30. Weiss, R. A.; Agarwal, P. K.; Lundberg, R. D. *J. Appl. Polym. Sci.* **1984**, *29*, 2719–2734.

31. Bagrodia, S.; Pisipati, R.; Wilkes, G. L.; Storey, R. F.; Kennedy, J. P. *J. Appl. Polym. Sci.* **1984**, *29*, 3065.

32. Bazuin, C. G.; Eisenberg, A. *J. Polym. Sci. B Polym. Phys.* **1986**, *24*, 1121–1135.

33. Agarwal, P. K.; Duvdevani, I.; Peiffer, D. G.; Lundberg, R. D. *J. Polym. Sci. B Polym. Phys. Ed.* **1987**, *25*, 839–854.

34. Greener, J.; Gillmor, J. R.; Daly, R. C. *Macromolecules* **1993**, *26*, 6416–6424.

35. Kim, J.-S.; Wu, G.; Eisenberg, A. *Macromolecules* **1994**, *27*, 814–824.

36. Hird, B.; Eisenberg, A. *Macromolecules* **1992**, *25*, 6466–6474.

37. Tomita, H.; Register, R. A. *Macromolecules* **1993**, *26*, 2791–2795.

38. Gauthier, M.; Eisenberg, A. *Macromolecules* **1990**, *23*, 2066–2074.

39. Lundberg, R. D.; Makowski, H. S. In *Ions in Polymers*; Eisenberg, A., Ed.; Advances in Chemistry 187; American Chemical Society: Washington, DC, 1980; Chapter 2.

40. Rigdahl, M.; Eisenberg, A. *J. Polym. Sci. Polym. Phys. Ed.* **1981,** *19,* 1641–1654.

41. Weiss, R. A.; Fitzgerald, J. J.; Kim, D. *Macromolecules* **1991,** *24,* 1071–1076.

42. Hara, M.; Jar, P.; Sauer, J. A. *Polymer* **1991,** *32,* 1622–1626.

43. Lefelar, J. A.; Weiss, R. A. *Macromolecules* **1984,** *17,* 1145–1148.

44. Clas, S.-D.; Eisenberg, A. *J. Polym. Sci. B Polym. Phys.* **1986,** *24,* 2743–2756.

45. Clas, S.-D.; Eisenberg, A. *J. Polym. Sci. B Polym. Phys.* **1986,** *24,* 2757–2766.

46. Clas, S.-D.; Eisenberg, A. *J. Polym. Sci. B Polym. Phys.* **1986,** *24,* 2767–2777.

47. Gauthier, S.; Duchesne, D.; Eisenberg, A. *Macromolecules* **1987,** *20,* 753–759.

48. Wollmann, D.; Williams, C. E.; Eisenberg, A. *Macromolecules* **1992,** *25,* 6775–6783.

49. Smith, P. Ph.D. Dissertation, McGill University, 1985.

50. Douglas, E. P.; Waddon, A. J.; MacKnight, W. J. *Macromolecules* **1994,** *27,* 4344–4352.

51. Gauthier, M.; Eisenberg, A. *Macromolecules* **1989,** *22,* 3756–3762.

52. Peyser, P. In *Polymer Handbook,* 3rd ed.; Bandrup, J.; Immergut, E. H., Eds.; Wiley: New York, 1988; Section VI.

53. Smith, P.; Eisenberg, A. *J. Polym. Sci. B Polym. Phys.* **1988,** *26,* 569–580.

54. Kim, J.-S.; Yoshikawa, K.; Eisenberg, A. *Macromolecules* **1994,** *27,* 6347–6357.

55. Navratil M.; Eisenberg, A. *Macromolecules* **1974,** *7,* 84–89.

56. Kim. J.-S.; Eisenberg, A. *J. Polym. Sci. B Polym. Phys.* **1995,** *33,* 197–209.

57. Dulac, L.; Bazuin, C. G. *Acta Polym.* **1997,** *48,* 25–29.

58. Hara, M.; Jar, P.; Sauer, J. A. *Macromolecules* **1990,** *23,* 4964–4969.

59. Register, R. A.; Cooper, S. L. *Macromolecules* **1990,** *23,* 310–317.

60. O'Connell, E. M.; Root, T. W.; Cooper, S. L. *Macromolecules* **1994,** *27,* 5803–5810.

61. Leibler, L.; Rubinstein, M.; Colby, R. H. *Macromolecules* **1991,** *24,* 4701–4707.

62. Mikeš, F.; Morawetz, H.; Bendnár, B. *Macromolecules* **1994,** *27,* 6577–6580.

63. Yano, S.; Fujiwara, Y.; Aoki, K.; Yamauchi, J. *Colloid Polym. Sci.* **1980,** *258,* 61–69.

64. Arai, K.; Eisenberg, A. *J. Macromol. Soc. Phys.* **1980,** *B17,* 803–832.

65. Yano, S.; Fujiwara, Y.; Kato, F.; Aoki, K.; Koizumi, N. *Polymer J.* **1981,** *13,* 283–291.

66. Weiss, R. A.; Lundberg, R. D.; Turner, S. R. *J. Polym. Sci. Polym. Chem. Ed.* **1985,** *23,* 549. Miles, I. Eisenberg, A.; unpublished results.

67. Hodge, I. M.; Eisenberg, A. *Macromolecules* **1978,** *11,* 283–288.

68. Boiteux, G. In *Structure and Properties of Ionomers;* Pineri, M; Eisenberg, A., Eds.; NATO ASI Series C: Mathematical and Physical Sciences 198; Reidel:, Dordrecht, 1987; pp. 227–245.

69. Tsatsas, A. T.; Risen, W. M. Jr. *Chem. Phys. Lett.* **1970,** *7,* 354–356.

70. Rouse, G. B.; Risen, W. M. Jr.; Tsatsas, A. T.; Eisenberg, A. *J. Polym. Sci. Polym. Phys. Ed.* **1979,** *17,* 81–85.

71. Mattera, V. D. Jr.; Risen, W. M. Jr. *J. Polym. Sci. Polym. Phys. Ed.* **1984,** *22,* 67–77.

72. MacKnight, W. J.; McKenna, L. W.; Read, B. E.; Stein, R. S. *J. Phys. Chem.* **1968,** *72,* 1122–1126.

73. Alberty, R. A.; Silbey, R. J. *Physical Chemistry;* Wiley: New York, 1997.

74. Ogura, K.; Sobue, H.; Nakamura, S. *J. Polym. Sci. Polym. Phys. Ed.* **1973,** *11,* p 2079–2088.

75. Zundel, G. *Hydration of Intermolecular Interaction*; 1969 Academic: New York, 1969.

76. Fitzgerald, J. J.; Weiss, R. A. In *Coulombic Interactions in Macromolecular Systems*; Eisenberg, A.; Bailey, F. E., Eds.; ACS Symposium Series 302; American Chemical Society: Washington, DC, 1986; Chapter 3.

77. Fan, X.-D.; Bazuin, C. G. *Macromolecules* 1995, *28*, 8209–8215.

78. Neppel, A.; Butler, I. S.; Eisenberg, A. *J. Polym. Sci. Polym. Phys. Ed.*, **1979**, *17*, 2145–2150.

79. Neppel, A.; Butler, I. S.; Eisenberg, A. *Can. J. Chem.* **1979**, *57*, 2518–2519.

80. Neppel, A.; Butler, I. S.; Brockman, N.; Eisenberg, A. *J. Macromol. Sci. Phys.* **1981**, *19(B)*, 61–73.

81. Neppel, A.; Butler, I. S.; Eisenberg, A. *Macromolecules* **1979**, *12*, 948–951.

82. Tsujita, Y.; Hsu, S. L.; MacKnight, W. J. *Macromolecules* **1981**, *14*, 1824–1826.

83. Dickinson, L. C.; MacKnight, W. J.; Connolley, J. M.; Chien, J. C. W. *Polymer Bull.* **1987**, *17*, 459–464.

84. Park, J.-K.; Lim, J.-C.; Kim, C.-H.; Ryoo, B.-K.; Yoo, I.-S. *Polym. Eng. Sci.* **1993**, *33*, 353–357.

85. Pineri, M.; Meyer, C.; Levelut, A. M.; Lambert, M. *J. Polym. Sci. Polym. Phys. Ed.* **1974**, *12*, 115.

86. Yamauchi, J.; Yano, S. *Makromol. Chem.* **1978**, *179*, 2799–2802.

87. Weiss, R. A.; Lefelar, J.; Toriumi, H. *J. Polym. Sci. Polym. Lett. Ed.* **1983**, *21*, 661–667.

88. Toriumi, H.; Weiss, R. A.; Frank, H. A. *Macromolecules* **1984**, *17*, 2104–2107.

89. Fitzgerald, J. J.; Weiss, R. A. *J. Polym. Sci. B Polym. Phys.* **1990**, *28*, 1719–1736.

90. Teo, B. K.; Joy, P. A. *EXAFS Spectroscopy;* Plenum: New York, 1981.

91. Yarusso, D. J.; Cooper, S. L.; Knapp, G. S.; Georgopoulos, P. *J. Polym. Sci. Polym. Lett. Ed.* **1980**, *18*, 557–562.

92. Yarusso, D. J.; Ding, Y. S.; Pan, H. K.; Cooper, S. L. *J. Polym. Sci. Polym. Phys. Ed.* **1984**, *22*, 2073–2093.

93. Ding, Y. S.; Yarusso, D. J.; Pan, H. K. D.; Cooper, S. L. *J. Appl. Phys.* **1984**, *56*, 2396–2403.

94. Ding, Y. S.; Register, R. A.; Nagarajan, M. R.; Pan, H. K.; Cooper, S. L. *J. Polym. Sci. B Polym. Phys.* **1988**, *26*, 289–300.

95. Register, R. A.; Sen, A.; Weiss, R. A.; Cooper, S. L. *Macromolecules* **1989**, *22*, 2224–2229.

96. Galambos, A. F.; Stockton, W. B.; Koberstein, J. T.; Sen, A.; Weiss, R. A.; Russell, T. P. *Macromolecules* **1987**, *20*, 3091–3094.

97. Grady, B. P.; Cooper, S. L. *Macromolecules* **1994**, *27*, 6627–6634.

98. Grady, B. P.; Cooper, S. L. *Macromolecules* **1994**, *27*, 6635–6641.

99. Zhao, Y.; Bazuin, C. G.; Prud'homme, R. E. *Macromolecules* **1989**, *22*, 3788–3793.

100. Fan, X.-D.; Bazuin, C. G. *Macromolecules* **1993**, *26*, 2508–2513.

101. Fan, X.-D.; Bazuin, C. G. *Macromolecules* **1995**, *28*, 8216–8223.

102. Bazuin, C. G.; Fan, X.-D.; Lepilleur, C.; Prud'homme, R. E. *Macromolecules* **1995**, *28*, 897–903.

103. Liu, C.; Martin, C. R. *Nature* **1991**, *352*, 50–52.

104. Chen, W.-J.; Martin, C. R. *J. Membr. Sci.* **1994**, *95*, 51–61.

105. Escoubes, M.; Pineri, M.; Gauthier, S.; Eisenberg, A. *J. Appl. Polym. Sci.* **1984**, *29*, 1249–1266.

106. Brockman, N. L.; Eisenberg, A. *J. Polym. Sci. Polym. Phys. Ed.* **1985**, *23*, 1145–1164.

107. Yasuda, M.; Tsujita, Y.; Takizawa, A.; Kinoshita, T. *Kobunshi Ronbunshu* **1989**, *46*, 347–351.

108. *Perfluorinated Ionomer Membranes*; Eisenberg, A.; Yeager, H. L.; Eds.; ACS Symposium Series 180; American Chemical Society: Washington, DC, 1982.

109. *Structure and Properties of Ionomers*; Pineri, M.; Eisenberg, A., Eds.; NATO ASI Series C, Mathematical and Physical Sciences 198; Reidel: Dordrecht, 1987.

110. *Ionomers: Characterizations, Theory, and Applications*; Schlick, S., Ed.; CRC: Boca Raton, FL, 1996.

111. Hara, M.; Jar, P.-Y.; Sauer, J. A. *Macromolecules* **1988**, *21*, 3183–3186.

112. Hara, M.; Jar, P.; Sauer, J. A. *Macromolecules* **1990**, *23*, 4465–4469.

113. Hara, M.; Jar, P. *Polymer Commun.* **1987**, *28*, 52–54.

114. Bellinger, M.; Sauer, J. A.; Hara, M. *Macromolecules* **1994**, *27*, 1407–1412.

115. Hara, M.; Jar, P.-Y. *Macromolecules* **1988**, *23*, 4465–4469.

116. Eisenberg, A.; Trepman, E. *J. Polym. Sci. Polym. Phys. Ed.* **1978**, *16*, 1381–1387.

117. Weiss, R. A. *J. Polym. Sci. Polym. Phys. Ed.* **1982**, *20*, 65–72.

118. Petrie, S. E. B. *J. Polym. Sci. A2* **1972**, *10*, 1255.

119. Lundberg, R. D.; Phillips, R. R. *J. Polym. Sci. Polym. Lett. Ed.* **1984**, *22*, 377–384.

120. Bukin, I. I. *Plast. Massy* **1977**, *6*, 33–34.

121. Suchocka-Gałas, K. *Polimery Warsaw* **1982**, *27*, 383–386.

122. Suchocka-Gałas, K. *J. Therm. Anal.* **1989**, *35*, 1423–1431.

123. Rozwadowska, B.; Gronowski, A.; Wojtczak, Z. *J. Therm. Anal.* **1990**, *36*, 939–946.

124. Suchocka-Gałas, K. *J. Therm. Anal.* **1987**, *32*, 315.

125. Charlier, P.; Jérôme, R.; Teyssié, P; Prud'homme, R. E. *J. Polym. Sci. A Polym. Chem.* **1993**, *31*, 129–134.

126. Krishnan, M.; White, J. R.; Fox, M. A.; Bard, A. J. *J. Am. Chem. Soc.* **1983**, *105*, 7002–7003.

127. Mau, A. W.-H.; Huang, C.-B.; Kakuta, N.; Bard, A. J.; Campion, A.; Fox, M. A.; White, J. M.; Webber, S. E. *J. Am. Chem. Soc.* **1984**, *106*, 6537–6542.

128. Kuczynski, J. P.; Milosavljevic, B. H.; Thomas, J. K. *J. Phys. Chem.* **1984**, *88*, 980–984.

129. Wang, Y.; Suna, A.; Mahler, W.; Kasowski, R. *J. Chem. Phys.* **1987**, *87*, 7315–7322.

130. Mahler, W. *Inorg. Chem.* **1988**, *27*, 435–436.

131. Hilinski, E. F.; Lucas, P. A.; Wang, Y. *J. Chem. Phys.* **1988**, *89*, 3435–3441.

132. Wang, Y.; Suna, A.; McHugh, J.; Hilinski, E. F.; Lucas, P. A.; Johnson, R. D. *J. Chem. Phys.* **1990**, *92*, 6927–6939.

133. Risen, W. M. Jr. In *Ionomers: Characterizations, Theory, and Applications*; Schlick, S., Ed.; CRC: Boca Raton, FL, 1996; Chapter 12.

134. Shim, I. W.; Mattera, V. D. Jr.; Risen, W. M. Jr. *J. Catal.* **1985**, *94*, 531–542.

135. Barnes, D. M.; Chaudhuri, S. N.; Chryssikos, G. D.; Mattera, V. D. Jr.; Peluso, S. L.; Shim, I. W.; Tsatsas, A. T.; Risen, W. M. Jr. In *Coulombic Interactions in Macromolecu-*

lar Systems; Eisenberg, A.; Bailey, F. E., Eds.; ACS Symposium Series 302; American Chemical Society: Washington, DC, 1986; Chapter 5.

136. Risen, W. M. Jr. In *Structure and Properties of Ionomers*; Pineri, M.; Eisenberg, A., Eds.; NATO ASI Series C, Mathematical and Physical Sciences 198; Reidel: Dordrecht, 1987; pp 87–96.

137. Moffitt, M.; Eisenberg, A. *Chem. Mater.* **1995,** *7*, 1178–1184.

138. Weller, H.; Schmidt, H. M.; Koch, U.; Fojtik, A.; Baral S.; Henglein, A.; Kunath, W.; Weiss, K. *Chem. Phys. Lett.* **1986,** *124*, 557–560.

139. Spanhel, L.; Haase, M.; Weller, H.; Henglein, A. *J. Am. Chem. Soc.* **1987,** *109*, 5649–5655.

CHAPTER 6

PARTLY CRYSTALLINE IONOMERS

In view of the large amount of information available on the properties of random ionomers other than polystyrene, the goal of the next two chapters, which deal with various families of materials, is to present only the most important results, with special emphasis on those that are significantly different from the styrene ionomers. Because we now understand, in broad terms, the structure–morphology–property relationships in the styrenes, these ionomers can be regarded as a baseline system. We will stress those areas in which much more work has been done in nonstyrene systems than in the styrene ionomers. However, wherever recent reviews are available on specific families, e.g., the perfluorosulfonates, we will attempt to avoid duplication, and the reader will be referred to the appropriate review.

Rather than attempting an encyclopedic coverage, which is considerably beyond the scope of this book, we selected 11 ionomer families for extended treatment. In this chapter we cover polyethylene, polytetrafluoroethylene, and the polypentenamers. Ionomeric elastomers, ethylene–propylene–diene terpolymers (EPDM), polyurethanes, acrylates (and methacrylates), and zwitterionomers are discussed in the first part of Chapter 7. In addition, Chapter 7 covers block copolymers, monochelics, telechelics, stars, ionenes, and poly(propylene oxide)s (PPrO) with lithium perchlorate ($LiOCl_4$).

These materials are divided into four major groups. The first group is the partly crystalline ionomers, such as those based on polyethylene, polytetrafluoroethylene, and the polypentenamers. The second group contains rubbery ionomers such as those based on EPDM, the polyurethanes, and the polysiloxanes. The third group consists of materials of relatively high glass transition temperature; two families are considered here: the acrylates and methacrylates. Finally, miscellaneous ionomers are covered in the fourth group, which includes materials such as PPrO with $LiOCl_4$ and acrylate-base zwitterionomers.

6.1 POLYETHYLENE

The ethylene ionomers were the first family of ionomers to bring manufacturers commercial success. The applications of these materials will be discussed briefly in Chapter 10. In this section, we treat the properties from a fundamental point of view. The first part is devoted to a discussion of stress–relaxation, followed by a description of dynamic mechanical properties and melt rheology. We will then discuss other physical properties and spectroscopy of the ethylene ionomers. Throughout the discussion of polyethylene (PE) ionomers, it should be borne in mind that they differ from styrene ionomers because of the presence of crystallinity. As the ion content increases, the degree of crystallinity decreases slightly; but more important, the crystalline morphology changes from that of a spherulitic material to that of a nonspherulitic one. Thus, even at relatively low ion contents, one can obtain highly transparent ethylene ionomers that are still highly crystalline. The trends in the crystallinity are reflected in the mechanical properties. The properties of the ethylene ionomers depend more on sample history than do those of the styrene ionomers (1) because of the effect of ions on the crystallinity and crystalline morphology.

The total number of papers on ethylene ionomers is now large, and their conclusions are sometimes contradictory; e.g., no comprehensive picture has as yet emerged of the details of the morphology of the salts of ethylene–methacrylic acid–based materials. Historically, the ethylene ionomers were developed in the laboratories of E. I. du Pont de Nemours & Co., and the word *ionomer* was coined in connection with these materials, which were first marketed under the trade name of Surlyn (2,3). Several publications are associated with early studies of this ionomer family (4–8); and extensive studies were performed in the laboratories of MacKnight (9,10) and Otocka (11,12). The more recent and on-going research of Yano's group (13–26) is probably the most extensive body of work on these materials.

6.1.1. Stress–Relaxation

The stress–relaxation curves of a copolymer containing ethylene and methacrylic acid are shown in Figure 6.1 (1), which compares the salt (47% neutralized with Na^+) and the acid for the 8 mol % ionomer. It is seen clearly that the relaxation for the salt is slower than for the acid at low temperatures. For example, a 100-s modulus of 10^8 N/m^2 is reached at ~10°C for the acid but at ~25°C for the salt. The mechanical properties of these partly neutralized ionomers are a function not only of the degree of crystallinity, which dominates the behavior, but also of the presence of acid groups, which increase the stress–relaxation rate considerably over that of completely neutralized systems by facilitating ion hopping between multiplets.

Quenching has a major effect on the properties, in that the stress–relaxation behavior of the quenched sodium salt (47% neutralized 8 mol % acid sample) is not substantially different from that of the quenched acid sample, indicating that the degree of crystallinity plays a major role in both systems. If the samples are annealed, however, further significant effects are observed (Fig. 6.1c). For example, the 100-s

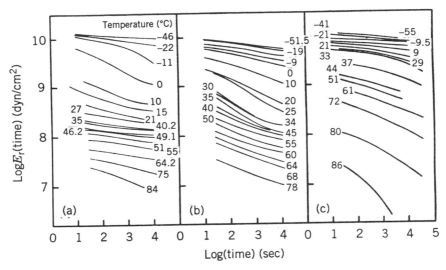

Figure 6.1. Stress–relaxation curves for ethylene-based carboxylated ionomer or precursor. (a) 8 mol % COOH, quenched; (b) Na⁺, 47% neutralized, quenched; (c) Na⁺, 47% neutralized, quenched, annealed. Numbers indicate temperature (°C). Modified from Ward and Tobolsky (1).

modulus reaches a value of 10^7 N/m² for the quenched acid sample at ~50°C, for the quenched sodium salt sample at ~60°C, and for the annealed sample at ~80°C. However, once the temperature exceeds 80°C, stress–relaxation becomes quite rapid, because the effect of crystallinity is minimized and ion hopping is facilitated.

6.1.2. Dynamic Mechanical Properties

A number of studies have been performed on the dynamic mechanical properties of the ethylene ionomers. In the first study, MacKnight et al. (9) found four loss peaks (Fig. 6.2): In order of decreasing temperature, they are the α peak (>50°C), which has been ascribed to motions involving the ionic phase, the β' peak (~50°C), which has been ascribed to the micro-Brownian motions of chain segments occurring in the amorphous phase cross-linked with hydrogen-bonded acid dimers; the β peak (~0°C), which has been assigned to a relaxation occurring in the amorphous branched polyethylene phase from which most of the ionic groups have been excluded; and the γ peak, the behavior of which is only indirectly related to the presence of ionic groups and is essentially the same as that found in ethylene homopolymers. It was found that the α peak position shifts to higher temperatures with increasing degree of neutralization, the β' peak position is independent of degree of neutralization, and the β peak position shifts to lower temperatures as the degree of neutralization is raised. In a subsequent investigation that dealt with ethylene ionomers containing various amounts of acrylic acid (11), the authors found that the results generally

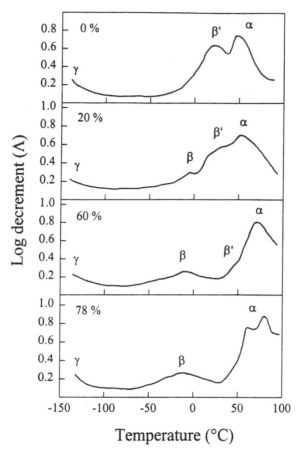

Figure 6.2. Log decrement versus temperature for P(E-4.1-MANa). 0% neutralized; 20% neutralized; 60% neutralized; 78% neutralized. Modified from MacKnight et al. (9).

paralleled those of the study by MacKnight et al. (9), except that the position of the β peak was found to shift to higher temperatures with increasing ion content (Fig. 6.3). The reason for this difference in the position of the β peak is not completely clear. However, MacKnight (27) suggested that it may be related to the sample history. Most recently, Tachino et al. (23) reinvestigated the mechanical relaxations in ethylene ionomers and found that, in addition to the α, β', and β peaks, an α' peak was also present. The α' peak appears at ~50°C, and its position is independent of frequency. The authors assigned this peak to a first-order transition.

6.1.3. Melt Rheology

A study of the melt rheology of ethylene-based ionomers was carried out in 1970 by Sakamoto et al. (28), who found that time–temperature superposition was operative for the acid copolymer but not for the sodium or calcium salt. A comparison of plots of the apparent viscosity η_a versus shear rate γ and the complex viscosity $|\eta^*|$ versus angular frequency ω for the acid and sodium copolymer of 4.1 mol % methacrylic acid (59% ionized), is shown in Figure 6.4. The authors concluded that the copolymer salts have a different structure at different temperatures and suggested that flow takes place by the interchange of ionic groups from one domain to another. A subsequent investigation by Earnest and MacKnight (29) on a copolymer of ethylene–methacrylic acid (3.5 mol % acid content) showed that it was possible to construct a storage modulus super master curve for the ethylene ionomers comparing the acid, the ester, and the 70% neutralized sodium salt derivatives, along the lines suggested by Shohamy and Eisenberg (30) (Fig. 6.5). The complex viscosities are shown as a function of temperature in the insert, with temperature shift factors for superposition of the three master curves. Superposition of G' data is not followed for G'' data, so that it is impossible to construct a comprehensive master curve. Attempts to do so show significant deviations in some regions of each curve.

6.1.4. Other Mechanical Properties

An early example that clearly illustrates the advantages of ionomers over nonionic materials in terms of stress–strain behavior is shown in Figure 6.6 for an ethylene ionomer (3), which also shows elongation at rupture versus degree of ionization.

6.1.5. Dielectric Properties

Dielectric studies of the ethylene-based ionomers began in 1969 (10,31). In general, it was found that the dielectric results support the mechanical results in terms of the

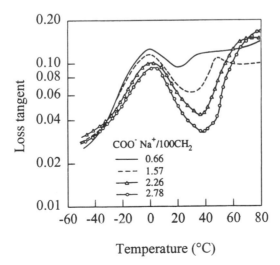

Figure 6.3. Tan δ versus temperature for P(E-*co*-ANa) of various ion contents. *Solid line*, 0.66 COO$^-$Na$^+$/100CH$_2$, *dashed line*, 1.57; $\triangle$ 2.26; $\bigcirc$, 2.78. Modified from Otocka and Kwei (11).

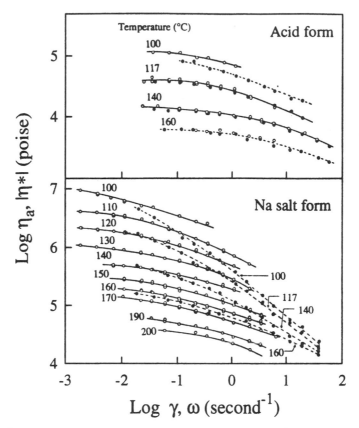

Figure 6.4. Log apparent viscosity (○) and complex viscosity (●) versus log shear rate or frequency for P(E-5.9-MAA) acid (top) and its sodium salt (bottom). Numbers indicate temperature (°C). Modified from Sakamoto et al. (28).

assignment of individual loss peaks. The α peak, which has been assigned to motions occurring in the ionic phase, increases in intensity with increasing degree of neutralization, and its position shifts to higher temperatures. The intensity of the β' peak, assigned to micro-Brownian segmental motion of the chains in the amorphous phase cross-linked with hydrogen-bonded acid dimers, decreases with increasing degree of neutralization, although its position remains constant. In contrast, the β peak, which has been assigned to motion of the amorphous polyethylene phase, increases in intensity and also shifts to lower temperature with increasing degree of neutralization, again in agreement with mechanical results. A significant difference was found in the activation energies of this dispersion. Mechanically, the activation energy was found to be in excess of 150 kJ/mol, whereas the dielectric activation energy was <60 kJ/mol. A possible explanation is the presence of lone salt groups or free acid groups, which may remain in the amorphous phase of polyethylene. The reorientation of these unassociated groups, because of their high dipole moment, would be a major

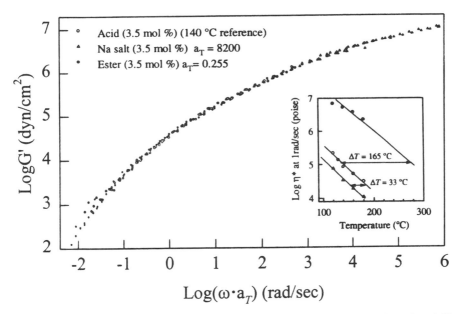

Figure 6.5. Master curve of log G' versus log ωa_T for ($\bigcirc$) P(E-3.5-MAA) acid (3.5 mol %), ($\triangle$) Na$^+$ salt (3.5 mol %; $a_T = 8200$), and ($\bullet$) ester (3.5 mol %; $a_T = 0.255$). $T_{ref} = 140°C$ for the acid. **Insert**, Complex viscosity versus temperature in the range of 10^4 P $< \eta^* <$ 10^7 P. Horizontal temperature shifts are indicated. Modified from Shohamy and Eisenberg (29).

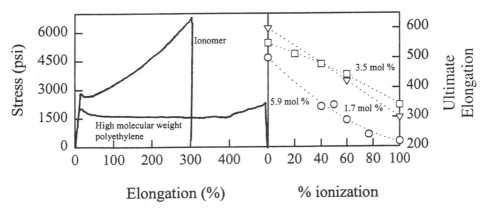

Figure 6.6. Left: Stress versus elongation plots for high molecular weight polyethylene and an ethylene ionomer containing carboxylate groups. Right: Ultimate elongation versus percent ionization for ethylene ionomers of various ion contents and levels of ionization. $\square$, 3.5 mol %; ∇; 1.7 mol %; $\bigcirc$, 5.9 mol %. Modified from Rees and Vaughan (3).

contributor to the dielectric β relaxation; thus, because these isolated pairs are quite small, the dielectric activation energy is smaller than the mechanical activation energy.

A much more recent series of dielectric studies was performed by Yano et al. (20). These studies essentially confirm the previous results and also provide much additional information on a range of counterions that had not been studied before. For example, it was shown that Na^+, K^+, Ca^{2+}, Mg^{2+}, and Co^{2+} salts of ethylene–methacrylic acid (5.4 mol %) start forming clusters at ~30% neutralization; however, Cu^{2+} and Mn^{2+} salts do not form clusters until a degree of neutralization >60% is reached.

A sudden increase in the intensity of the dielectric loss as a function of temperature was found (20), which coincides with an apparent order–disorder transition from mechanical results. However, this order–disorder transition in ionomer clusters is a matter of controversy (32).

Also, it was found that the incorporation of 1,3-bis(aminomethyl)cyclohexane (BAC) induces major changes in the type of aggregates that are found. The addition of BAC results in the formation of complexes with transition metals, which promote ionic cluster formation. Extensive dielectric results on this system have been acquired by Yano's group (17–19,33). Although their results are important, a discussion of the details is beyond the scope of the present treatment.

The effect of water has also been investigated in some detail (20). A peak was found at $\sim -3°C$ at 1 kHz, which seems to be quite independent of the water content over a wide range. It was suggested that approximately three water molecules can be absorbed by a $-COO^- \ Na^+$ group, which can thus be considered to be the primary hydration shell, with excess water forming a more extended shell around it.

Yano's group also studied the dielectric relaxations of ethylene–acrylate ionomers (13,14,17) and found differences in the behavior of the dielectric increment $\Delta\epsilon$ for styrene and ethylene ionomers (14). For the styrene ionomers, $\Delta\epsilon$ is essentially independent of the degree of neutralization. However, in the ethylene ionomers, $\Delta\epsilon$ has a maximum at approximately 20% neutralization. This difference can probably be ascribed to the difference in the number of unassociated polar groups in the two systems.

6.1.6. Spectroscopy

6.1.6.1. Infrared. The earliest study of the far-infrared spectra of ethylene ionomers was performed by Risen's group (34,35). The trends are the same as those seen in the styrene ionomers (Fig. 5.34), i.e., the cation-related band moves progressively to higher energies with decreasing atomic number of the alkali cation. The width of this band probably indicates that several cationic environments are present, which exert different forces on the cation. It was also suggested that the vibrations of the ionic assemblies are anharmonic (34). Not unexpectedly, the presence of alkali ions perturbs the vibrational motions of the carboxylate group, and the motion of the polyethylene backbone is restricted as a result of the presence of the ionic groups (35).

In a study of the near infrared spectra of the ethylene-based ionomers, MacKnight et al. (36) showed that the relative intensities of the 1700 cm^{-1} un-ionized carbonyl stretching band could be related to the degree of neutralization. Furthermore, they showed that a monomer–dimer equilibrium exists among the acid groups that favors the hydrogen-bonded dimers in the unneutralized material. In a subsequent paper, Andreeva et al. (37) found that the 1560 cm^{-1} band (asymmetric stretching mode of the carboxylate group) was split into two bands at high temperatures: one at 1550 cm^{-1} and the other at 1565 cm^{-1}. These two bands were assigned to the asymmetric stretching vibration of the carboxylate groups in the ionic multiplets and in the ionic clusters, respectively. However, it should be stressed that the definition of the multiplet and cluster is different in that paper from the one given here in Chapter 3. In any case, these two bands represent ions in different environments. This cautionary note should be kept in mind generally when reading the literature published before 1990.

An extended series of studies on the infrared spectra of ethylene ionomers was performed by Coleman and Painter's group (38–42). These authors found that the intensity of the band near 1550 cm^{-1} increases with increasing temperature (40). They ascribed this band as originating from simple ion pairs. The authors suggested that for the Na$^+$ system the doublet at 1568 and 1547 cm^{-1} is associated with local structure found in the ionic aggregates. At 70°C, cation–related behavior was observed; the K$^+$ and Cs$^+$ salts showed only a single band at 1550 cm^{-1}, while the Li$^+$ and Na$^+$ salts showed a doublet. Subsequent studies (41) found that soaking the ionomer in water for 45 days converted the well resolved doublet into a broad band centered at 1540 cm^{-1}. Also, a new band appeared at 1700 cm^{-1}, which was assigned to hydrogen-bonded carboxylic acid dimers. Further bands were observed at 1680 and 1590 cm^{-1}, which had previously been assigned to acid–salt complexes (40). When the ethylene ionomers are soaked in water, a fraction of the metal cations is exchanged with protons to yield partly neutralized salts and dimerized acid groups. The soaking also results in a more rapid appearance of crystallinity compared to annealing at room temperature for comparable time periods. It was suggested that the origin of this phenomenon lies in the decreasing strength of ionic interactions in the presence of water. The authors concluded that infrared spectroscopic studies in the carboxylate stretching region may not be used to differentiate between multiplets and clusters. They also suggested that multiplets retain their structural integrity within the clusters upon cluster formation.

Coleman et al. (42) showed that the immersion of an ethylene–methacrylic acid copolymer (10.3 mol % of acid groups) in a dilute solution of diethylzinc in hexane results in the formation of a material that has a "skin-core" morphology. The skin of the Zn^{2+}-neutralized material reduces the diffusion coefficient for further penetration of the Zn^{2+} into the core of the ethylene–methacrylic acid copolymer. However, at elevated temperatures, ion hopping is sufficiently rapid to allow the material to become homogeneous. It was suggested that the fully neutralized zinc ionomer is tetra-coordinated, which yields a highly hydrolytically stable local structure. Immersion of the film in water did not affect the IR spectrum, and the Zn^{2+}-coordinated species appeared to be stable. However, in partly neutralized systems, a mixture of species was found, such as tetra- and hexa-coordinated Zn^{2+} salts, Zn^{2+} acid salt, and carboxylic acid dimers.

Ishioka (43) showed that three bands at 1539, 1560, and 1625 cm^{-1} that were present in dry, zinc-neutralized ethylene-co-methacrylate ionomers (5.8 mol % of acid groups) of 47% neutralization collapsed into a single band at 1587 cm^{-1} upon water absorption. From the integrated intensities of the bands, Ishioka calculated the number of H_2O molecules associated with a COO^- group in the water-swollen state to be 3.56. In an investigation of the absorption of deuterium oxide D_2O, it was shown that the asymmetric stretching COO^- band appears at the same position as in H_2O, from which the authors concluded that hydration takes place preferentially around the Zn^{2+} cation. More recently, Ishioka (44) found that there is no correlation between the temperature dependence of the elastic constant of the polymer and the local structural changes in the ionic aggregates or the acid monomer–dimer equilibrium, indicating that it is the presence of aggregates that determines the properties and not the detailed structure of the aggregates.

6.1.6.2. *Raman.*

In the Raman spectrum of ethylene-*co*-potassium methacrylate ionomers (45), two bands were identified in the 200–400 cm^{-1} range and one was found at a much lower wave number, ~40 cm^{-1}. Although some salt group sensitivity was detected, none of the three bands was thought to originate in the ionic regions but was thought to be related to crystallinity in polyethylene. Therefore, these bands are not discussed here further.

In a recent Raman study (44), it was found that the temperature dependence of the conformational order, or the relative amount of a short *trans* polymethylene sequence in the amorphous phase, is correlated with that of the elastic constant. This finding is corroborated by the temperature dependence of the intensity of the longitudinal acoustic mode band (reflecting the relative fraction of polymethylene chain segments having liquidlike conformations).

6.1.6.3. *Electron Spin Resonance.*

An early study of the electron spin resonance (ESR) spectrum of a Mn^{2+} salt of an ethylene–acrylic acid copolymer containing 9.3 mol % acrylic acid was performed by Yamauchi and Yano (16), who found that at 40% neutralization, the spectrum showed six peaks, characteristic of Mn^{2+} ions in the isolated state. However, at 62% neutralization, a single broad peak was detected, suggesting the presence of an Mn^{2+}–Mn^{2+} exchange interactions within the Mn^{2+} complexes. Infrared spectra were also monitored; it was shown that there is no dependence of the peak positions in the IR spectra on the degree of neutralization, only changes in the intensities. Therefore, these spectral changes were ascribed to a Mn^{2+}–Mn^{2+} exchange rather than the formation of aggregates or complexes. It should be recalled, however, that other evidence suggests that aggregation may already be occurring at the degrees of neutralization investigated in this study, so that the change in the spectra may possibly the result of aggregate formation.

A subsequent study by Takei et al. (46), involving both Cu^{2+} and Mn^{2+} neutralized ethylene–methacrylic acid copolymers (5.4 mol %), confirmed the earlier results of Yamauchi and Yano (16) by investigating a much larger number of ionomer samples. From 17 to 99% neutralization, a progressive disappearance of the six-line spectrum of Mn^{2+} in favor of one broad line was seen. A plot of the line width as a function of temperature showed an inflection point at around 70°C, which is associated with a change in a mobility of Mn^{2+} ions in the ionic aggregates. The Cu^{2+} salt showed similar behavior in terms of line width. The complete reversibility of the line width on heating and cooling suggests that it is not induced by crystallinity of the sample, because some type of hysteresis would be associated with crystallinity.

6.1.6.4. Nuclear Magnetic Resonance.

A number of nuclei have been used in the NMR study of ethylene ionomers, including ^{13}C, ^{1}H, and alkali metals. Belfiore et al. (47,48), using solid-state ^{13}C NMR, showed that the peak width of the amorphous CH_2 units in the ionomers was three times wider than that in low-density polyethylene, which indicates that the amorphous domains in the ionomer are more rigid than in the nonionic ethylene. It was also found that the intensity of the crystalline CH_2 peak decreased with increasing degree of neutralization. In a subsequent study, Belfiore et al. (49) suggested that the crystalline/amorphous weight ratios are sensitive to the acid content in the acid copolymer but insensitive to the type of cation.

Read et al. (10) studied the proton spin-lattice relaxation times T_1 (at 30 MHz) and $T_{1\rho}$ (at 10–100 kHz) of ethylene–methacrylic acid and its 53% neutralized Na^+ salt. They found that the magnetization decay in the T_1 measurement is exponential, with a single time constant, while the decay in $T_{1\rho}$ is a double exponential. The authors suggested that both the crystalline and amorphous components may contribute to this double exponential decay. In the acid sample, T_1 and $T_{1\rho}$ minima were found at the same point as the dielectric β' and γ processes. In addition, a δ relaxation was found below -100°C, which was not found in the dielectric or mechanical results, which was attributed to the rotation of methyl groups. In the Na^+ salt, both the δ and γ relaxations were found in addition to a process at high temperature (~ 50°C).

In an early study, Otocka and Davis (12) observed that the ^{7}Li-NMR line width of the lithium-neutralized ionomer changes, corresponding to the γ and β transitions, at -60 and 55°C, respectively. These transition regions were also found in the plot of the ^{1}H-NMR line widths versus temperature, suggesting that the increased motion at these two temperatures corresponded to an increase in the matrix mobility. It was suggested that if lithium nuclei were present in exclusive domains, then the matrix mobility would not be transferred effectively to the aggregates. The authors also found that a finite ^{7}Li line width persists up to 150°C, which may be owing to either quadrupolar broadening or to weak restrictions in the mobility of the salt groups. A study of solid-state ionomers using ^{23}Na-NMR showed that the line shape at room temperature is broad and featureless (50). However, exposure to water-saturated air at 95°C for 24 h resulted in the appearance of a narrow signal at -12 ppm, suggesting exchange among water and the carboxylate groups but incomplete hydration of the Na^+ ions.

In a subsequent study by Connolly (51), a broad stepwise increase was seen in the plot of the spin–spin relaxation time T_2, as a function of temperature of the Na^+ salt sample, which had been saturated with water or ethylene glycol. It was suggested that this increase in T_2 is correlated with a relaxation process, attributed to the fast exchange of water around the Na^+ nuclei. In the dry Na^+ salt, the broad ^{23}Na line was found to narrow and to shift to a lower field with increasing temperature, indicating an increase in the mobility of the polymer backbone and in the exchange rate of the carboxylates.

6.1.7. Water Uptake

A detailed study of water absorption in ethylene ionomers was performed by Yano's group (21,22), who observed Langmuir-type absorption behavior at a degree of hydration of <0.5 water molecules per ion pair. They suggested that the water molecules in that region interact primarily with unassociated ion pairs and with the surfaces of ionic aggregates (22). At intermediate degrees of hydration (0.5–3 water molecules per ion pair), the mechanism changes to one that resembles solution behavior, in that the water molecules are incorporated into the ionic aggregates themselves. It was suggested that at three water molecules per ion pair, the hydration of the ion pairs becomes complete. Because this number is smaller than the four water molecules per ion that is expected for free Na^+ ions, it appears that incomplete dissociation occurs, in that the ions are present at least as ion pairs. For degrees of hydration higher than three water molecules per ion pair, a solution-like mechanism is clearly operative, with water molecules present in the environment of the ionic aggregates. Inverted micelle structures are thought to exist under those conditions, with the number of water molecules in excess of three per ion pairs placed outside the primary hydration shell.

The effect of annealing at various relative humidities on the mechanical properties of ethylene ionomers was also investigated (25). Figure 6.7 shows the effect of storage at various levels of relative humidity on the stiffness of a sample containing 5.4 mol % acrylic acid, 60% neutralized with either Na^+ or Zn^{2+}. For Na^+ ions, the effect is considerable; the stiffness after 100 days of storage at 100% relative humidity is more than a factor of two lower than that of the dry material. By contrast, the Zn^2-neutralized material shows a much smaller effect, which illustrates the appreciable difference in the interaction of the metal cations with water. The modulus of the Na^+ material increases with storage time at 0% relative humidity, which is thought to be related to the increase in crystallinity of the sample.

6.1.8. Crystallinity

One of the major complications in the study of the ethylene ionomers compared to those based on styrene is the presence of crystallinity. Crystallinity in polyethylene by itself, i.e., in the absence of an ionic comonomer, is a complicated topic, partly because of the architectural complexity of the polyethylene chain (e.g., degree of branching and length of branches). The presence of ionic groups increases the degree

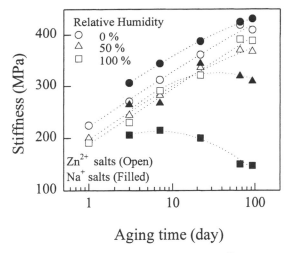

Figure 6.7. Stiffness versus aging time of 60% Na+- and Zn²⁺-neutralized P(E-5.4-MAA) ionomers at various relative humidities. *Open symbols*, Zn²⁺ salts; *solid symbols*, Na+ salts; *circles*, 0% relative humidity; *triangles*, 50% relative humidity; *squares*, 100% relative humidity. Modified from Tachino et al. (25).

of complexity considerably; thus relatively few detailed studies have been performed on crystallinity in ethylene ionomers. In one early study, Marx and Cooper (52) showed that annealing leads to the appearance of a secondary endotherm in the thermogram, accompanied by a shift to smaller angles of the x-ray scattering peak owing to crystallinity. The data were interpreted in terms of a model that suggested that the hydrocarbon part of the ionomer crystallizes into folded-chain lamellae.

More recent studies from Tsujita's group (53,54) examined the topic of ethylene ionomer crystallization in considerable detail. They found that both the degree of crystallinity and the dynamic modulus of the ionomers (neutralized up to 60% with Na+ or Zn²⁺) increase gradually with aging time (54). In a highly neutralized Zn²⁺ ionomer (90%), while the crystallinity again increases gradually, the modulus at 100°C increases for aging times of 40–100 h and then approaches an asymptotic value. This pattern indicates that clustering is progressing up to 100 h of aging. The authors suggested that ionic aggregation takes place much more rapidly than does crystallization and that only after the ionomer aggregates have developed does crystallization proceed more fully. In an accompanying paper (53), the authors suggested that two different types of crystallites are present: chain-folded lamellae and bundle-like crystals, which melt at higher and lower temperatures, respectively. The lower melting temperature shifts to higher temperatures with increasing annealing time, even in the presence of ionic aggregates, confirming that the presence of ionic aggregates does not immobilize the chain. This mobility of chains was also seen in the work of Orler et al. (55) on the syndiotactic styrene ionomers, which suggests that the kinetics of crystallization in that system can be used as a measure of the internal viscosity or of chain mobility in ionomers. The modulus, in general, also increases with increasing annealing time as a result of the development of crystallinity.

More recently, the degree of crystallinity as a function of degree of neutralization was investigated for the ethylene ionomers by Kutsumizu et al. (26). The results of their study for the ethylene–methacrylate (5.4 mol %) system with various cations are shown in Figure 6.8. Generally, the degree of crystallinity decreases with increasing neutralization or temperature (56).

6.1.9. Permeation

Itoh et al. (57) studied the permeation of water through ethylene-based ionomers. Their results confirmed that the Zn^{2+} salt absorbs much less water than does the Na^+ salt. For example, a 5.4 mol % ionomer, 60% neutralized with Zn^{2+} exhibited a degree of hydration 0.0035 g/g, whereas in the comparable Na^+ salt the degree of hydration was 0.102 g/g. The steady-state diffusion coefficient $\bar{D}$ of water through membranes prepared from these materials showed a $\bar{D}$ value of ~6 × 10^{-9} cm²/s for the Zn^{2+} salt, compared to 4 × 10^{-10} cm²/s for the corresponding Na^+ salt. By contrast, the steady-state permeability coefficient of water vapor through the same membranes showed only a small difference. The value for Zn^{2+} was ~8 × 10^{-9} (cm³ stp cm/cm² s cm Hg), independent of relative vapor pressure, whereas the value for Na^+ ranged from 6 × 10^{-9} to 14 × 10^{-9}. The same authors showed that in the Na^+ ionomer above 6% of water (w/w) some of the water was freezable; but

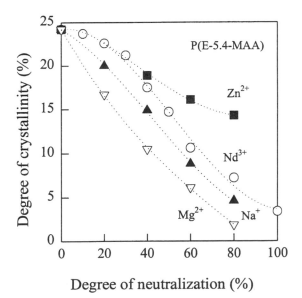

Figure 6.8. Degree of crystallinity versus degree of neutralization for P(E-5.4-MAA) neutralized with various cations. ■, Zn^{2+}; ○, Nd^{3+}; ▲, Na^+; ▽, Mg^{2+}. Modified from Kutsumizu et al. (26).

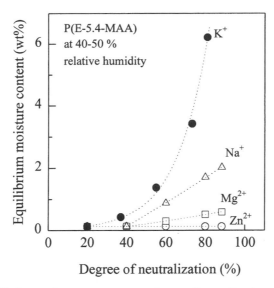

Figure 6.9. Equilibrium moisture content versus degree of neutralization with various cations at 40–50% relative humidity for P(E-5.4-MAA). ●, K$^+$, △, Na$^+$; □, Mg^{2+}; ○, Zn^{2+}. Modified from Tachino et al. (24).

below that point, all the water was nonfreezable or bound. The equilibrium moisture contents for ethylene ionomers containing 5.4 mol % of methacrylate salt (40–50% relative humidity) neutralized with various cations are shown in Figure 6.9 (24).

6.1.10. Orientation

The birefringence and orientation function at 50% elongation f^{50} were studied as a function of temperature by Uemura et al. (56) for various wavelengths in a 55% neutralized P(E-4.1-MANa) ionomer. The behavior of the acid copolymer is quite straightforward; both the orientation function f^{50} and the birefringence Δ at 50% elongation generally decrease with increasing temperature. This behavior is in marked contrast to the ionomer (55% ionization). The value of f^{50} decreases initially between ~15 and ~30°C, then goes through a maximum at ~45°C. Generally Δ shows similar behavior. The maximum is related to the α relaxation mechanism observed in dynamic mechanical studies, which is associated with the softening of the cluster regions of the material. In those regions, the orientation is at a maximum, because all of the material can participate, i.e., not only the nonionic or crystalline regions but also the ionic aggregates themselves. The authors suggested that there is effectively a parallel connection between the ionic regions and nonionic regions, because the birefringence increases in both regions simultaneously. A plot of the strain-optical coefficient from a companion study by Kajiyama et al. (58) is shown in Figure 6.10. Whereas the crystalline contribution decreases throughout the range, the amorphous

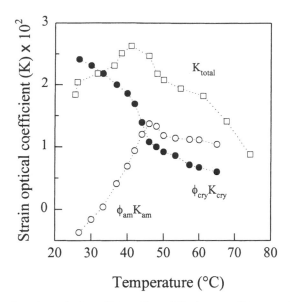

Figure 6.10. Total strain optical coefficient K_{total} ($\square$), the crystalline contribution K_{cry} ($\bullet$), and the amorphous contribution K_{am} ($\bigcirc$). K_{cry} and K_{am} are multiplied by their relative volume fraction. The sample is P(E-4.1-MANa) 78% neutralized. Modified from Kajiyama et al. (58).

contribution goes through a maximum, again, at ~45°C, coincident with the α peak maximum in the mechanical loss tangent plot (59).

In a small-angle x-ray scattering (SAXS) study of the deformation of highly amorphous ethylene ionomer, Roche et al. (60) found that at elongations above ~ 60%, the SAXS pattern becomes significantly azimuthally dependent. For example, at 300% elongation, a completely neutralized P(E-6.1-MACs) sample showed a scattering peak at an azimuthal angle θ of 90° of dramatically increased intensity, whereas no peak was seen at a θ of 0°. After applying several models to fit the data, the authors concluded that the model that best fit the results is one involving lamellar aggregates rather than spherical or core-shell-type systems (Section 3.1.3.4).

6.1.11. Concluding Comments

Of the 570 papers and 2000 patents dealing with ethylene ionomers published up to 1996, about 60 papers were mentioned here, thus only a small portion of the literature has been covered. A much more extended treatment would be useful but is considerably beyond the scope of this book.

6.2. POLYTETRAFLUOROETHYLENE

Polytetrafluoroethylene (PTFE) systems are an important family of ionomers, because they are used extensively as membranes in the chlor-alkali industry and in fuel cells, among other areas. Because of their many industrial applications, the literature on the perfluorinated ionomers is large. The 152nd National Meeting of the Electrochemical Society held in October 1977 hosted the first symposium on perfluorocarbon ion exchange membranes (61). The research up to 1981 was summarized in a collection of review papers (62). The first material in this family to be commercialized has the trade name of Nafion (E. I. du Pont de Nemours & Co.)

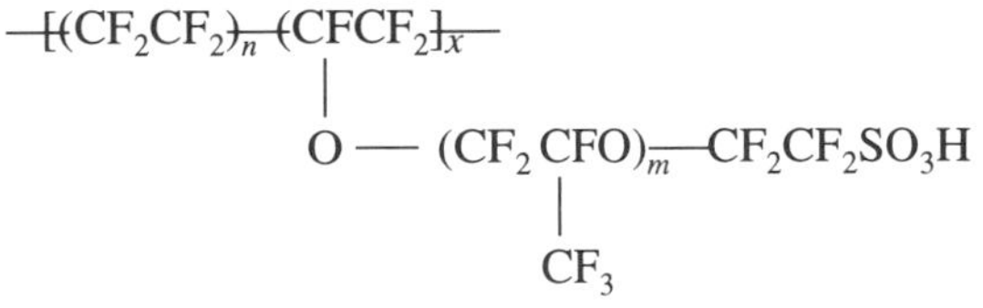

$$-[(CF_2CF_2)_n-(CFCF_2)_x-$$
$$|$$
$$O-(CF_2CFO)_m-CF_2CF_2SO_3H$$
$$|$$
$$CF_3$$

Nafion

Several other materials exist, including Flemion (Asahi Glass) and a short-side chain system perfluorosulfonate ionomer (PFSI; Dow Chemical). Both sulfonates and carboxylates are available. Several reviews of perfluorinated ionomers have appeared recently (63) that deal with small-angle scattering, infrared spectroscopy, ESR, and NMR studies. The topics that will allow the reader to compare these materials with other ionomers are discussed here.

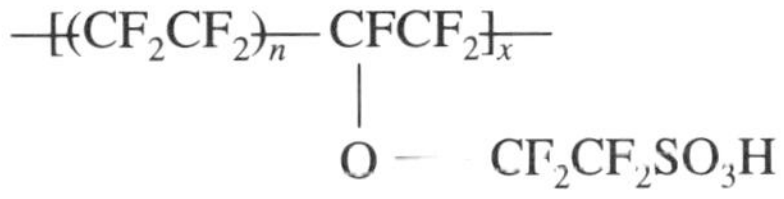

$$-[(CF_2CF_2)_n-CFCF_2]_x-$$
$$|$$
$$O-CF_2CF_2SO_3H$$

PFSI

6.2.1. Crystallinity

The crystallinity of the perfluorosulfonic acid copolymers as a function of the equivalent weight (EW) of the acid (Section 1.8) was studied by Starkweather (64). The heat of fusion was found to be proportional to the EW. At an EW of 1100 the heat of fusion was about 4 J/g, whereas at an EW of 1800 it was 30J/g. Similarly, the melting point was found to be proportional to -ln X, where X is the mol fraction of crystallizable comonomer, i.e., tetrafluoroethylene. Thus, at an EW of ~1000, the crystalline melting point was ~250°C, whereas for an EW of 2000, it was ~290°C. Most interesting, the size of crystallites was found to be greater than the average separation between ionic groups. It was suggested that the polymer is arranged in bilayers, with the ionizable side groups extending on both sides of the crystallites.

In a subsequent study of the annealing of solution cast perfluorosulfonate ionomer films, (which will be discussed more extensively below), Gebel et al. (65) found that crystalline order increases with temperature. Before thermal treatment, the cast perfluorosulfonate membranes contain a large fraction of material with small crystal-

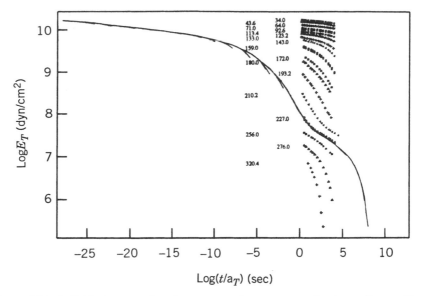

Figure 6.11. Individual stress relaxation curves and master curve for the perfluorosulfonate Nafion -K). Modified from Yeo and Eisenberg (67).

lites, together with the amorphous material. Upon annealing, however, the size of lamellar crystallites increases; at the same time their internal order appears to improve. Long-range order also develops.

A study of the crystallinity of the short side chain Perfluorosulfonate Ionomers (Dow PFSI) showed that the melting points vary only slightly with EW. Moore and Martin (66) suggested that the copolymers must be blocky in nature.

6.2.2. Mechanical Properties

The first study of the stress–relaxation of PTFE ionomers was performed by Yeo and Eisenberg (67), who showed that although the acid precursor obeys time–temperature superposition the K^+ salt does not <180°C; time–temperature superposition seems to be reestablished >180°C (Fig. 6.11). The 10-s modulus versus temperature plots for the perfluorosulfonates in the PTFE acid, K^+-PTFE ionomer, and PTFE precursor, along with those for styrene and the P(S-9.0-MANa) ionomer are shown in Figure 6.12 (68–70). In the PTFE precursor, it is seen that the 10-s tensile modulus curves are relatively broad, the breadth in this case being undoubtedly related to the presence of crystallinity in the system. In nonionic styrene, the modulus decreases rapidly with temperature because crystallinity is absent. However, in the styrene ionomer, the curve is also relatively broad; the width in this case is the result of the ionic interactions, since crystallinity is absent. As was pointed out before, the effect of crystallinity on mechanical properties resembles the effect of ionic interactions. In the PTFE ionomer, the 10-s modulus curve is broad because of the presence of both crystallinity and ionic interactions. It is noteworthy that in the PTFE acid, which does show time–temperature superposition, the addition of a small amount of water leads to a breakdown in this superposition, presumably by introducing another relaxation mechanism with an activation energy different from that of the primary relaxation mechanism (67) but effective on similar time scale.

The modulus and loss tangent of the acid and Cs^+ salt at 1 Hz are given as a function of temperature in Figure 6.13. It is seen that the position of both the primary transition (α peak), owing to the glass transition in the ionic regions, and the secondary transition (β peak), owing the glass transition of the matrix regions, increase in temperature as the acid is converted to the salt. The third peak in the loss tangent plot (γ peak) is attributed to short-range motions of the fluorocarbon backbone. The effects of various parameters on the mechanical relaxation was explored, and it was shown that the effect of the degree of neutralization is similar to that seen in the styrene ionomers, i.e., the position of the peak owing to the primary relaxation shifts to higher temperature with increasing degree of neutralization (69,71).

Kyu and Eisenberg (69, 72) studied the effect of crystallinity by investigating samples in which crystallinity is absent as a result of quenching from above the crystalline melting point. The modulus–temperature plots of amorphous and crystalline samples are similar, except that the position of the curve for the amorphous material is lower by ~40°C. Soaking in water has a major effect on the mechanical properties, with the modulus in the range of 20–50°C dropping by one order of magnitude. A smaller change in the modulus of the amorphous material relative to that of the crystalline material is found in studies in which the samples were allowed to relax under water.

The dynamic mechanical properties of perfluorocarboxylate ionomers have been investigated by Nakano and MacKnight (73). The glass transition temperature of the ionic phase was found to be related to Z/r, where Z is the charge of the cation,

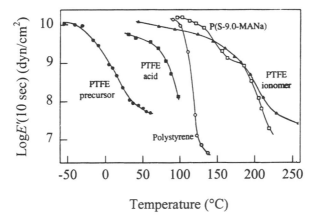

Figure 6.12. Comparison of log E' (10s) versus temperature for various ionomers and parent polymers. Modified from Hodge and Eisenberg (68).

and r is the distance between the centers of charge for the anion and cation at closest approach. Two different curves were obtained: one for alkali ions, as well as Ag^+ and Ba^{2+}, and the other for the Zn^{2+} and Mn^{2+} ionomers. The difference is possibly the result of the partial covalent character of the bonds involving the latter ions. Another noteworthy feature of the perfluorocarboxylates is that they absorb far less water than the perfluorosulfonates. Dynamic mechanical loss peaks for the perfluoro-carboxylates are the α, β', β, γ', and γ, in order of decreasing temperature. The assignments of the α, β, and γ relaxations to molecular mechanisms are identical to those for the perfluorosulfonates, whereas the mechanism of the β' relaxation is not clear. The γ' relaxation is assigned to local motions of the polar side groups, which are not phase separated from the matrix. The perfluorosulfonates do not show any distinction between the γ' and γ relaxations, but the perfluorocarboxylates do.

Tant et al. (74) investigated short side chain perfluorosulfonate ionomers in considerable detail. They studied the precursor, sulfonic acid, and its salt. As might be expected, the equivalent weight of the polymer chain, which controls crystallinity, exerts a major effect on the mechanical properties of the precursor. The higher the equivalent weight, the greater the crystallinity. The length of the side chain at constant mole percent of ionic groups influences the glass transition temperature, as has been found for a number of nonionic systems (75) and for ionomers (76). Here, the longer the side chain, the lower the glass transition temperature. As in the long side chain systems, the short side chain systems show a dramatic increase in the

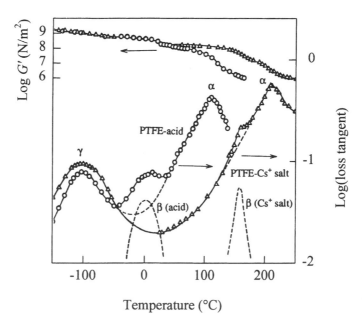

Figure 6.13. Log G' and log tan δ versus temperature for a perfluorosulfonate in the acid and Cs$^+$ salt form. *Dashed lines* represent the subtracted β peaks. Modified from Yeo and Eisenberg (67).

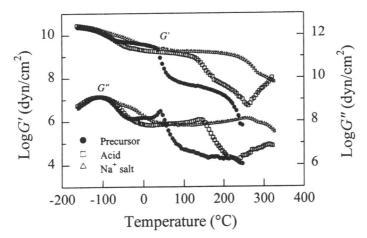

Figure 6.14. Log G' and log G'' versus temperature for short side chain perfluorosulfonate with EQ = 1000. ●, precursor; □, acid; and △, Na⁺ salt. Modified from Tant et al. (74).

glass transition temperature (~100°C) when the precursor is converted to the acid and a further increase when the acid is neutralized (~150°C for the sodium salt). An example of the effect on the shear modulus of conversion from precursor to acid to the sodium salt of a short side chain ionomer with an equivalent weight of 1000 is shown in Figure 6.14 (74). The upswing for the acid at high temperatures in both storage and loss moduli is probably owing to decomposition of the samples.

Starkweather and Chang (77) investigated the relaxation behavior of water in perfluorosulfonate ionomers. They found that the dynamic mechanical relaxation of water is seen near − 100°C. The presence of the relaxation is confirmed by dielectric and ¹H-NMR measurements. Neutralization of the acid groups with K⁺ and Na⁺ cations increases the temperature of the relaxation and lowers its activation energy. The authors assigned these phenomena to the glass transition of aqueous domains in the water swollen ionomer. The mobility of water molecules in the perfluorosulfonates was studied using high-resolution quasielastic neutron scattering by Volino et al. (78), who found that on the 10 Å scale the water molecules move essentially as freely in the water soaked PTFE acid membrane as in bulk water. However, in contrast to bulk water, the long range motion of water molecules is greatly restricted. This type of behavior is completely consistent with the most common models for perfluorosulfonate morphology (discussed below). In a study of the low temperature properties of water in perfluorinated membranes by Boyle et al. (79), the glass transition temperature of water in the ionic regions of the PTFE ionomers was taken as − 105°C, in close agreement with the results of Starkweather and Chang.

6.2.3. Other Physical Properties

The densities and expansion coefficients of various perfluorosulfonates were investigated by Takamatsu and Eisenberg (80). Graphs of density as a function of the ratio

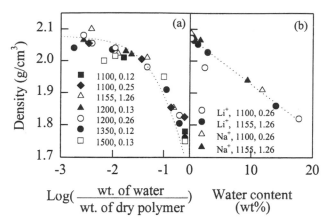

Figure 6.15. Density versus water content for perfluorosulfonates of various equivalent weights (1100–1500) and sample thickness (0.12–1.26 mm). Modified from Takamatsu and Eisenberg (80).

of the weight of water to the weight of dry acid polymer and as a function of water content by weight for the Li$^+$ and Na$^+$ salts are shown in Figure 6.15. The expansion coefficients of the annealed salts were also investigated; Figure 6.16 shows the linear expansion of Li$^+$ and Na$^+$ salts versus temperature. Two kinks in the plot are clearly seen. The linear expansion coefficients derived from these plots are given in Table

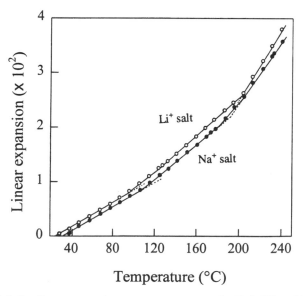

Figure 6.16. Relative linear expansion versus temperature for Na$^+$ (●) and Li$^+$ (○) salt, of Nafion with EW = 1100. Modified from Takamatsu and Eisenberg (80).

TABLE 6.1. Linear Expansion Coefficients of PTFE Polymers ($\times 10^4$) in Various Temperature Ranges

Acid form	$R.T.^a{-}80\,°C$	$> 100\,°C$	
α	1.47	2.98	
d	0.03	0.12	
σ	0.04	0.25	
Li⁺ salt	$R.T.{-}90\,°C$	$120{-}190\,°C$	$>190\,°C$
α	1.23	1.98	3.3
d	0.10	0.21	
σ	0.13	0.29	
Na⁺ salt	$R.T.{-}100\,°C$	$120{-}180\,°C$	$>200\,°C$
α	0.95	1.40	2.8
d	0.10	0.13	
σ	0.15	0.21	
K⁺ salt	$R.T.{-}120\,°C$	$130{-}190\,°C$	$>230\,°C$
α	1.23	1.44	2.5
d	0.11	0.16	
σ	0.16	0.22	
Cs⁺ salt	$R.T.{-}120\,°C$	$130{-}190\,°C$	$>200\,°C$
α	0.94	1.20	2.6
d	0.09	0.13	
σ	0.14	0.17	

[a] R.T., room temperature; α, draw direction; d, 45° direction; σ, transverse direction.

6.1. The two intersections of the straight line segments in the curve (the glass transitions of the matrix and cluster regions) coincide approximately with the loss tangent maxima.

6.2.4. Dielectric Properties

A dielectric relaxation study of perfluorosulfonate ionomers of various water contents was performed by Yeo and Eisenberg (67). Two peaks in the tan δ_ϵ versus temperature curve were seen above a water content of 0.4 H_2O/SO_3H group. The peak positions shift to lower temperatures as the water content increases. For example, the major peak shifts from ~100°C for the dry sample to -70°C for the 4 H_2O/SO_3H group. By contrast, the mechanical tan δ peak shifts from ~20° to -70°C over the same range of water contents. The activation energy for the dielectric relaxation of the major peak is ~100 kJ/mol. The K⁺-neutralized perfluorinated sample shows similar results, in that above a water content of 0.67 H_2O/SO_3K, two peaks are observed; these shift to lower temperatures with increasing water content. A subsequent study of the PTFE precursor was performed by Hodge and Eisenberg (68), who identified three peaks. The β (at ~-70°C) and γ (at ~-190°C) peaks were identified as fluorocarbon backbone and/or ether side chain motions and the motion of the SO_3F groups, respectively. The α peak was identified as the glass transition temperature (at ~20°C, at 100 Hz).

Mauritz's group (81–85) performed a series of studies on the dielectric relaxation attributed to ion motions. Perfluorosulfonate ionomers containing NaOH, NaCl, CH_3COONa, KCl, KI, $ZnSO_4$, and $CaCl_2$ were investigated. The authors suggested that the mechanisms involve the relaxation of interfacial polarization, originating from the accumulation of mobile ionic charge, and its subsequent dissipation along the direction of the external applied field. Evidence for another mechanism (i.e., a long-range ion motion) was also found, which could be attributed to migration of ions between clusters.

The perfluorocarboxylate ionomers were investigated by Perusich et al. (86), who found five dielectric relaxations: α, α', β, γ, and δ. The α relaxation, occurring between -20 and $100°C$, represents the motion of the polar vinyl ether side chains, α' is not well defined, and the β relaxation (between -80 and $-10°C$) is assigned to the ester rotation. The γ relaxation is seen at $\sim -100°C$ and is assigned to crankshaft rotational motion of the linear $-CF_2-$ segments along the PTFE backbone, whereas the δ relaxation (-200 to $-150°C$) is attributed to motions of the highly polar O—CH_3, O—K, or O—H dipoles at the end of the side groups.

6.2.5. Spectroscopy

6.2.5.1. Nuclear Magnetic Resonance. A number of NMR studies have focused on various perfluorosulfonates. Komoroski and Mauritz (87) investigated the ^{23}Na NMR spectra of the salt as a function of water content in the range of 1–30 wt %. At 1%, a broad peak is seen, which narrows with increasing water content. A large chemical shift change of approximately 130 ppm and a decrease in the line width with increasing water content can be reversed when the temperature is decreased. The data have been interpreted in terms of a fast equilibrium between bound and unbound, but loosely associated, cations in the hydrophilic regions of the material. A smaller fraction of cations is bound as the water content and temperature increase. The chemical shift data, in the Na^+ case, are sensitive only to the immediate environment, suggesting that the first hydration sphere of the Na^+ ion consists of three to four water molecules. Much of the NMR data can be explained by a three-phase model such as that proposed by Rodmacq et al. (88), who suggested that there are microcrystallites of a few hundred angstroms in diameter, regions of an aqueous ionic cluster phase with sizes of 20–35 Å, and finally an intermediate phase that contains few hydrated ions. A similar model was suggested by Yeager and Steck (89), who propose the existence of two different ion environments in the cation-neutralized membranes observed by Boyle et al. (90,91).

6.2.5.2. Infrared. The far infrared (IR) study of the cation sensitive bands in the perfluorosalts was performed by Risen's group (92). As in the styrene ionomers, a band in the 100–200 cm^{-1} range was found, the position of which shifted with the mass of the counterion. Thus, for Cs^+, the peak occurs near 100 cm^{-1}; whereas for Na^+, it is close to 180 cm^{-1}. Divalent cations show a similar trend. A plot of the peak position against $(mass)^{-1/2}$ shows two linear relationships: one for the alkali ions and the other for the alkaline earth ions. This observation suggests that in any one family the vibrational force constants are approximately the same for each ion in a given group.

In view of the importance of water and its transport in the perfluorosulfonate membranes, a large amount of research has been devoted to the IR study of water in the perfluorosulfonate copolymers. An assignment of the IR bands of perfluorinated membranes was made first by Heitner-Wirguin (93) and more recently by Cable et al. (94). In an early study of partly deuterated water in the perfluorosulfonates, Falk (95,96) showed that the O—D stretching region exhibited two distinct absorption bands. It was suggested that these bands, in view of their relative intensities, correspond to two different environments of the —OH groups: O—H $\cdots$ O and O—H $\cdots$ CF$_2$. It was concluded that the sizes of the hydrated regions must be small because a substantial fraction of the hydroxyl groups is exposed to a fluorocarbon environment. Because small-angle x-ray estimates of the aggregate dimensions from SAXS are approximately 40 Å (97,98), whereas those from the infrared study are ~12 Å, Falk suggested that the geometry of the aggregates could not be spherical. Quezado et al. (99) studied the IR spectra of perfluorinated membranes in various cationic forms to characterize the water–anion–cation interactions at low water contents.

Hydration effects in the perfluorosulfonate ionomers were investigated for various counterion salts by Lowry and Mauritz (100). The band corresponding to the SO$_3^-$ symmetric stretching mode was observed to shift to higher frequency and to broaden with decreasing water content. These changes were interpreted as an increase in the interaction between the polyanion and the cation, resulting from the decreased shielding accompanying the removal of water. The effect is cation dependent; the shifts are largest for the smallest cation (Li$^+$), and disappear for large cations (Rb^{2+}).

6.2.6. Morphology

Because of the importance of the perfluorinated ionomers in a wide range of electrochemical and other applications (Chapter 10), extensive studies have been devoted to the morphology of the perfluorosulfonate and perfluorocarboxylate ionomers. Because the presence of crystallinity introduces a high level of complexity into the morphological picture (Section 6.1.2), it is not surprising that many morphological studies have resulted in a wide range of proposed models for the structure of the perfluorinated ionomers.

In an early investigation of the morphology of perfluorinated systems by Fujimura et al. (101), SAXS and wide-angle x-ray studies were performed simultaneously. Two small-angle features were found. The one at the lower angle ($s = 2\sin\theta/\lambda \approx 0.1$ nm^{-1}) was ascribed to crystallinity in the system, and attributed to the long period in lamellar crystals, i.e., the interlamellar spacing. Another feature at an s of ~0.3 nm^{-1} was associated with the existence of ionic clusters. The size of these clusters was found to be a function of the number of pendent ionic groups in the polymer. These two features are seen in Figure. 6.17, which shows typical SAXS peaks for several different functional groups, including carboxylate and sulfonate. The EW of all the polymers was 1100. In a subsequent paper, the same authors

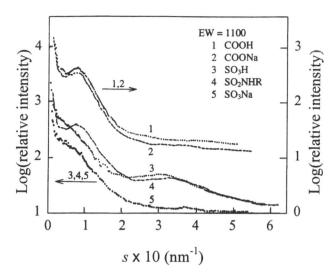

Figure 6.17. SAXS profile of perfluorosulfonates and perfluorocarboxylates of 1100 EW in various forms 1, COOH; 2, COONa; 3, SO$_3$H; 4, SO$_2$NHR; 5, SO$_3$Na. Modified from Fujmura et al. (101).

(102) attempted to fit various ionomer models to the SAXS profiles. The fits are marginally better for the core-shell model than for the other models.

In another investigation of the morphology of PTFE ionomers by x-rays, Gierke et al. (98,103) found that a SAXS peak is observed in the water-swollen, neutralized material; the Bragg spacing corresponds to a real space distance of 40–50 Å, which can be attributed distances between scattering regions. The intensity increases with decreasing EW; and the spacing is proportional to the ion exchange capacity, which, in turn, is related to the ion content. Thus, for an ion exchange capacity of 0.5 mEq/g (EW = 2000 g/mol), the spacing was 40 Å, whereas for an ion exchange capacity of 1 mEq/g (EW = 1000 g/mol) it was close to 50 Å. It was also found that the spacing and relative intensity of the peak increase with water content. These authors suggested a model based on the existence of nearly spherical clusters, consisting of water and ion-rich domains connected by narrow channels. The ionic domains become larger as the water content increases. This change is accompanied by a decrease in the number of water-rich ionic regions and an increase in the number of ionic groups per aggregate. The superpermselectivity of the membranes is related to the existence of narrow channels.

Extensive additional studies have been performed on the morphology of the perfluorinated ionomers, but these are far too numerous to be treated in detail here. A useful review was written by Mauritz (104). The various models are listed in Table 6.2 (104); naturally, the more general models (such as the core-shell model and the EHM model), although not listed here specifically, should be kept in mind also.

Because it is possible to dissolve perfluorinated ionomers at high temperatures and high pressures in a range of solvents (109), membranes can be prepared by casting the polymers from these solvents. An extensive study of properties of cast films was performed by Moore and Martin (66,110,111). Because ionic aggregate formation occurs rapidly in these materials, the nature of the aggregates in the films is independent of the method of treatment of the samples. However, no crystallinity was found for films cast at low temperature. Under these conditions, the mechanical properties of the films (without heat treatment) are relatively poor. However, the properties improve dramatically upon annealing of the films at elevated temperatures, as the crystallinity develops and as the polymer chains interpenetrate extensively.

The perfluorinated ionomers in solution have also received some attention. In one study, Aldebert et al. (108,112) proposed a hexagonal packing of rodlike micellar structures in perfluorinated ionomer solutions and gels. The rods, with a radius of between 18 and 31 Å, depending on the solvent, consist of a perfluorinated core with charges on the surface (108).

6.2.7. Water Uptake and Diffusion

Takamatsu et al. (113) examined the kinetics of neutralization of perfluorosulfonate films. The sorption (or weight gain) of the acid form of the membrane is shown in Figure 6.18 during neutralization with 0.01 N CsOH solution at room temperature. Curve 1 is the total weight gain as a function of log t, curve 2 shows the part of the weight gain owing to the cation alone, curve 3 shows the water uptake, curve 4 indicates degree of neutralization ($\times$ 0.1), and curve 5 is the weight gain in pure water for comparison. The maximum in the water uptake (curve 3) can be understood in terms of the neutralization process. Water diffusion through the membranes is rapid, so they take up a considerable amount of water in a short time. However, as the degree of neutralization increases, the total amount of water absorbed by the membranes decreases, because the Cs^+ salt form of the membrane at equilibrium absorbs less water than does the acid form.

The same study explored a range of other cations. Figure 6.19 shows the number

TABLE 6.2. Various Models of Perfluorinated Ionomers

Model	Reference
Three-phase model	88
Three-region structural model	89
Ion–dipole hydration shell cluster model	104
Four-state model of hydration-mediated counterion dissociation equilibrium	100
Cluster-network model	97
Elastic theory of cluster swelling	105
Spherical surface micelle geometry model	106
Water sorption isotherm model	107
Redistribution of ion exchange sites upon hydration	98
Rodlike micellar structures in solutions	108
Two environments of sorbed water	95

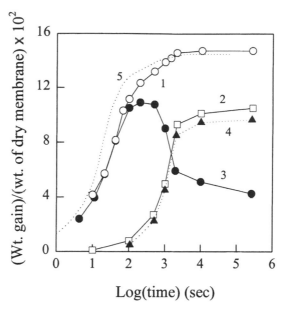

Figure 6.18. Relative weight gain versus time and individual contribution to the weight gain for perfluorosulfonate Cs^+ salt. See text for details. Modified from Takamatsu et al. (113).

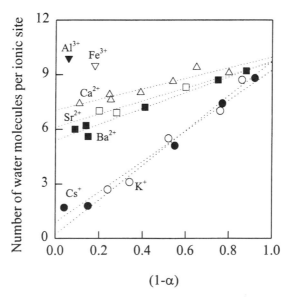

Figure 6.19. Number of water molecules per ionic site for water-swollen perfluorosulfonate membrane versus $1 - \alpha$. Modified from Takamatsu et al. (113).

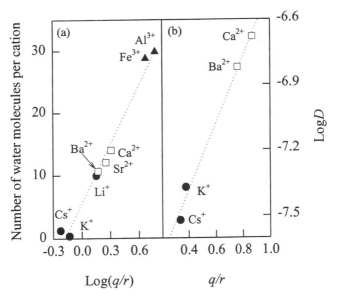

Figure 6.20. a, Number of water molecules per cation versus log q/r. **b,** Log of the diffusion coefficient D versus q/r for a perfluorosulfonate membrane neutralized with various cations. Modified from Takamatsu et al. (113).

of water molecules per ion exchange site plotted against the fraction of unneutralized acid groups $1-\alpha$ (where α is degree of neutralization) for several salts. It is clear that the number of water molecules per ionic group differs drastically from salt to salt. This finding has been correlated with the ratio of the charge to the radius of the cation (q/r); the diffusion coefficient at room temperature seems to follow a similar trend (Fig. 6.20).

The diffusion of various ions through the perfluorosulfonate membranes was investigated by Yeager and Steck (89). The results are plotted as the log of the self-diffusion coefficient $\bar{D}$ versus the reciprocal of absolute temperature (1000/T) for Na^+ and Cs^+ salts in Figure 6.21. The Na^+ ion clearly diffuses more rapidly.

The first study of the diffusion coefficient of water through perfluorosulfonate membranes was published in 1977 (67). The diffusion coefficient was found to obey the following equation:

$$D = 6.0 \times 10^{-3} \exp(-20 \text{ kJ}/RT) \text{ cm}^2/\text{s} \qquad (6.1)$$

It is interesting to note that at room temperature the diffusion coefficient for water through the Na^+ salt of the perfluorosulfonate polymer is only about a factor of three lower than the self-diffusion coefficient of water, despite the water content in the polymer being only ~ 10 vol %. A comparison of the diffusion of water through the acid and the salt forms was published by Takamatsu et al. (113). The diffusion coefficient could be described in a normal Arrhenius form:

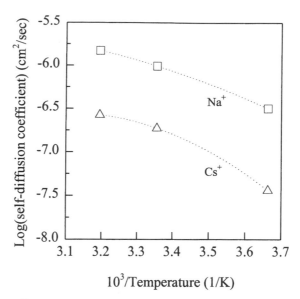

Figure 6.21. Log $\bar{D}$ versus 1000/temperature for Cs^+-neutralized ($\triangle$) and Na^+-neutralized ($\square$) perfluorosulfonate membranes. Modified from Yeager and Steck (89).

$$D = D_0 \exp(-E_d/RT) \qquad (6.2)$$

where $D_0 = 8 \times 10^{-3}$ cm^2/s and E_d is the activation energy, which equals 20 kJ/mol for water through the acid membrane. By contrast, D_0 was ~1.9×10^2, and E_d was 55 kJ/mol for diffusion of water through the K^+ salt membrane (EW = 1100). The value of D_0 changes only slightly if pure water is replaced by 1 N KOH (aq.) (1.9×10^2 versus 1.6×10^2).

A percolation treatment of water diffusion through perfluorosulfonate membranes in the acid form (EW = 1050, 1100, 1350, and 1500) was given by Hsu et al. (105). It was shown that percolation concepts are applicable (the critical exponent being 1.5 ± 0.2) and the percolation threshold is 0.10 (lower than the ideal value of 0.15 found in many other random systems). From this difference, the authors concluded that the ion-containing clusters or aggregates are not randomly distributed and suggested that there is an extended conductive network of aggregates connected by narrow channels.

A more qualitative study of water sorption in acid PTFE membranes was published by Duplessix et al. (114), who showed that the energy of hydration is a function of the water content. At low overall water contents (between zero and five water molecules per sulfonate group), the hydration energy is constant at ~50 kJ/mol; between five and eight water molecules per ionic group it drops to ~10 kJ/mol. Eight water molecules per sulfonate group represents the absorption limit. There is a slight difference between the samples dried at room temperature and those dried at 220°C: for the samples dried at elevated temperatures, the drop starts at a slightly lower H_2O/SO_3^- ratio (i.e., three instead of five).

It was found that vacuum drying at room temperature for 24 h does not desorb all the water. Interestingly, the binding energies of the last trace of the water are not the same as those at higher water contents. It was, therefore, suggested that drying of PTFE is a kinetic phenomenon, and the water molecules are trapped kinetically rather than thermodynamically. In water sorption, the drop in the hydration energy has been attributed to rheological effects, i.e., the removal of hydrophobic chains from the hydration zone. Water molecules entering the polymer after a certain amount of water had already been absorbed need to push chains out of the way to make room, which reduces the apparent binding energy.

Rodmacq et al. (88) performed SAXS and small-angle neutron-scattering (SANS) studies to investigate the morphology of water-swollen PTFE ionomers. No major change in the microstructure between the acid and salt forms was seen. They also found that there was only a small change in the sizes of the clusters resulting from the substitution of various counterions. From a Guinier analysis (115) of the SAXS data, the authors noted that up to ~83% relative humidity the radius of gyration of the particles is ~8 Å. Soaking and boiling the samples raises the size of the regions to 15 and 24 Å, respectively. On the basis of these results, the authors proposed a three-phase model, consisting of crystalline material, ionic aggregates, and an intermediate ionic phase of lower ion content. Direct connectivity between the phases is not made explicit in this model, in contrast to the models of Gierke (97,98,103) and Yeager and Steck (89) (Table 6.2).

Morris and Sun (116) investigated water sorption and transport in Nafion-117 perfluorosulfonates. They found that water sorption isotherms obey Henry's law with a nonzero intercept, suggesting that there is some retention of water. The mutual diffusion coefficient of water was measured as a function of the molar ratio of water to ion exchange sites (Fig. 6.22) and the electrical conductivity was measured as a function of water content for similar systems (Fig. 6.23). The conductivity was interpreted as a percolation phenomenon, which is similar to the treatment of Hsu et al. (105), and is given by the following equation:

$$\sigma = 0.125 \, (f_v - 0.06)^{1.95} \qquad (6.3)$$

where f_v is the volume fraction of the aqueous phase. It should be mentioned that the critical exponent and critical concentration at the percolation threshold (1.95 and 0.06) are different from those of Hsu et al. (1.5 and 0.1).

The conductivity of Nafion-117 membranes was also measured by Zawodzinski et al. (117) (Fig. 6.24). The swelling behavior of the perfluorosulfonates in a range of solvents was investigated by Yeo and Cheng (118). Solvent uptake of the perfluorosulfonate membranes is given in Figure 6.25 as a function of the solubility parameters for the acid, Li^+ salt, Na^+ salt, and K^+ salt.

6.2.9. Permeation

Gas permeation through perfluorosulfonate membranes was investigated by Sakai et al. (119,120), who found that the permeability coefficients are a strong function

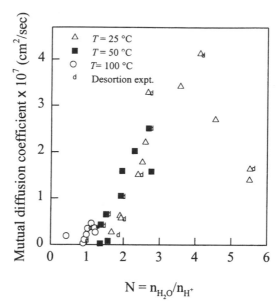

Figure 6.22. Water diffusion coefficient versus degree of swelling for perfluorosulfonate membrane. △, at 25°C; ■, at 50°C; ○, at 100°C; d, desorption experiment. Modified from Morris and Sun (116).

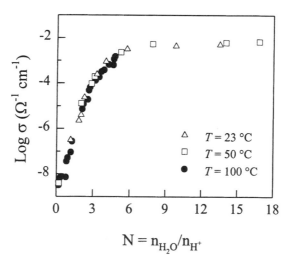

Figure 6.23. Log of conductance versus degree of swelling at various temperatures for perfluorosulfonate membranes. △, at 23°C; □, at 50°C; ●, at 100°C. Modified from Morris and Sun (116).

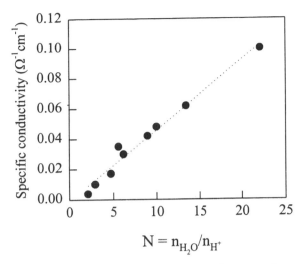

$$N = n_{H_2O}/n_{H^+}$$

Figure 6.24. Specific conductivity versus degree of swelling at various temperatures for perfluorosulfonate membranes. Modified from Zawodzinski et al. (117).

of the water content, the cation, and the ion exchange capacity. Naturally, the permeation rates also vary with temperature, pressure, and membrane thickness. Examples of the permeability coefficients for hydrogen and oxygen for a Nafion-117 K^+ salt membrane at various water contents are shown in Figure 6.26. In a follow-up study (120), the same group found that the gas diffusivity in the dry membrane was similar to that of polytetrafluoroethylene.

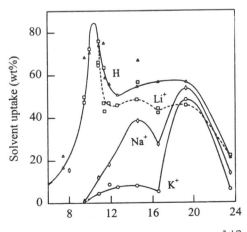

Figure 6.25. Solvent uptake versus solubility parameter for perfluorosulfonate. Modified from Yeo and Cheng (118).

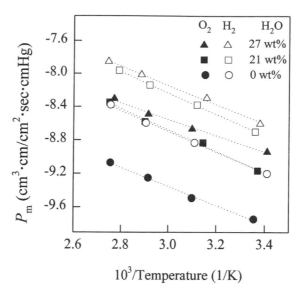

Figure 6.26. Permeability coefficients of H_2 (*open symbols*) and O_2 (*solid symbols*) for Nafion-117 membrane with various water contents. *Triangles,* 27 wt % H_2O; *squares,* 21 wt % H_2O; *circles,* ○ wt % H_2O. Modified from Sakai et al. (119).

Chiou and Paul (121) investigated the transport of a wide range of gases through Nafion-117 at 35°C (including He, H_2, O_2, Ar, N_2, CH_4, and CO_2). They found that for all but CO_2, the permeability coefficients were independent of the upstream pressure. For CO_2 there was a pressure dependence, with the permeability coefficient changing from ~2.4 at 0.1 MPa to ~4.2 at 2 MPa. The authors found that while no productivity–selectivity advantages relative to other polymers were seen in perfluorosulfonates for mixtures such as CO_2/CH_4 and O_2/N_2, the membrane was as an effective separator for mixtures such as He/CH_4, N_2/CH_4, and He/H_2. The pressure dependence for CO_2 is typical for rubbery polymers (122,123).

6.2.10. Chemistry

Olah's group (124–126) found that the PTFE acid resins can be used as a superacid catalyst for various reactions, e.g., alkylation of benzene and transalkylation of alkylbenzenes, the pinacol rearrangement, and deacetylation and decarboxylation of aromatic molecules. Of the other extensive studies that have been performed on chemical reactions in the perfluorosulfonates, only one example will be mentioned here. Barns et al. (127) investigated reactions of transition metal ions in PTFE ionomers with various gases; CO, NO, NH_3, C_2H_4, N_2H_4, and H_2. For example, it was found that the Ag^+-PTFE ionomer reacts with CO to form a silver carbonyl. However, the other transition metal-PTFE ionomers do not react with CO.

6.3. POLYPENTENAMER

6.3.1. Synthesis and Thermal Properties

Polypentenamer is a noncrystalline polymer, whereas hydrogenated polypentenamer is partly crystalline. A series of studies on hydrogenated, functionalized polypentenamers were performed by MacKnight's group using materials containing carboxylate, sulfonate, and phosphonate groups. The synthesis of the carboxylated polypentenamer and the hydrogenation procedure were described by Sanui et al. (128). The structure of the carboxylated polypentenamer is shown in Scheme **6.1**. This effort

$$-(CH_2CH_2CH_2CH_2CH_2)_x-(CH_2CH_2CH_2CHCH_2)_y-$$
$$|$$
$$SCH_2COO^-Na^+$$

Scheme 6.1

was followed by an investigation of polypentenamer and its hydrogenated derivatives (129,130). The carboxylation of the polypentenamer—which was performed by thioglycolation using methyl thioglycolate ($HSCH_2COOCH_3$) as a reagent and subsequent hydrogenation using p-toluenesulfonhydrazide ($CH_3C_6H_4SO_2NHNH_2$)—introduces some thermal instability to the material in that, in contrast to most other ionomers, the carboxylates decompose considerably below the temperatures at which the polypentenamer itself decomposes. The origin of the instability was suggested to be the thioglycolate side chain.

The preparation of the phosphonylated polypentenamer was described by Azuma and MacKnight (131). Dimethylphosphite [$HPO(OCH_3)_2$] was used as a reagent for phosphonylation. The authors found that an increase in the phosphonate side group content decreases the melting point in the case of hydrogenated derivatives and increases the glass transition temperature for unhydrogenated derivatives. As might be expected, the free acid derivatives are thermally unstable; and hydrogenated derivatives were found to be generally more stable than unhydrogenated systems.

The sulfonated polypentenamers and their hydrogenated derivatives were prepared by Rahrig et al. (132). A complex of liquid sulfur trioxide and triethyl phosphate [SO_3-$OP(OC_2H_5)_3$] was used as the reagent for sulfonation. As before, the melting point decreases with the degree of substitution for the sodium sulfonate; the melting point of the hydrogenated derivatives is ~130°C but drops linearly to ~100°C for the 20 mol % substituted material. Similar trends were seen for the thioglycolate and phosphonate groups; in the thioglycolate, the T_m of the 20 mol % material is ~90°C, whereas in the phosphonate it is ~70°C.

6.3.2. Mechanical Properties

In the unhydrogenated phosphonylated derivatives, a major relaxation, called β, was seen by mechanical testing between -160 and $+100$°C (133). It was suggested that this relaxation arises from micro-Brownian segmental motion accompanying the glass transition. In addition to the β relaxation, α and γ relaxations were also observed. As for the ethylene-based ionomers, these two relaxations were believed to be the result of the melting of crystals containing low levels of *trans* double bonds and the local motions of the three methylene groups, respectively. α, β, and γ relaxations were also observed in the hydrogenated derivatives. It was found that the temperatures of the α and β relaxations strongly depend on the chemical nature and concentration of the substituents, as well as the thermal history of the samples. As was found in the previous study on ethylene ionomers, the γ relaxation is independent of these factors.

The mechanical properties of the sulfonated polypentenamer were also investigated extensively (134,135). The phosphonylated and sulfonated polypentenamer derivatives are similar in their mechanical properties. Again α, β and γ peaks were observed in the unsaturated system (134). It was suggested that the α relaxation is the result of a relaxation mechanism in a separate ionic phase, the β relaxation is related to micro-Brownian segmental motion in the amorphous hydrocarbon matrix, and the γ relaxation to crankshaft motion of methylene groups between double bonds. In the hydrogenated system, the behavior is complicated by the presence of crystallinity (135). In this case, it was found that the α relaxation peak increases in magnitude and decreases in temperature with increasing sulfonation. In addition, annealing results in an increase of the temperature and magnitude of the α process. Thus it was suggested that this relaxation is probably the result of an ionic-phase mechanism. The magnitude of the β relaxation increases with decreasing degree of crystallinity and was thus assigned to the apparent glass transition of semicrystalline part of the sample. The assignment of the γ relaxation was the same as in the other systems.

6.3.3. Morphology

A morphological study on sulfonated polypentenamers was performed using SAXS and small-angle neutron scattering (136). For the 17% Cs^+ salt in the dry state, the SAXS ionomer peak is seen at $q = 0.2$ Å^{-1}, but no SANS ionomer peak was observed in this sample. It was found, however, that when several D_2O molecules per ionic group are added ($<$ six D_2O molecules/ionic group), a SANS ionic peak is seen ($q = 0.2$ Å^{-1}), corresponding to the SAXS peak in dry state. If more D_2O is added to the sample, the SANS peak shifts to lower angles (Chapter 8). It was found that the trend in the shift of the SANS peak as a function of the amount of D_2O added is similar to that in the perfluorosulfonates (137). It was suggested that phase separation may occur at that high water content.

6.4. REFERENCES

1. Ward, T. C.; Tobolsky, A. V. *J. Appl. Polym. Sci.* **1967,** *11,* 2403–2415.
2. Rees, R. W.; Vaughan, D. J. *Polym. Prepr. Am. Chem. Soc. Div. Polym. Chem.* **1965,** *6*(1), 287–295.

3. Rees, R. W.; Vaughan, D. J. *Polym. Prepr. Am. Chem. Soc. Div. Polym. Chem.* **1965,** *6*(1), 296–303.

4. Rees, R. W. U.S. Patent 3 264 272, 1966.

5. Wilson, F. C.; Longworth, R.; Vaughan, D. J. *Polym. Prepr. Am. Chem. Soc. Div. Polym. Chem.* **1968,** *9*(1), 505–512.

6. Davis, H. A.; Longworth, R.; Vaughan, D. J. *Polym. Prepr. Am. Chem. Soc. Div. Polym. Chem.* **1968,** *9*(1), 515–524.

7. Longworth, R.; Vaughan, D. J. *Polym. Prepr. Am. Chem. Soc. Div. Polym. Chem.* **1968,** *9*(1), 525–533.

8. Bonotto, S.; Bonner, E. F. *Macromolecules* **1968,** *1*, 510–515.

9. MacKnight, W. J.; McKenna, L. W.; Read, B. E. *J. Appl. Phys.* **1967,** *38*, 4208–4212.

10. Read, B. E.; Carter, E. A.; Connor, T. M.; MacKnight, W. J. *Br. Polym. J.* **1969,** *1*, 123–131.

11. Otocka, E. P.; Kwei, T. K. *Macromolecules* **1968,** *1*, 401–405.

12. Otocka, E. P.; Davis, D. D. *Macromolecules* **1969,** *2*, 437.

13. Yano, S.; Fujiwara, Y.; Aoki, K.; Yamauchi, J. *Colloid Polym. Sci.* **1980,** *258*, 61–69.

14. Yano, S.; Fujiwara, Y.; Kato, F.; Aoki, K.; Koizumi, N. *Polym. J.* **1981,** *13*, 283–291.

15. Yano, S.; Yamashita, H.; Matsushita, K.; Aoki, K.; Yamauchi, J. *Colloid Polym. Sci.* **1981,** *259*, 514–521.

16. Yamauchi, J.; Yano, S. *Macromolecules* **1982,** *15*, 210.

17. Yano, S.; Yamamoto, H.; Tadano, K.; Yamamoto, Y.; Hirasawa, E. *Polymer* **1987,** *28*, 1965–1970.

18. Yano, S.; Tadano, K.; Hirasawa, E.; Yamauchi, J. *Polym. J.* **1991,** *23*, 969.

19. Tsunashima, K.; Kutsumizu, S.; Hirasawa, E.; Yano, S. *Macromolecules* **1991,** *24*, 5910.

20. Yano, S.; Nagao, N.; Hattori, M.; Hirasawa, E.; Tadano, K. *Macromolecules* **1992,** *25*, 368–376.

21. Yano, S.; Tadano, K.; Nagao, N.; Kutsumizu, S.; Tachino, H.; Hirasawa, E. *Macromolecules* **1992,** *25*, 7168–7171.

22. Kutsumizu, S.; Nagao, N.; Tadano, K.; Tachino, H.; Hirasawa, E.; Yano, S. *Macromolecules* **1992,** *25*, 6829–6835.

23. Tachino, H.; Hara, H.; Hirasawa, E.; Kutsumizu, S.; Tadano, K.; Yano, S. *Macromolecules* **1993,** *26*, 752–757.

24. Tachino, H.; Hara, H.; Hirasawa, E.; Kutsumizu, S.; Yano, S. *Polym. J.* **1994,** *26*, 1170.

25. Tachino, H.; Hara, H.; Hirasawa, E.; Kutsumizu, S.; Yano, S. *J. Appl. Polym. Sci.* **1995,** *55*, 131–138.

26. Kutsumizu, S.; Ikeno, T.; Osada, S.; Hara, H.; Tachino, H.; Yano, S. *Polym. J.* **1996** *28* 299–308.

27. MacKnight, W. J. In *Structure and Properties of Ionomers*; Pineri, M.; Eisenberg, A., Eds.; NATO ASI Series C, Mathematical and Physical Sciences 198; Reidel: Dordrecht, 1987; pp. 267–277.

28. Sakamoto, K.; MacKnight, W. J.; Porter, R. S. *J. Polym. Sci. A2* **1970,** *8*, 277–287.

29. Earnest, T. R. Jr.; MacKnight, W. J. *J. Polym. Sci. Polym. Phys. Ed.* **1978,** *16*, 143–157.

30. Shohamy, E.; Eisenberg, A. *J. Polym. Sci. Polym. Phys. Ed.* **1976,** *14*, 1211–1220.

31. Phillips, P. J.; MacKnight, W. J. *J. Polym. Sci. A2* **1970,** *8*, 727–738.

32. Goddard, R. J.; Grady, B. P.; Cooper, S. L. *Macromolecules* **1994,** *27,* 1710–1719.

33. Hirasawa, E.; Yamamoto, Y.; Tadano, K.; Yano, S. *Macromolecules* **1989,** *22,* 2776–2780.

34. Tsatsas, A. T.; Risen, W. M. Jr. *Chem. Phys. Lett.* **1970,** *7,* 354–356.

35. Tsatsas, A. T.; Reed, J. W.; Risen, W. M. Jr. *J. Chem. Phys.* **1971,** *55,* 3260–3269.

36. MacKnight, W. J.; McKenna, L. W.; Read, B. E.; Stein, R. S. *J. Phys. Chem.* **1968,** *72,* 1122–1126.

37. Andreeva, E. D.; Nikitin, V. N.; Boyartchuk, Yu. K. *Macromolecules* **1976,** *9,* 238–243.

38. Painter, P. C.; Brozoski, B. A.; Coleman, M. M. *J. Polym. Sci. Polym. Phys. Ed.* **1982,** *20,* 1069–1080.

39. Brozoski, B. A.; Coleman, M. M.; Painter, P. C. *J. Polym. Sci. Polym. Phys. Ed.* **1983,** *21,* 301–308.

40. Brozoski, B. A.; Coleman, M. M.; Painter, P. C. *Macromolecules* **1984,** *17,* 230–234.

41. Brozoski, B. A.; Painter, P. C.; Coleman, M. M. *Macromolecules* **1984,** *17,* 1591–1594.

42. Coleman, M. M.; Lee, J. Y.; Painter, P. C. *Macromolecules* **1990,** *23,* 2339–2345.

43. Ishioka, T. *Polym. J.* **1993,** *25,* 1147–1152.

44. Ishioka, T. *Macromolecules* **1995,** *28,* 1298–1305.

45. Tsujita, Y.; Hsu, S. L.; MacKnight, W. J. *Macromolecules* **1981,** *14,* 1824–1826.

46. Takei, M.; Tsujita, Y.; Shimada, S.; Ichihara, H.; Enokida, M.; Takizawa, A.; Kinoshita, T. *J. Polym. Sci. B. Polym. Phys.* **1988,** *26,* 997–1008.

47. Belfiore, L.; Shah, R. J. *Polym. Mat. Sci. Eng.* **1986,** *54,* 490.

48. Belfiore, L.; Patwardhan, A. A. *Polym. Mat. Sci. Eng.* **1986,** *54,* 638.

49. Belfiore, L. A.; Shah, R. J.; Cheng, C. In *Contemporary Topics in Polymer Science,* Vol. 6; Culbertson, B. M., Ed.; Plenum: New York, 1989.

50. Dickinson, L. C.; MacKnight, W. J.; Connolly, J. M.; Chien, J. C. W. *Polym. Bull.* **1987,** *17,* 459–464.

51. Connolly, J. M. Ph.D. Dissertation University of Massachusetts at Amherst, 1990.

52. Marx, C. L.; Cooper, S. L. *J. Macromol. Sci. Phys.* **1974,** *B9,* 19–33.

53. Tsujita, Y.; Shibayama, K.; Takizawa, A.; Kinoshita, T.; Uematsu, I. *J. Appl. Polym. Sci.* **1987,** *33,* 1307–1314.

54. Kohzaki, M.; Tsujita, Y.; Takizawa, A.; Kinoshita, T. *J. Appl. Polym. Sci.* **1987,** *33,* 2393–2402.

55. Orler, E. B.; Yontz, D. J.; Moore, R. B. *Macromolecules* **1993,** *26,* 5157.

56. Uemura, Y.; Stein, R. S.; MacKnight, W. J. *Macromolecules* **1971,** *4,* 490–494.

57. Itoh, K.; Tsujita, Y.; Takizawa, A.; Kinoshitu, T. *J. Appl. Phys. Sci.* **1986,** *32,* 3335–3343.

58. Kajiyama, T.; Oda, T.; Stein, R. S.; MacKnight, W. J. *Macromolecules* **1971,** *4,* 198.

59. Kajiyama, T.; Stein, R. S.; MacKnight, W. J. *J. Appl. Phys.* **1970,** *41,* 4361.

60. Roche, E. J., Stein, R. S., Russell, T. P., MacKnight, W. J. *J. Polym. Sci.: Polym. Phys. Ed.* (1980) *18,* 1497.

61. *J. Electrochem. Soc.* **1977,** *124,* 318C–319C.

62. *Perfluorinated Ionomer Membranes*; Eisenberg, A.; Yeager, H. L., Eds.; ACS Symposium Seriers 180; American Chemical Society: Washington, DC, 1982.

63. *Ionomers: Characterizations, Theory, and Applications*; Schlick, S., Ed.; CRC: Boca Raton, FL, 1996.

64. Starkweather, H. W. Jr. *Macromolecules* **1982**, *15*, 320–323.

65. Gebel, G.; Aldebert, P.; Pineri, M. *Macromolecules* **1987**, *20*, 1425–1428.

66. Moore, R. B. III; Martin, C. R. *Macromolecules* **1989**, *22*, 3594–3599.

67. Yeo, S. C.; Eisenberg, A. *J. Appl. Polym. Sci.* **1977**, *21*, 875–898.

68. Hodge, I. M.; Eisenberg, A. *Macromolecules* **1978**, *11*, 289–293.

69. Kyu, T.; Eisenberg, A. In *Perfluorinated Ionomer Membrane*; Eisenberg, A., Yeager, H. L., Eds.; ACS Symposium Seriers 180; American Chemical Society: Washington, DC, 1982; Chapter 6.

70. Kyu, T. In *Materials Science of Synthetic Membranes*; Lloyd, D. R., Ed.; ACS Symposium Series 269; American Chemical Society: Washington, DC, 1985; Chapter 18.

71. Kyu, T.; Hashiyama, M.; Eisenberg, A. *Can. J. Chem.* **1983**, *61*, 680–687.

72. Kyu, T.; Eisenberg, A. *J. Polym. Sci. Polym. Symp.* **1984**, *71*, 203–219.

73. Nakano, Y.; MacKnight, W. J. *Macromolecules* **1984**, *17*, 1585.

74. Tant, M. R.; Darst, K. P.; Lee, K. D.; Martin, C. W. In *Multiphase Polymers: Blends and Ionomers*; Utracki, L. A., Weiss, R. A., Eds.; ACS Symposium Series 395; American Chemical Society: Washington, DC, 1989; Chapter 15.

75. Vincent, P. I. In *The Physics of Plastics*; Ritchie, P. D., Ed.; Iliffe Books: London, 1965.

76. Gauthier, M.; Eisenberg, A. *Macromolecules* **1990**, *23*, 2066–2074.

77. Starkweather, H. W. Jr.; Chang, J. J. *Macromolecules* **1982**, *15*, 752–756.

78. Volino, F.; Pineri, M.; Dianoux, A. J.; De Geyer, A. *J. Polym. Sci. Polym. Phys. Ed.* **1982**, *20*, 481–496.

79. Boyle, N. G.; Coey, J. M. D.; McBrierty, V. J. *Chem. Phys. Lett.* **1982**, *86*, 16–19.

80. Takamatsu, T.; Eisenberg, A. *J. Appl. Polym. Sci.* **1979**, *24*, 2221–2235.

81. Mauritz, K. A.; Fu, R.-M. *Macromolecules* **1988**, *21*, 1324–1333.

82. Mauritz, K. A.; Yun, H. *Macromolecules* **1988**, *21*, 2738–2743.

83. Mauritz, K. A.; Yun, H. *Macromolecules* **1989**, *22*, 220–225.

84. Mauritz, K. A. *Macromolecules* **1989**, *22*, 4483–4488.

85. Deng, Z. D.; Mauritz, K. A. *Macromolecules* **1992**, *25*, 2369–2380.

86. Perusich, S. A.; Avakian, P.; Keating, M. Y. *Macromolecules* **1993**, *26*, 4756–4764.

87. Komoroski, R. A.; Mauritz, K. A. *J. Am. Chem. Soc.* **1978**, *100*, 7487–7489.

88. Rodmacq, B.; Coey, J. M. D.; Escoubes, M.; Roche, E.; Duplessix, R.; Eisenberg, A.; Pineri, M. In *Water in Polymers*; Rowland, S. P., Ed.; ACS Symposium Series 127; American Chemical Society: Washington, DC, 1980; Chapter 29.

89. Yeager, H. L.; Steck, A. *J. Electrochem. Soc.* **1981**, *128*, 1880–1884.

90. Boyle, N. G.; McBrierty, V. J.; Douglass, D. C. *Macromolecules* **1983**, *16*, 75–80.

91. Boyle, N. G.; McBrierty, V. J.; Douglass, D. C.; Eisenberg, A. *Macromolecules* **1983**, *16*, 80–84.

92. Mattera, V. D. Jr.; Peluso, S. L.; Tsatsas, A. T.; Risen, W. M. Jr. In *Coulombic Interactions in Macromolecular Systems*; Eisenberg, A.; Bailey, F. E., Eds.; ACS Symposium Series 302; American Chemical Society: Washington, DC, 1986; Chapter 4.

93. Heitner-Wirguin, C. *Polymer* **1979,** *20,* 371–374.

94. Cable, K. M.; Mauritz, K. A.; Moore, R. B. *J. Polym. Sci. B Polym. Phys.* **1995,** *33,* 1065–1072.

95. Falk, M. *Can. J. Chem.* **1980,** *58,* 1495–1501.

96. Falk, M. In *Perfluorinated Ionomer Membranes*; Eisenberg, A.; Yeager, H. L., Eds.; ACS Symposium Seriers 180; American Chemical Society: Washington, DC, 1982; Chapter 8.

97. Gierke, T. D. *J. Electrochem. Soc.* **1977,** *124,* 319(C).

98. Gierke, T. D.; Munn, G. E.; Wilson, F. C. *J. Polym. Sci. Polym. Phys. Ed.* **1981,** *19,* 1687–1704.

99. Quezado, S.; Kwak, J. C. T.; Falk, M. *Can J. Chem.* **1984,** *62,* 958–966.

100. Lowry, S. R.; Mauritz, K. A. *J. Am. Chem. Soc.* **1980,** *102,* 4665–4667.

101. Fujimura, M.; Hashimoto, T.; Kawai, H. *Macromolecules* **1981,** *14,* 1309–1315.

102. Fujimura, M.; Hashimoto, T.; Kawai, H. *Macromolecules* **1982,** *15,* 136–144.

103. Gierke, T. D.; Munn, G. E.; Wilson, F. C. In *Perfluorinated Ionomer Membranes*; Eisenberg, A.; Yeager, H. L., Eds.; ACS Symposium Series 180; American Chemical Society: Washington, DC, 1982; Chapter 10.

104. Mauritz, K. A.; Hora, C. J.; Hopfinger, A. J. In *Ions in Polymers*; Eisenberg, A., Eds.; Advanced Chemistry Series 187; American Chemical Society: Washington, DC, 1980; Chapter 8.

105. Hsu, W. Y.; Barley, J. R.; Meakin, P. *Macromolecules* **1980,** *13,* 198–200.

106. Datye, V. K.; Taylor, P. L.; Hopfinger, A. J. *Macromolecules* **1984,** *17,* 1704.

107. Mauritz, K. A.; Rogers, C. E. *Macromolecules* **1985,** *18,* 483–491.

108. Aldebert, P.; Dreyfus, B.; Gebel, G.; Nakamura, N.; Pineri, M.; Volino, F. *J. Phys. France,* **1988,** *49,* 2101–2109.

109. Martin, C. R.; Rhodes, T. A.; Ferguson, J. A. *Anal. Chem.* **1982,** *54,* 1639–1641.

110. Moore, R. B. III; Martin, C. R. *Anal. Chem.* **1986,** *58,* 2569–2570.

111. Moore, R. B. III; Martin, C. R. *Macromolecules* **1988,** *21,* 1334–1339.

112. Aldebert, P.; Dreyfus, B.; Pineri, M. *Macromolecules* **1986,** *19,* 2651–2653.

113. Takamatsu, T.; Hashiyama, M.; Eisenberg, A. *J. Appl. Polym. Sci.* **1979,** *24,* 2199–2220.

114. Duplessix, R.; Escoubes, M.; Rodmacq, B.; Volino, F.; Roche, E.; Eisenberg, A.; Pineri, M. In *Water in Polymers*; Rowland, S. P., Ed.; ACS Symposium Series 127; American Chemical Society: Washington, DC, 1980; Chapter 28.

115. Guinier, A.; Fournet, G. *Small Angle Scattering of X-Rays*, Wiley: New York, 1965.

116. Morris, D. R., Sun, X. *J. Appl. Polym. Sci.* **1993,** *50,* 1445–1452.

117. Zawodzinski, T. A. Jr.; Derouin, C.; Radzinski, S.; Sherman, R. J.; Smith, V. T.; Springer, T. E.; Cottesfeld, S. *J. Electrochem. Soc.* **1993,** *140,* 1041–1047.

118. Yeo, R. S.; Cheng C.-H. *J. Appl. Polym. Sci.* **1986,** *32,* 5733–5741.

119. Sakai, T.; Takenaka, H.; Wakabayashi, N.; Kawami, Y.; Torikai, E. *J. Electrochem. Soc.* **1985,** *132,* 1328–1332.

120. Sakai, T.; Takenaka, H.; Torikai E. *J. Electrochem. Soc.* **1986,** *133,* 88–92.

121. Chiou, J. S.; Paul, D. R. *Ind. Eng. Chem. Res.* **1988,** *27,* 2161–2164.

122. El-Hibri, M. J.; Paul, D. R. *J. Appl. Polym. Sci.* **1986,** *31,* 2533.

123. Chiou, J. S.; Barlow, J. W.; Paul, D. R. *J. Appl. Polym. Sci.* **1985,** *30,* 1173.

124. Olah, G. A.; Kaspi, J.; Bukala, J. *J. Org. Chem.* **1977,** *42,* 4187.

125. Olah, G. A.; Meidar, D. *Synthesis* **1978,** *358.*

126. Olah, G. A.; Laali, K.; Mehrotra, A. K. *J. Org. Chem.* **1983,** *48,* 3360.

127. Barnes, D. M.; Chaudhuri, S. N.; Chryssikos, G. D.; Mattera, V. D. Jr.; Peluso, S. L.; Shim, I. W.; Tsatsas, A. T.; Risen, W. M. Jr. In *Coulombic Interactions in Macromolecular Systems.* Eisenberg, A.; Bailey, F. E., Eds.; ACS Symposium Series 302; American Chemical Society: Washington, DC, 1986; Chapter 5.

128. Sanui, K.; Lenz, R. W.; MacKnight, W. J. *J. Polym. Sci. Polym. Chem. Ed.* **1974,** *12,* 1965–1981.

129. Sanui, K.; MacKnight, W. J.; Lenz, R. W. *Macromolecules* **1974,** *7,* 101–105.

130. Earnest, T. R. Jr.; MacKnight, W. J. *Macromolecules* **1977,** *10,* 206–210.

131. Azuma, C.; MacKnight, W. J. *J. Polym. Sci. Polym. Chem. Ed.* **1977,** *15,* 547–560.

132. Rahrig, D.; MacKnight, W. J.; Lenz, R. W. *Macromolecules* **1979,** *12,* 195–203.

133. Rahrig, D.; Azuma, C.; MacKnight, W. J. *J. Polym. Sci. Polym. Phys. Ed.* **1978,** *16,* 59–80.

134. Rahrig, D.; MacKnight, W. J. In *Ions in Polymers.* Eisenberg, A., Eds.; Advanced Chemistry Series 187; American Chemical Society: Washington, DC, 1980; Chapter 6.

135. Rahrig, D.; MacKnight, W. J. In *Ions in Polymers*; Eisenberg, A., Eds.; Advanced Chemistry Series 187; American Chemical Society: Washington, DC, 1980; Chapter 7.

136. Earnest, T. R. Jr.; Higgins, J. S.; MacKnight, W. J. *Macromolecules* **1982,** *15,* 1390–1395.

137. Roche, E. J.; Pineri, M.; Duplessix, R.; Levelut, A. M. *J. Polym. Sci. Polym. Phys. Ed.* **1981,** *19,* 1–11.

CHAPTER 7

OTHER IONOMERS

7.1. IONOMERS WITH STATISTICAL ION PLACEMENT

7.1.1. Elastomers

7.1.1.1. Butadiene-Based Elastomers. The carboxylated elastomers were the first series of synthetic ionomeric materials. As early as 1957, Brown (1) wrote a review of carboxylated rubber, which describes the most important features that result from the introduction of acidic or ionic groups into polymers. Of particular interest is the section dealing with metal salt vulcanization of carboxylated rubbers. It was recognized that cations can act as cross-linking agents in these systems and thus influence a wide range of physical properties. The metal ion-containing compounds that were used included zinc oxide, lead oxide, magnesium oxide, calcium hypochlorite, dibutyl tin oxide, and cadmium hydride. The first two were more effective cross-linking agents than most of the others, such as $Ca(OH)_2$, CaS, $ZnCO_3$, and HgO. An example of one study involving the effect of the amount of zinc oxide on the tensile properties of a vulcanized butadiene–acrylonitrile–methacrylic acid (55/35/10 w/w/w) is shown in Figure 7.1.

Cooper (2) studied the mechanical properties of butadiene-based ionomers of various molecular weights. He found that for high molecular weight samples, the rate of creep in the ionomers is significant and is greater for zinc or lead ionomers than for magnesium or calcium systems. He suggested that the ionic interactions in the zinc and lead vulcanizates are weaker than those of the alkaline earth vulcanizates. Ionic bond interchange is, therefore, easier with zinc and lead than with the alkaline earths. Low molecular weight copolymers were found to be elastic when subjected to a rapid impact, but plastic under slow applied stress. This effect is most significant for zinc and cadmium salts; materials based on other cations are hard at room temperature.

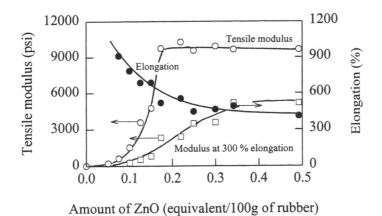

Figure 7.1. Effect of ZnO content on mechanical properties of butadiene-acrylonitrile-metha-crylic acid (55/35/10 w/w/w) rubber. Modified from Brown (1).

More recent studies of carboxylated rubbers include the work of Tobolsky et al. (3) and Otocka and Eirich (4,5). Tobolsky et al. (3) investigated the mechanical properties of ZnO and sulfur-vulcanized butadiene–acrylonitrile–methacrylate co-polymers. They found that a sulfur cross-linked sample showed a true rubbery pla-teau, whereas the zinc salt showed a decrease in the modulus with increasing temper-ature above the primary glass transition temperature T_g. This effect illustrates the difference between chemical and physical cross-links. It was proposed that the high strength of the elastomers cured with ZnO is probably the result of the presence of ionic aggregates, which act both as filler and as time-dependent physical cross-links. Otocka and Eirich studied the effect of ionic bonding on the mechanical properties of lithium poly(butadiene-*co*-methacrylate) and poly(butadiene-*co*-2-methyl-5-vi-nylpyridinium methyl iodide) ionomers. They found that the curves of log E' as a function of temperature show pseudo-equilibrium rubbery plateaus at ~10^7 N/m², with the height of the plateau increasing with ion content (4). They also suggested that at high temperatures the decrease in the modulus is related to bond interchange (5).

Morphological studies of the butadiene rubber ionomers have been carried out by electron microscopy as well as small-angle x-ray scattering (SAXS) techniques. When a series of methylvinylpyridine-*co*-butadiene copolymers was quaternized with α,ω-dibromo alkanes, it was found that the amount of curing agent (bromoal-kane) and the chain length have no effect on the ionic aggregate dimensions, which were ~45 Å (6). Another study on the morphology of butadiene-*co*-methacrylate ionomers was performed by Pineri et al. (7) who found the size of aggregates to be ~6 Å and the intermultiplet distance ~70 Å. They also noted no apparent change in the size of aggregates and in the intermultiplet distance with the degree of neutraliza-tion. The structure of the ionic aggregates was investigated using electron paramag-netic resonance (EPR) (7). From the EPR results, the authors postulated that the ionic aggregate consists of 2 Cu^{2+}, 4 RCOO$^-$, and 2 H$_2$O or 2 RCOOH units. As the temperature increases, the signal for Cu^{2+}-Cu^{2+} pairs disappears because of the thermally activated ion exchange.

7.1.1.2. Ethylene–Propylene–Diene Terpolymers. An extensive series of

studies was performed on the ethylene–propylene-based ionomers, notably sulfo-nated ethylene–propylene terpolymer with norbornadiene serving as a site for sulfo-nation (8). The chemical structure of the third monomer, 5-ethylidene-2-norbornene (ENB), is shown in Scheme **7.1**. The copolymerization was achieved through the

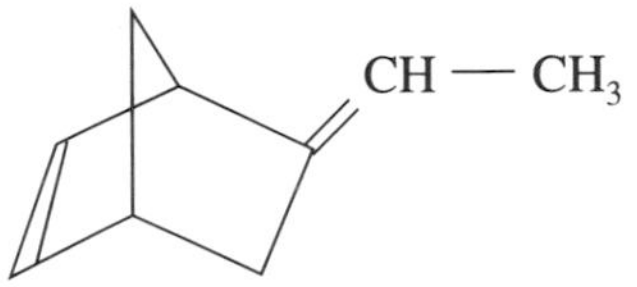

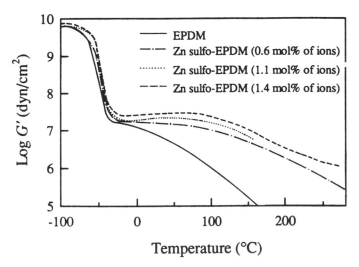

Scheme 7.1

endocyclic double bond of the ENB, whereas the ions were introduced by sulfonation of the exocyclic double bond of the ENB. The zinc ion was found to be the most useful as a counterion, as it gave excellent cross-link properties at low temperatures while preserving a relatively low viscosity at the processing temperature, this is in contrast to the sodium ion, which had a much higher viscosity under processing conditions.

The effect of the zinc sulfonate content on the modulus of ethylene–propy-lene–diene (EPDM) is shown in Figure 7.2, which compares the modulus of the

Figure 7.2. Log G' versus temperature for zinc sulfonate EPDM ionomers of various sulfonate content; *solid line*, EPDM; *dashed and dotted line*, zinc sulfonate-EPDM at 0.6 mol % of ions; *dotted line*, zinc sulfonate-EPDM at 1.1 mol %; *dashed line*, zinc sulfonate-EPDM at 1.4 mol %. Modified from Agarwal et al. (8).

unfunctionalized EPDM with that of the 0.6, 1.1 and 1.4 mol % zinc sulfonate. It is clear that the rubbery plateau is extended considerably, without a significant increase in the glass transition temperature, by the presence of the ionic groups. In view of the low ion content, the glass transition temperature should not be expected to increase extensively. It was found that the cross-link density is higher than theoretically expected by a factor of three to four, which suggests that topological constraints on the nonionic sections of the polymer have a strong influence on the elastic behavior in the presence of ionic cross-links. For this particular system, if we add the value of 5×10^{-4} mol/cm^3 of physical entanglements to the theoretically expected values of the inter-cross-link molecular weight calculated from stoichiometry, the sum of two values yields a modulus value close to that obtained from experiment. As might be expected, the use of a sulfonate ion as the ionic group results in a rubbery modulus that extends to relatively high temperatures. Among other property changes, it was found that the T_g value increases by approximately 5 °C/mol % in the range of 0–1.4 mol % of ionic groups.

Dynamic mechanical studies were also performed. It was found that time–temperature superposition is operative for the 1.4 mol % zinc salt. Again, in the light of the behavior of the styrene system, this result is not surprising, because at such low ion contents, clustering would not be expected. It was found that the ammonium ion, like Zn^{2+}, yields a polymer that flows easily; e.g., whereas the ammonium salt shows a substantial drop in the rubbery modulus at 50°C, the lithium, cesium, and barium salts show a constant rubbery modulus up to ~200°C. The persistence of the rubbery modulus is also seen in the lead and magnesium salts. The T_g values were found to be independent of the specific cation and to depend only on the ion content. This behavior is indicative of a system in which the multiplets remain intact far above the glass transition.

The effect of ion content on other properties was also studied (9). The melt viscosity increases markedly from ~8 $\times$ 10^5 P for 0.7 mol % sulfonation to ~35 $\times$ 10^5 P for 1.6 mol % sulfonation. Both tensile strength at 25°C and stress at 300% were found to increase with sulfonation level, whereas elongation decreased steadily. The effect of the type of cation was also investigated. No melt flow was observed with Mg^{2+}, Ca^{2+}, Co^{2+}, Li$^+$, Ba^{2+}, and Na$^+$ cations, but samples containing Pb^{2+} and Zn^{2+} cations did show it. These two cations also show lower melt viscosity and melt-fracture at higher shear rates than do the other cations.

7.1.1.3. Polyurethane Ionomers.

The polyurethane-based ionomers make up a large family of materials that shows much more complex behavior than do the styrene ionomers because of the heterogeneity of the backbone, i.e., the presence of both soft and hard segments. The properties of backbone materials can range all the way from those of homogeneous systems if the components are short chains, to phase-separated systems if the individual components have a relatively high molecular weight. For example, if poly(ethylene glycol) is used as the soft segment, one obtains materials that are either phase separated or homogeneous, depending entirely on the molecular weight of the diol. The same phenomenon can be obtained if the length of the hard segments is raised. The introduction of ionic interactions further increases the level of complexity.

7.1.3.1.1. Physical Properties.

The first extensive study of polyurethane ionomers was performed by Dieterich et al. (10). A wide series of materials based on various methods of placing ionic groups on the polymer chain was investigated. These methods include using diols containing nitrogen (which, after quaternization, yield cationic chains) and phenyl rings with carboxylate anions (which, in the ionic form, can yield polyanionic chains). Wide structural variability was demonstrated. For example, materials were prepared with sulfonate groups, which were attached to the chain through a nitrogen atom by postpolymerization methods using propanesultone. Another example is the preparation of materials with sulfur linkages in the diol, which may be converted to sulfonium ions. The authors suggested a number of other interesting possibilities, including the cross-linking of the polyurethane chains by using a dihalide, which would bridge nitrogen atoms on different chains, thus providing cross-linking. Another linkage that introduces ionic groups is produced by the reaction of diazabicyclooctane with a urethane oligomer terminated by bromine groups, forming materials similar to the ionenes.

Dieterich et al. (10) also investigated the mechanical properties of the materials and suggested, in complete analogy with previous work on ionomers, that ionic species can aggregate in these materials. This aggregation results in a considerable increase in the viscosity and obviously parallels a wide range of previous investigations on ionomers. The reduction in the effectiveness of ionic association in the presence of water was also noted. As in the nonionic polyurethanes, hydrophobic interactions can play an important role and can lead to a high degree of complexity in the viscosity–concentration plots of some materials. Thus, at high water contents, hydrophobic association is postulated to lead to the formation of colloidal particles in aqueous solution. It was suggested that the most technically interesting property of the polyurethane ionomers is their ability to form stable dispersions in water. The polyurethane phase is the discontinuous phase dispersed in water; the diameters of the particles range between 20 and 5000 nm.

In addition to the methods mentioned above a number of other methods of introducing ionic segments into polyurethanes have been used; two examples are shown in Scheme **7.2** (11).

An extensive series of investigations of the ionic polyurethanes was carried out in Cooper's laboratory. A wide range of systems was investigated, based on a number of different materials and ionic groups. In the first of these studies, polyether-polyurethanes based on 4,4'-diphenylmethane diisocyanate (MDI), N-methyldiethanolamine (MDEA), and poly(tetramethylene oxide) (PTMO) containing various hard segment contents were investigated (12). The MDEA was used as the ionic site by reaction with 1,3-propanesultone, which converts the nitrogen in the polymer backbone into a quaternary ammonium ion and a sulfonate anion at the end of the side chain. The synthetic scheme is shown in Scheme **7.3**.

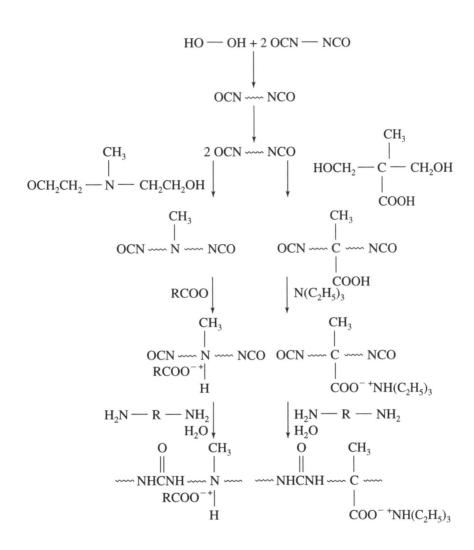

Cationic Urethane   Anionic Urethane

Scheme 7.2

The effects of changes in the chemical composition and the degree of sulfonation were investigated. As might be expected, for a polymer of constant MDI/PTMO/ MDEA ratio, the higher the degree of sulfonation, the higher the rubbery modulus or inflection region. Thus the position of the inflection point can range from $\sim 10^7$ to $\sim 10^8$ N/m^2 for degrees of sulfonation from 0 to 6 wt %. The T_g is also affected; it is interesting that there is a shift of the peak position from $-17°C$ for the unfunctionalized materials to $-40°C$ for the 6 wt % sulfonated material containing 37 wt % MDI. This suggests that both the degree of sulfonation and the chemical composition of the chain influence microphase separation, with the degree of separation increasing with ion content.

A subsequent study compared polybutadiene– and polyether–polyurethane zwitterionomers; it was shown that the ionic interactions in polybutadiene–polyurethane ionomers, which are more highly phase separated, have a much smaller effect on dynamic mechanical, tensile, and orientation properties than in polyether-based ionomers (13). It was suggested that the difference in the properties of these two systems arises from the interconnectivity of the hard domain of the polyether system and the possible strain-induced crystallization of the polyether soft segment.

In a subsequent paper (14), the zwitterion system was converted to a pendent anion by reacting the zwitterion with a metal acetate in DMF at 90°C under N_2. Two different series were investigated: one is based on PTMO of molecular weight (MW) 1000 and the other contains PTMO of MW 2000. In the first series, neutralization with different counterions (such as Na^+, Zn^{2+}, and Fe^{3+}) increases the moduli of the anionic systems as the charge of the neutralizing species increases. It was suggested that that system has an interconnected hard domain morphology. By contrast, the series based on 2000-MW PTMO exhibits an isolated hard domain morphology and is largely unaffected by a variation of the cation charge or the degree of neutralization in terms of thermal and mechanical properties. This behavior is illustrated in Figure 7.3, which shows the modulus–temperature plots of these two systems. Tensile properties were also investigated. For the material with the interconnected domain morphology, the type of cation has a major effect. The materials could be converted from a soft rubbery material to a material that resembles a hard thermoplastic elastomer. The difference between the interconnected domain morphology and the isolated domain morphology is obviously an important one in the urethane ionomers. In addition, it was observed that in the well phase-separated system, T_g of the soft segment is unaffected by the degree of sulfonation.

In one study involving a material of low hard segment content (20 wt % MDI), the nonionic control material exhibited a single-phase morphology (15). Upon ionization, phase separation was observed, with subsequent development of hard segment ordering. This hard segment ordering was shown to lead to a distinct endotherm at 105–110°C and the presence of a high-temperature feature in the log E' curve, occurring around 100°C (the precise position depends on the composition). Figure 7.4 shows the effect of the hard segment content and the degree of ionization on the stress–strain properties for two series of polyurethane anionomers based on MDI

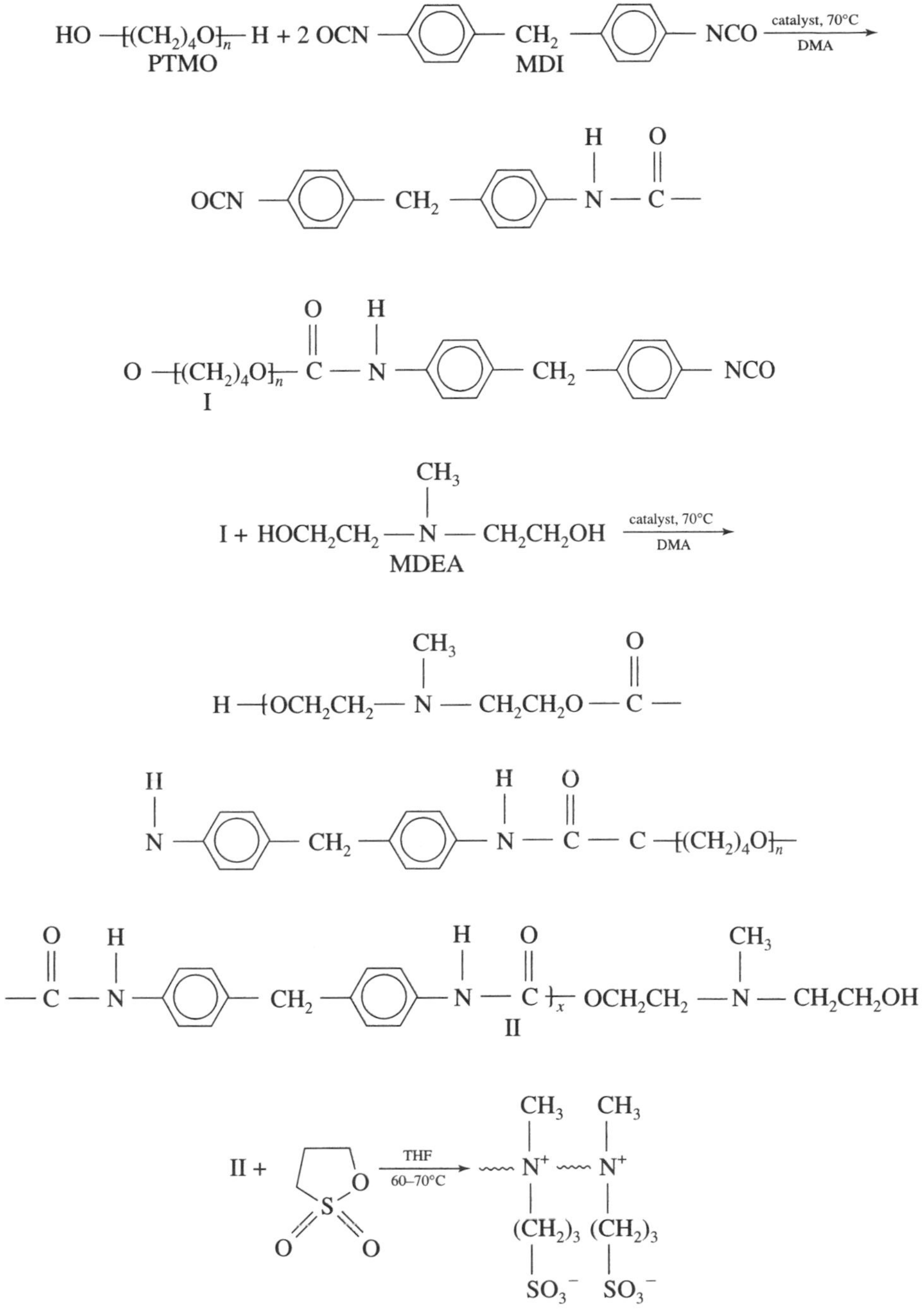
HO ⟨(CH2)4O⟩n H + 2 OCN CH2 NCO
PTMO
MDI
catalyst, 70°C
DMA
I
I + HOCH2CH2 N CH2CH2OH
MDEA
II
II +
THF
60–70°C
SO3−

Scheme 7.3

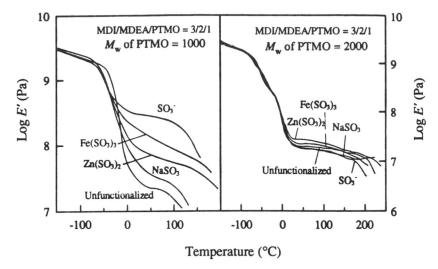

Figure 7.3. Log E' versus temperature for MDI: MDEA: PTMO in a ratio of $3:2:1$ with (left) 1000 MW PTMO and (right) 2000 MW PTMO. Modified from Miller et al. (14).

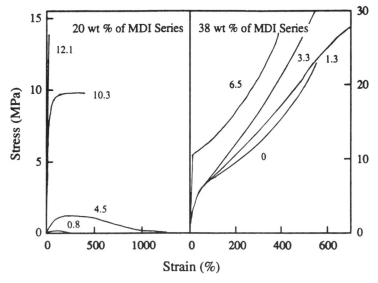

Figure 7.4. Stress–strain curves for the (left) PTMO-20 series and the (right) PTMO-38 series of polyurethanes. Modified from Hwang et al. (15).

and PTMO with different degrees of ionic substitution. The effects of ionization are large for the material with the lower hard segment content (PTMO 20) and much less dramatic for the material of higher hard segment content.

In another investigation (16), the properties of four series of polyurethanes containing zwitterions were compared. The series are based on poly(ethylene oxide) (PEO), poly(propylene oxide) (PPrO), PTMO, and polybutadiene (PBD) as the soft segments. For the PBD-based zwitterionomers, the un-ionized sample shows a high degree of phase separation. However, ionization leads to an increase in the degree of domain cohesion. As expected, the Young's modulus of the materials increases with increasing ion content, whereas the degree of elongation decreases. In the un-ionized polyether materials, the phases are not well separated. However, ionization leads to an increase in the phase separation as well as in the hard segment domain cohesion. It was found that the tensile properties of the ionized PEO- and PTMO-based zwitterions are more affected by an increase in ion content than are those of the PPrO- and PBD-based materials. This effect is probably owing to the fact that PEO and PTMO segments can crystallize under stress.

Another extended series of investigations on urethane ionomers originated from Frisch's laboratory. In an early study, the influence of various parameters on the mechanical properties of the polyurethane ionomers was examined (17). A polyure-thane ionomer based on MDI and PTMO was used with chain extension with a diol containing a tertiary amine. The MDEA chain extender was shown to give the most interesting physical properties. These properties improved progressively with decreasing chain length of the soft segment (PTMO), with increasing concentration of quaternary ammonium centers, and with increasing degree of quaternization for partly quaternized systems (Fig. 7.5).

The effect of monovalent and divalent metal salts on the properties of a prepoly-mer containing MDI and poly(caprolactone) glycol with dimethylolpropionic acid as the chain extender was also investigated (18). In the glass transition studies, it was found that the increase in T_g was linearly related to the charge/distance (between the centers of charge for the anion and cation at closest approach) ratio (q/a) of the ions. The effect of various cations on water absorption was also investi-gated. It was shown that water absorption in these anionic systems changes in the following order: $Mg^{2+} > Ba^{2+} > Ni^{2+} > Mn^{2+} > Cd^{2+} > Zn^{2+} > Hg^{2+} > Pb^{2+} > Cu^{2+}$.

A subsequent investigation was devoted to the properties of polyurethane cat-ionomers (19), prepared by quaternizing a tertiary amine on the backbone, with various polymeric quaternizers containing acid groups. It was found that the mechan-ical properties of the cationomers were improved with decreasing molecular weight of the polyol and with increasing degree of quaternization. The mechanical properties were also improved with decreasing carbon number in the alkyl groups of the amine chain extenders or by the use of rigid diisocyanates.

Zielinski and Rutkowska (20) conducted a study involving both ionic interactions and cross-linking. The urethane elastomer was based on a mixture of 3-bromo-1,2-

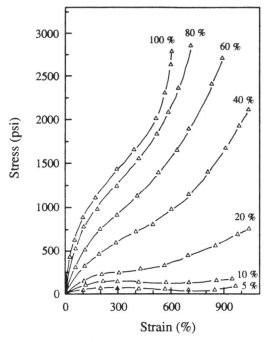

Figure 7.5. Stress–strain curves for polyurethane ionomers of various degrees of neutralization. Modified from Hsu et al. (17).

propanediol, 1,4-butanediol, and triethanolamine, which was both a chain extender and a cross-linking agent. They showed that hydrocarbon chains with two pendent bromines could cross-link at the nitrogen atom, forming quaternary ammonium and bromide ions. Ionic interactions were thought to dominate at higher ion contents, and the trifunctional triethanolamine was thought to be the main chemical cross-linker at low ion contents. In addition, the authors found that at low degrees of ionic cross-linking, the material was phase separated, but that phase separation decreased with increasing ion content. In another investigation (21), involving *N*-methyldiethanolamine as the nitrogen carrier, dibromohexane was used as the cross-linking agent. Again, cross-linking had a significant effect.

A study of tensile properties in a series of model ionomers (22) suggested that the ultimate tensile properties of the ionomer improve as the sulfonate content increases. Mixing the sulfonates and carboxylates shows a mixed ion effect, i.e., the modulus of a mixed system is higher than that of the fully sulfonated ionomer. It was, therefore, suggested that the modulus enhancement arises from a combination of effects, including aggregate packing, immobilization of the ionic chain segments, and the degree of phase separation. In contrast, the temperature of the onset of flow seems to be related to the carboxylate ion content and decreases with increasing carboxylate ion content, as expected.

One series of ionic urethanes was prepared in the complete absence of chain

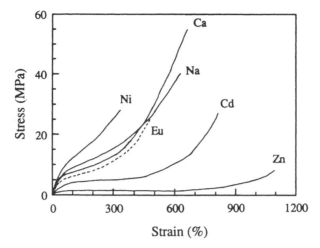

Figure 7.6. Stress–strain curves for sulfonated polyurethanes with various counterions. Modified from Ding et al. (23).

extenders (scheme **7.3**) (23). The ionogenic group was based on a urethane nitrogen, which was converted to a pendent sulfonate group with propanesultone. The soft segment consisted of tetramethylene oxide or similar materials. In this series of polyurethanes, the elastomeric properties depend completely on the presence of ionic groups, which provide the cross-linking, because there are no chain extenders. The T_g value was found to be independent of ion content or type. In contrast, the stress–strain properties were very much a function of the type of ion; Figure 7.6 shows an example for the M1 series (1000 molecular weight polyol). The casting solvent was also found to influence the mechanical properties.

Lee and Kim (24) studied polyurethane ionomer diisocyanates and found that the T_g of the mixed phase polyurethane or of the hard segment–rich phase increases with increasing extender functionality. This effect is the result of the increase in the cross-link density. The tensile strength also increases with increasing extender functionality, whereas the degree of elongation at break decreases.

Frisch's group investigated a range of urea anionomer dispersions. They found that as the 2,2-bis(hydroxymethyl)propionic acid (DMPA)/polyol ratio increases, the T_g of the soft segment increases, because of the miscibility enhancement between the hard and soft segments. It was also found that with an increasing NCO/OH ratio, the T_g of the soft segment decreases, which arises from phase separation of the hard and soft segments. With an increasing DMPA/polyol or NCO/OH ratio, the tensile strength and modulus increases, while the elongation decreases (25). In a subsequent study (26), it was found that there is partial mixing between the hard and soft segments. The T_g of the soft segment changes with the hard segment structure.

One study explored the difference in behavior between sulfonate and carboxylate ionomers based on the same backbone structure shown in Scheme **7.4** (27). Carboxyl-

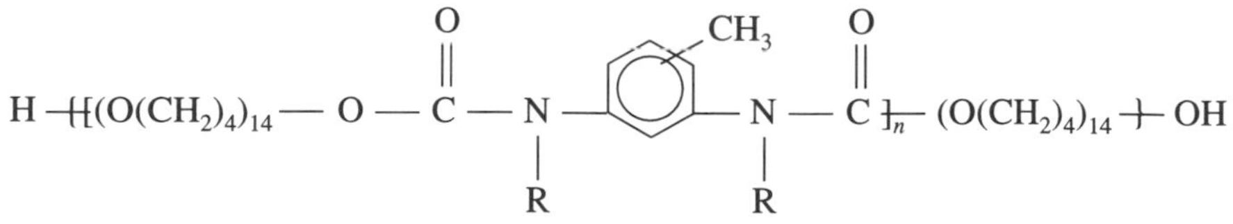

$$H \text{---} [(O(CH_2)_4)_{14} \text{---} O \text{---} \overset{\overset{\displaystyle O}{\|}}{C} \text{---} \underset{\underset{\displaystyle R}{|}}{N} \text{---} \overset{CH_3}{\bigcirc} \text{---} \underset{\underset{\displaystyle R}{|}}{N} \text{---} \overset{\overset{\displaystyle O}{\|}}{C}]_n (O(CH_2)_4)_{14} \text{---} OH$$

$$R = \text{---}(CH_2)_3 \text{---} SO_3^- Na^+ \text{ or } \text{---}(CH_2)_2 \text{---} COO^- Na^+$$

Scheme 7.4

ation or sulfonation was introduced on the nitrogen of the tolylene diisocyanate (TDI), which formed the linking group, with a PTMO with molecular weight of either 1000 or 2000 being the chain extender. A change of the countercation shows a relatively small effect on the tensile properties, whereas a change in anion shows major effects (27,28). For polymers of relatively high ion contents with the 1000 molecular weight soft segment, it was shown that the T_g of the carboxylate was 9–10°C higher than that of the sulfonate ionomers. It was suggested on the basis of the narrower temperature range of the glass transition (by dynamic mechanical thermal analysis; DMTA) that this difference in T_g is seen because the sulfonated ionomers showed a higher degree of phase separation than did the carboxylates. By contrast, the ionomers based on the 2000 MW soft segment showed equal T_g values and a similar degree of phase separation regardless of anion type. The identity of the glass transitions can be attributed to the fact that at 2000 MW, the parent polymers are already highly phase separated, so the added effect of ionic groups is not as pronounced as in the absence of phase separation in the parent polymer. As might be expected, the onset of flow occurred at higher temperatures in the sulfonated species than in the carboxylated systems, indicating a stronger ionic interaction in the sulfonates, as has been found in other ionomers. Thus carboxylate groups decrease the tensile strength (relative to sulfonate groups) and increase the degree of phase-mixing and cross-linking efficiency in this polyurethane.

7.1.1.3.2. Morphology. A series of studies from Cooper's group was devoted to an investigation of the morphology of ionic polyurethanes. Model polyurethanes were prepared, based on PTMO of various molecular weights using MDI as the connecting unit (29). A liquidlike core-shell model, similar to one proposed by MacKnight et al. (30), was invoked on the basis of SAXS profiles (Chapter 3). In this, the core unit is connected to the shell units by a molecular springlike arrangement (Fig. 7.7). The model is known as the ''liquidlike core-shell model.'' However, it was found that the postulated bead-spring micronetwork model, in which the multiplets are the beads and PTMO coils are the springs, fits the profiles better. In the same series of materials, the mechanical properties were investigated as a function of concentration of ions (31). The ionic spacing was changed by varying the chain length of the PTMO and by the degree of conversion of the nitrogen to a pendent sulfonate ion by methods described above. Not surprisingly, the rubbery modulus was shown to be a function of the weight percent of sodium ions (Fig. 7.8).

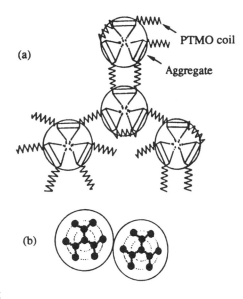

(a)

PTMO coil

Aggregate

(b)

Figure 7.7. (**a**) Postulated bead-spring model; the aggregates are the beads and the PTMO coils the springs. (**b**), Micronetworks distributed in the polymer matrix in a liquidlike manner. Modified from Lee et al. (29).

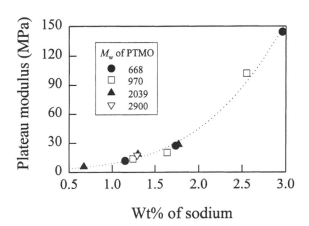

Figure 7.8. Relation between plateau modulus and ion content for polyurethane ionomers of various PTMO content. (●, M_w = 668; □, M_w = 970; ▲, M_w = 2039; ▽, M_w = 2900. Modified from Lee et al. (31).

In a subsequent morphology study, the SAXS patterns were investigated in the same sulfonated polyurethane ionomers based on TDI as a function of the soft segment length and degree of ionization (32). It was shown that with increasing soft segment length, the peak position moved to lower q ($= 4\pi\sin\theta/\lambda$), indicating a larger separation between aggregates. Similar behavior was observed in soft segments, which were based on PPrO and polybutadiene. The behavior of the PPrO materials reflects the solubilization of the ionic groups by the alkylene oxide chain, in that the sizes of the aggregates were somewhat smaller than in the other soft segment materials. The temperature dependence of the SAXS profiles was also studied for the cadmium and zinc salts. The cadmium salt showed no change in profile for temperatures between 25 and 225°C, while the zinc ionomer showed a slight shift of the peak to lower q value with increasing temperature. These results indicate that the ionic aggregates certainly persist at these elevated temperatures. However, it should be recalled that the SAXS peak does not address the kinetics of ion hopping. Thus, while flow (related to ion hopping) can proceed, the time-average aggregate size, at least in the cadmium case, is not affected.

An extended x-ray absorption fine structure spectroscopy (EXAFS) study probed the environment of the cations in the polyurethanes (33). It was found that for the calcium, nickel, and europium salts the first shell consists of six oxygen atoms, whereas for zinc the coordination shell consists of five oxygen atoms. Europium and nickel show a clear second coordination shell, cadmium and calcium a poorly ordered second shell, and zinc only a first shell. In the cesium case, only a first shell of low order was found. The degree of local order found from the EXAFS study correlates inversely with the strain at which stress crystallization begins. This indicates that the cohesiveness of ionic aggregates controls the large strain behavior, because it suppresses ion hopping. Hydration enhances ion hopping and causes the materials to lose their mechanical strength.

In one SAXS study (34), it was suggested that the carboxylate ionomers yield larger aggregates than do the sulfonate ionomers. A subsequent reinvestigation of the morphology of aggregates in the polyurethane ionomers suggested that the liquidlike polydisperse hard-sphere model gave a much better fit than did the adhesive hard-sphere model (35) and local-ordering model (36) investigated previously. Visser and Cooper (37) reanalyzed the interparticle versus intraparticle scattering, showing that an interparticle rather than intraparticle interference model reproduced the intensities better.

7.1.2. Acrylate and Methacrylate Ionomers

Matsuura and Eisenberg (38), investigated the T_g of the ethyl acrylate ionomers; the results were described briefly in the section on the glass transition. It should be recalled that by DSC only the matrix glass transition is detectable up to a cq/a value of ~0.05 (because the general shape of the curves is sigmoidal). The cluster glass transition can, of course, be seen in dynamic mechanical plots. It is also worth recalling that the cq/a effect is operative in these systems, indicating that ion hopping may already be of importance in connection with the matrix glass transition, possibly because of the larger number of multiplets in the matrix and the higher dielectric constant of the backbone polymer compared to the styrene-based ionomers.

The stress–relaxation properties of carboxylated ethyl acrylate ionomers were reported in a subsequent publication (39). It was shown that at low ion contents, time–temperature superposition was operative but broke down above a concentration of 16 mol %. It is noteworthy that the onset of thermorheological complexity in this system, 14 ± 2 mol %, is ~8 mol % higher than it is in polystyrene (40–42). This difference may be related to the dielectric constant of the backbone. Another contributing factor maybe the possibly shorter persistence length of the chain.

Duchesne (43) investigated the dynamic mechanical properties of carboxylated ethyl acrylate ionomers using a torsion pendulum and suggested that the behavior is quite similar to that of polystyrene ionomers, except that the cluster peak appears for ion contents as low as 4.1 mol %. The cluster peak is seen at ion concentrations in which stress–relaxation properties still show time–temperature superposition. This behavior parallels that found in the styrenes. The ethyl acrylate-*co*-vinylpyridine [P(EA-*co*-4VP)] ionomers were also investigated (43). In that system, it was shown that, in contrast to the sodium acrylate–based ionomers, no clustering was visible even for ion contents as high as 16 mol %. However, as will be discussed in Chapter 8, plasticization of the P(EA-*co*-4VP) system may induce clustering, just as it did in the P(S-*co*-4VP) case. Butyl acrylate–based ionomers with vinylpyridine (T_g matrix is about $-35°C$) do show clustering (44), unlike the P(EA-*co*-4VP) system (T_g matrix is about $-15°C$). It was concluded from that study that, for a constant electrostatic interaction strength, it is the position of the glass transition that determines whether the material is clustered or not. To prove this point, the P(EA-*co*-4VP) system, which is not clustered, was plasticized to reduce the T_g value. Clustering was, indeed, induced by plasticization, so that the plasticized ethyl acrylate ionomers behave, in some ways, similarly to the butyl acrylate ionomers. The same behavior was also found in the styrene–vinylpyridine case (45).

A comprehensive study of poly(methyl methacrylate) (PMMA) ionomers was published in several papers by Hara's group. In the first of these, the dynamic mechanical properties and morphology of PMMA were investigated (46). The glassy modulus was found to increase with ion content, in contrast to the behavior of styrene ionomers. The glassy modulus is $\sim2.35 \times 10^9$ N/m^2 for the pure PMMA and reaches a value of $\sim3.3 \times 10^9 N/m^2$ for the ~25 mol % (of sodium methacrylate) sample. The trend appears to be linear. It was also found that the rubbery inflection point increases with increasing ion content. Up to a mole fraction of ~0.1, the rubbery modulus increases, as expected from the kinetic theory of rubber elasticity. However, above that value, it increases rapidly, and at ~25 mol % it reaches a value of $\sim4 \times 10^8$ N/m^2. From rubber elasticity theory, a value of $\sim0.5 \times 10^8$ N/m^2 is expected. The T_g value increases linearly as a function of ion content. It is noteworthy that the matrix glass transition in this system increases at a rate of $\sim5.5°C/mol$ %, whereas in the styrene ionomer, it increases at $\sim3°C/mol$ %. The glass transition of the cluster phase increases only slightly with ion content over the same range of ion contents (6–13 mol %).

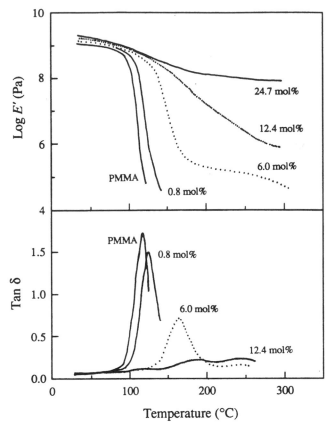

Figure 7.9. Log E' and loss tangent versus temperature for P(MMA-*co*-MANa) of various ion contents. Modified from Ma et al. (46).

The dynamic mechanical results are summarized in Figure 7.9 for the range of 0–20.4 mol % of ions. The loss tangent plot shows two peaks, especially for the 12 mol % sample. However, total area under the loss tangent curves decreases with increasing ion content, unlike polystyrene. A decrease in area with increasing cross-link density parallels, to some extent, the behavior of styrene-cross-linked PMMA polymers found by Chang et al. (47). The onset of cluster-dominated behavior (presumably related to the percolation threshold; Section 5.2.6) occurs at ~12 mol %. The ion content (12 mol %) at which thermorheological complexity appears is greater in PMMA ionomers than in the styrene-based materials (6 mol %). This difference may be related to a decrease in the persistence length of PMMA (7.2 ± 0.5 Å) relative to the value for polystyrene (9.1 ± 0.2 Å) (48), though the difference in dielectric constant may also contribute.

It is noteworthy that the β relaxation, which for pure PMMA occurs at ~ 20°C, is split in the ionomer into two relaxations. For example, at ~ 12 mol %, $T_{\beta 1}$ occurs

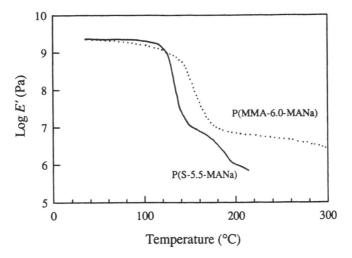

Figure 7.10. Log E' versus temperature for P(MMA-6.0-MANa) (*dotted line*) and for P(S-5.5-MANa) (*solid line*). Modified from Ma et al. (46).

at ~45°C and $T_{\beta 2}$ occurs at ~110°C. It was suggested that this splitting takes place because of the different environments of ionic groups. Thus $T_{\beta 1}$ is presumably related to the T_β of the unclustered phase, and $T_{\beta 2}$ is related to the β relaxation in the clustered regions. It is also noteworthy that the rubbery plateau extends over a much wider temperature range in the PMMA ionomers than in the styrene ionomers. The comparison is shown in Figure 7.10 (46) for a polystyrene ionomer of 5.5 mol % of ions and a PMMA ionomer of 6.0 mol % of ions. It is evident that the PMMA ionomer has some of the hallmarks of a composite system, i.e., the longer rubbery plateau and the decrease in the area under the loss tangent peak.

Another study of the same polymer revealed that the deformation mechanism changes as a function of ion content (49). PMMA deforms by crazing, which changes to crazing plus shear deformation for the material containing 6 mol % of sodium methacrylate, and changes further to shear deformation only for the material with 12 mol % of ions. These changes parallel the behavior of the styrene ionomers and arise from an increase in the strain density, which is related to the presence of ionic cross-links. In the calcium salt, the onset of shear deformation in addition to crazing occurs at ion contents as low as 0.8 mol %. This behavior is presumably owing to the increase in the strain density related to the stronger ionic cross-links in the calcium salt compared to the sodium salt.

The dynamic mechanical properties of the cesium ionomer of poly(methyl methacrylate-*co*-methacrylic acid) [P(MMA-*co*-MACs)] were investigated in connection with an ionic conductance study for water-swollen membranes (50). The dynamic mechanical properties are summarized in Figure 7.11. In contrast to the sodium salt, the area under the loss tangent peak does not change with ion content, and no second mechanical peak is visible in the loss tangent curve. Most interesting, SAXS reveals no ionic multiplets in the cesium salt, again, in contrast to the sodium salt where a SAXS peak related to multiplet scattering is seen. This work suggests that in the cesium ionomer, the ionic cross-links are weakened to the point at which the high value of dielectric constant, coupled with the high glass transition prevents multiplet formation (on a scale that would be detectable by SAXS). The delicate balance of forces that is involved in multiplet formation and clustering in some ionomer systems is illustrated by this behavior. The P(MMA-co-4VP)-based ionomers have also been investigated (43). Here, the dielectric constant and glass transition temperature of the backbone are also high, and thus no multiplets are encountered.

One possible explanation for the behavior of methyl methacrylate ionomers, especially the large change in the area under the loss tangent peak, comes from the suggestion that the nature of the multiplet is different in P(MMA-co-MANa) than in "classical" styrene ionomers. Methyl methacrylate behaves much more like a composite material than an ionomer. One can envisage the existence of regions of only mildly perturbed methyl methacrylate and other regions of a composite of methyl methacrylate and sodium–methacrylate. In the composite regions, the sodium–methacrylate does not form classical multiplets but, rather, forms regions in which the ionic groups interact with each other over larger distances. The ionic groups are perhaps coordinated to several carboxylate or ester groups from nonionic systems, with which they interact strongly. This idea would explain the smaller area under the loss tangent curve for the first transition. The strength of the interactions within the composite regions might explain the absence of a second glass transition.

The specific conductivity of a water-swollen P(MMA-co-MANa) system has been

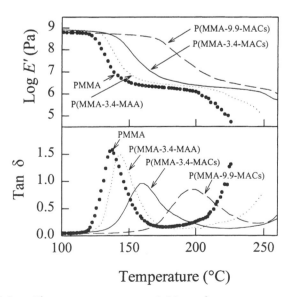

Figure 7.11. (a) Log E' versus temperature and (b) tan δ versus temperature for P(MMA-co-MACs) of various ion contents. Modified from Gronowski et al. (50).

studied (50). It was found that there is a sharp insulator–conductor transition at ~6.5 mol % of MANa content. For styrene ionomers, owing to the low polarity of polystyrene, a higher ion content (~18 mol %) is required for the observation of high conductivity.

7.1.3. Zwitterionic Systems

In the preceding sections, the materials were classified into families based on the backbone rather than the type of ionic group. The present section, dealing with zwitterionomers is handled somewhat differently, primarily because in the zwitter-ionomers the anion and cation are attached to each other and to the main chain. Thus the members of this family have many similarities in behavior, and backbone-based differences are smaller.

7.1.3.1. Siloxane-Based Di-Zwitterionomers. The first systematic study of a zwitterionomeric system was that of Graiver et al. (51–55), who investigated copolymers of dimethylsiloxane and 4,7-diazaheptyl-4,7-di(3-propane-sulfonate) methylsiloxane, (Section 2.1.2). These materials can have two zwitterions on one repeat unit, in which case they are called di-zwitterionomers. A wide range of techniques was used to investigate the system, including electron microscopy, SAXS, differential scanning calorimetry, torsion pendulum, dielectric measurements, and stress–strain measurements. The ion concentration ranged from 0.5–10 mol %. It was found that multiplets were lamellar in nature, with a thickness of ~7 Å (52). The proposed structure is given in Figure 7.12. The distance between the ionic lamellae is determined by the chain length of the siloxane polymer. Thus, not surprisingly, an increase in ion concentration, which decreases the chain length between ionic groups, decreases the spacing between lamellae. It was postulated that the

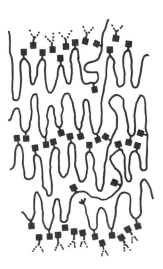

Figure 7.12. Model of microphase separation of di-zwitter-ions (■) in siloxane di-zwitterionomer. Modified from Graiver et al. (52).

ionic domains are monolayers of zwitterions. Changes in the ion content did not affect the domain shape or size. The dynamic mechanical loss peak (called α_z) was detected in the region of 350 K (54). This loss peak was attributed to a process occurring within the ionic domains. The peak is probably not a cluster glass transition because the interlamellar spacing is greater than twice the persistence length of the polydimethylsiloxane chain, and thus regions of restricted mobility near the lamellae cannot overlap. Another peak was found in the range of 175–225 K, which has been attributed to the motions of polar groups such as isolated ions or to chain segments near a di-zwitterion. The water sensitivity of this peak is markedly smaller than that of the α_z relaxation and is evident only at low ion content.

Thermal history was found to have a pronounced effect on the organization of zwitterions in the lamellae (54). When the sample was cooled slowly, the structure of the ionic aggregate was well organized, and the α_z relaxation was correspondingly large. It was suggested that, at temperatures above this relaxation region, the domain structure is still preserved but the lamellae can be shifted or broken easily under stress. In the light of current understanding of the viscoelasticity of ionomers, it seems likely that ion hopping becomes an important component of the relaxation mechanism in this temperature range. It was suggested that removal of the stress results in a reformation of the ionic morphology.

Not surprisingly, the elastic modulus shows a strong dependence on the ion concentration (55). At ~ 2 mol % of ions, the elastic modulus is ~ 20 kg/cm^2, at 5 mol % it jumps to ~ 150 kg/cm^2, and at 10 mol % to 750 kg/cm^2. This behavior is also explained qualitatively by the proposed morphology.

7.1.3.2. Hydrocarbon-Based Zwitterionomers.

An early study of zwitterionomers based on poly(vinyl imidazolium sulfobetaine) was performed by Salamone et al. (56), who investigated their aqueous solution properties. A structure of the polymer is shown in Scheme 7.5. The authors found that these ionomers are

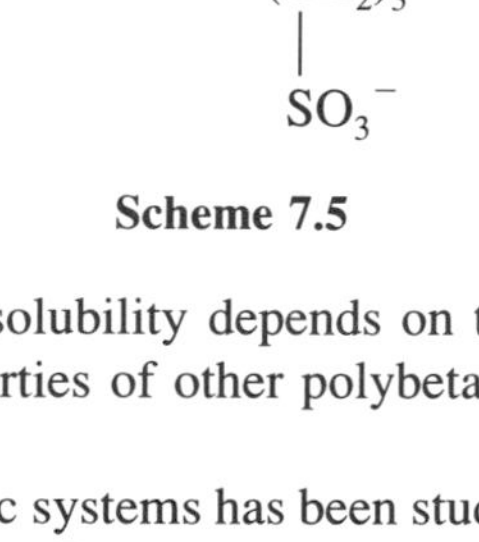

$$-(CH_2 - CH)-$$

Scheme 7.5

insoluble in water, and their solubility depends on the type and the concentration of added salts. Solution properties of other polybetaine zwitterionomers were also studied (57,58).

A wide range of zwitterionic systems has been studied by Galin's group (59–62); a selection is shown in Scheme 7.6. An extensive study was performed on the *n*-

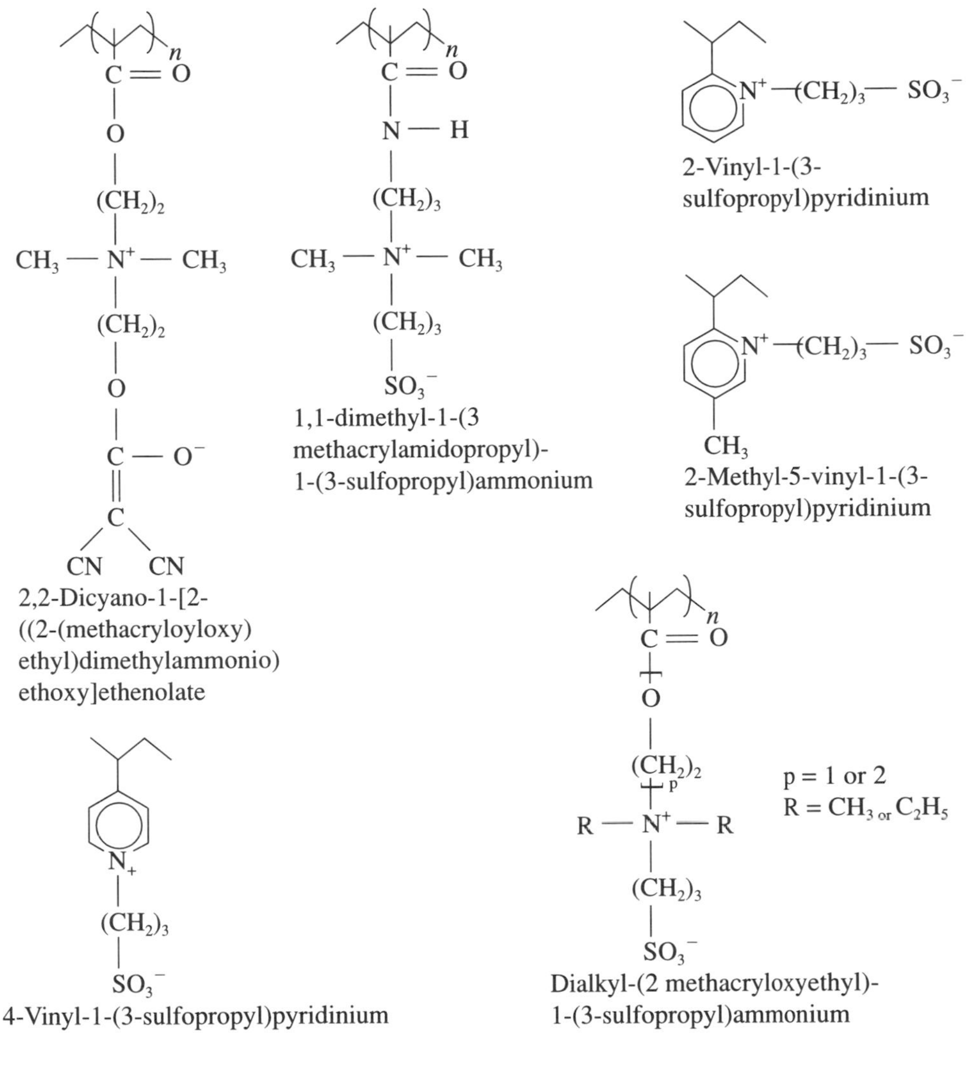

2-Vinyl-1-(3-
sulfopropyl)pyridinium

1,1-dimethyl-1-(3
methacrylamidopropyl)-
1-(3-sulfopropyl)ammonium

CH₃
2-Methyl-5-vinyl-1-(3-
sulfopropyl)pyridinium

2,2-Dicyano-1-[2-
((2-(methacryloyloxy)
ethyl)dimethylammonio)
ethoxy]ethenolate

$p = 1$ or 2
$R = CH_3$ or C_2H_5

4-Vinyl-1-(3-sulfopropyl)pyridinium

Dialkyl-(2 methacryloxyethyl)-
1-(3-sulfopropyl)ammonium

Scheme 7.6

butyl acrylate-sulfopropyl betaine copolymers, which in many ways are typical of
this family of materials. A wide range of techniques, including differential scanning
calorimetry (DSC), SAXS, and solid-state NMR spectroscopy, was used to investigate the copolymers with ion contents that ranged from 4 to 35 mol % (63).

DSC revealed the presence of two glass transitions in the *n*-butyl acrylate-sulfopropyl betain copolymer system (identified as matrix and cluster glass transition)
(63). The matrix T_g ranges from -46 to $-20°C$, whereas the cluster T_g lies between
30 and 220°C for the composition range investigated. A SAXS peak was observed
for these materials: the Bragg spacing d ranges from 4 to 7 nm and is proportional
to $\phi_B^{-0.27}$, where ϕ_B is the volume fraction of ionic repeat units. SAXS experiments

also suggested that the ionic aggregates are stable to temperatures 100°C above the cluster T_g.

The mobility of the polymers was investigated by solid-state NMR (63). Three regions of mobility were identified in which the polymer is rigid, of intermediate mobility, and mobile. The fractions of various mobilities as a function of temperature are given in Figure 7.13. The T_g values by DSC of the matrix and cluster phase for this composition (27.3 mol %) are -26 and 117°C, respectively.

In view of the distance of the ions from the backbone, it is not surprising that the multiplet sizes in these systems are considerably larger than those in ionomers containing sodium sulfonate or sodium carboxylate ion pairs. Ehrmann et al. (64) estimated the dimensions of the aggregates by solid-state NMR. The distances of spin diffusion were probed via $T_{1\rho}$ (spin-lattice relaxation time in the rotating frame) for ionomers containing 4 and 12 mol % of ions. It was shown that the dimensions of rigid microdomains were ~1 and 2.2 nm. These sizes are reasonable when compared to the dimensions of the sodium carboxylate multiplets (0.6–0.8 nm) and sodium sulfonate aggregates (~1 nm) (65,66) and illustrate the fact that the sulfopropyl betaine copolymers consist of multiplets that contain a substantial number of -CH_2- groups, unlike the other ionomers in which only contact ion pairs are located in the multiplet.

Given that the morphology is analogous to that of styrene ionomers discussed in Chapter 3, it is not surprising that the mechanical properties, which reflect the morphology, are also similar. E' and tan δ versus temperature are plotted in Figure 7.14 (67); the plots show considerable similarity to plots from other ionomers, especially in the variation of the modulus with ion content, decrease of the tan δ peak height of the matrix glass transition with ion content, and many other characteristics.

Melt rheology of these materials shows that time-temperature superposition is

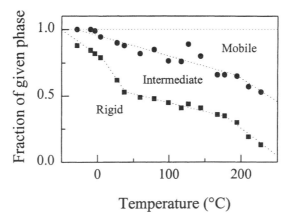

Figure 7.13. Relative amount of rigid, intermediate, and mobile material as a function of temperature for ethylene-based zwitterionomers of 27.3 mol %. Modified from Ehrmann et al. (63).

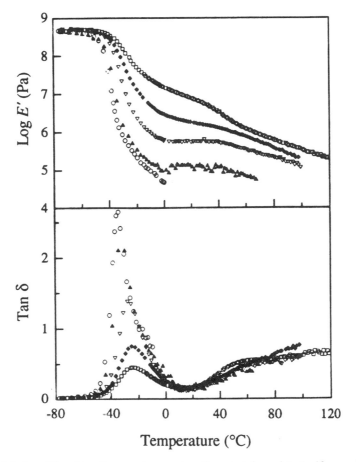

Figure 7.14. Log E' and tan δ versus temperature for *n*-butyl acrylate-(sulfopropyl)ammonium betaine for 0 (○); 2.0 (▲), 4.0 (▽), 7.1 (◆), and 12.8 (□) mol % of ions. Modified from Ehrmann et al. (67).

operative over the entire range of frequencies, concentrations and temperatures investigated (67). Otherwise the curves show considerable similarities to curves of other ionomers, especially in the plateau region of the modulus.

The ethyl acrylate zwitterionomers, which were investigated by Bazuin et al. (68), show a progressive shift of the tan δ peak to higher temperature with increasing ion content. Phase-separation in the sense of multiplet formation is clearly evident in the system, as seen from the presence of a SAXS peak (69). The tan δ curves, however, cannot be resolved into two components, as was done in some of the studies described before; therefore, clustering in this material at high ion contents cannot be investigated directly. In view of the similarity to the butyl acrylate zwitterionomers, it is likely that this system is also clustered.

7.2. IONOMERS WITH REGULAR ARCHITECTURES

The following sections concern themselves with the morphology and properties of ionomers other than the random ionomers discussed in Chapters 5, 6, and 7. Specifically, diblocks and triblocks, stars in which the ionic groups are placed at the ends of chains away from the junction points, monochelics and telechelics (see Section 2.1.1) as well as ionenes are discussed. The only feature that these ionomers have in common is the regularity in the positioning of the ionic groups.

7.2.1. Block Ionomers

We will discuss diblocks and triblocks of the classical type (i.e., materials in which each of the groups in the ionic block is actually ionized) and also a different type of block copolymer in which one block consists of completely nonionic material and the other segment or segments consist of a random ionomer of relatively low ion content.

7.2.1.1. Morphology A study of the morphology of compression-molded poly(4-vinylpyridinium methyl iodide)-*block*-polystyrene-*block*-poly(4-vinylpyridinium methyl iodide) triblock ionomers by SAXS and small-angle neutron scattering (SANS) was published in 1989 (70). The ionic endblocks were much shorter than the nonionic midblock. The general morphological features were found to be similar to those of other block copolymers of similar composition. Because the ionic blocks are short, the ionic aggregates or microdomains are expected to be mostly spherical. The relationship between the radius of the aggregate and the ionic block length for the triblock ionomer system is presented in Figure 7.15, which shows that the chains are highly extended in the phase-separated ionic regions. For example, for a block of 10 vinylpyridinium (VP) units, the radius was found to be ~50 Å, whereas for a 50-unit block, the radius was found to be >200 Å. This value is extremely large, considering that the length of a vinylpyridine chain of 50 units at full extension is only ~125 Å. Thus, at least for short ionic blocks, the radius is considerably larger than the fully stretched chain length, which, at first sight, seems unreasonable. However, as will be shown later, the large radii are possible because of chain length polydispersity.

The reason for the high degree of extension must be addressed. The interfacial energy is crucial in this connection. The large unfavorable interactions between the ionic and the nonionic regions provide a strong driving force for minimization of the total surface area of the aggregates. The total surface area minimization requires maximization of the radius of each aggregate, which will, therefore, contain the maximum possible number of chains. The only way to form such large aggregates is to stretch some of the chains in the ionic cores. It should be recalled that some of the chains must extend from the surface to the center of the core. Ionic interactions per se are not needed to observe chain stretching, as was proven by a study of polystyrene core micelles in an ionic continuum of poly(cesium methacrylate) (71).

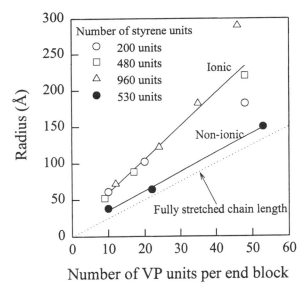

Figure 7.15. Radii of ionic aggregates versus 4-vinylpyridinium block length for PS-*b*-P4VP diblocks quaternized (*open symbols*) and unquaternized (*solid symbols*) for various styrene block lengths. ○, 200 styrene units; □, 480 units; △, 960 units; ●, 530 units. Modified from Gouin et al. (70).

Chain extension was observed in the styrene cores, of essentially the same extent as in the ionic cores. Thus it is the interfacial energy that is crucial for chain extension.

As mentioned above, the chain length polydispersity in the core-forming block can lead to a radius greater than the average fully stretched length (72). In a study of intentionally broadened molecular weight distributions, it was shown that when the core regions are prepared under near equilibrium conditions (e.g., by evaporation of a solvent rather than precipitation of the blocks), then the radius of the microdomains is linearly related to the ionic chain polydispersity index (PI) (Fig. 7.16). It is noteworthy that despite the high degree of stretching in the cores, it has been shown by neutron scattering that the polystyrene chains between the ionic regions have essentially random conformations (72).

The x-ray scattering profile of a diblock copolymer containing 18 units of cesium methacrylate and 440 units of polystyrene is shown in Figure 7.17 (71), along with the shape factor for spheres of radius 48 Å, which is a reasonable fit to the data. Polydispersity in sphere dimensions may cause the smoothing of the minima. Recently, it has been suggested that nonsphericity of the core may also explain this smoothing; our group is now investigating this in detail. At low q ($= 4\pi\sin\theta/\lambda$) values, the data deviate substantially from the fit for spheres because of the appearance in the scattering profile of features that arise from structure factor contributions, i.e., from the type of packing of the aggregates.

Fitting of the curves using shape factor information gives an idea of the sizes

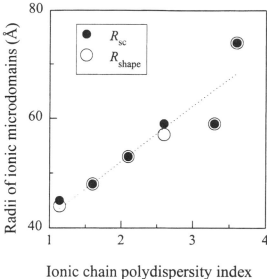

Figure 7.16. Radii of ionic aggregates versus ionic chain polydispersity index for solution cast PS-*b*-P4VPMeI calculated by different methods. ●, R_{sc} (radius calculated assuming simple cubic lattice); ○, R_{shape} (radius calculated from shape factor peak). Modified from Nguyen et al. (72).

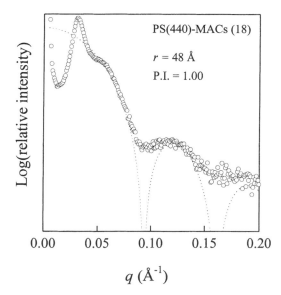

Figure 7.17. Log of the scattering intensity versus q for PS(440)-*b*-PMACs(18). ○, experimental; *dotted line*, shape factor fit for monodisperse sphere of $r = 48$ Å; PI = 1.00. Modified from Gouin et al. (71).

of phase-separated spherical regions. There is, however, another way of obtaining information about the sizes. The major peak in the structure factor component at low q (Fig. 7.17) originates from interparticle interference. Therefore, if the lattice structure of the aggregates is known, one can calculate their size from the stoichiometry and space filling considerations. Sometimes the shape factor information is not sufficient to decide precisely what the lattice type is. Frequently in the past, some type of cubic lattice has been postulated. If one assumes a simple cubic lattice, one obtains the sizes from the density of ionic groups and the space filling considerations mentioned above; these sizes are indicated by the solid symbols in Figure 7.16. The sizes from the shape factor fit are indicated by the open symbols. The values are similar, which means that the assumption may be reasonable.

A different type of block ionomer was investigated by Lu et al. (73). The material consisted of triblocks of styrene-*block*-(ethylene-*co*-butylene)-*block*-styrene, in which the styrene was lightly sulfonated. Phase separation on two different levels could be distinguished: (a) between polystyrene and the ethylene-*co*-butylene copolymers and (b) of the ionic aggregates within the styrene domains. Solution casting resulted in a well-ordered lamellar microstructure with a lamellar thickness of ~7 nm. In contrast, compression-molded samples exhibited an ellipsoidal microstructure. Ionic aggregates were identified as 2-nm-sized multiplets at a closest approach distance of ~3.5 nm. These results were obtained from a modified hard-sphere interparticle interference model for the ionic domains. The authors found that the ionic domain size was independent of the level of sulfonation; however, the aggregates for the sodium salt were somewhat larger than those for the corresponding zinc salt. The temperature dependence of the SAXS peak was also investigated. It was found that for the zinc salt, the ionic microstructure disappears in the range of 230–250°C. However, the alkali metal salts retain their SAXS features up to ~300°C, i.e., close to the decomposition temperature of the material.

In the same study, it was found that the degree of sulfonation did not affect the lamellar spacing, which suggests that, in this case, the polymer coil dimensions are independent of the degree of ionic character of the ionomer (Section 3.4). The solution-cast zinc salt exhibits a well-defined lamellar microstructure, in which four SAXS maxima were observed, with scattering vectors in the ratio of $1:2:3:4$, characteristic of a lamellar microstructure (74). This structure was not seen in the sodium salt. This copolymer system, incidentally, was the first in which an order–disorder transition has been observed in an ionomer (75). The order–disorder transition was found above the ionic aggregate dissociation temperature and was attributed to a destabilization of the ionic phase separation by the competing entropic driving force for block mixing at elevated temperatures.

Another type of block ionomer investigated was based on a *n*-hexyl methacrylate midblock with methacrylic acid end blocks (76). The methacrylic acid end blocks were short (8–42 units), whereas the *n*-hexyl methacrylate midblock had a molecular weight of ~46,000. SAXS and direct electron microscopic observations suggested strongly that the ionic regions phase separate (at least in some cases) to give a cylindrical morphology. Both ordered and disordered regions were found in the materials by transmission electron microscopy (TEM).

7.2.1.2. Mechanical Properties. In an early study of the mechanical properties of block ionomers, Gauthier and Eisenberg (77) investigated ABA triblock copolymers containing a midblock of ~5000 units of styrene with 4-VP methyl iodide end blocks of 12, 24, 35, and 46 units. A torsion pendulum study showed that, as the number of ionic groups in the end block increased, the modulus plot (above the glass transition temperature) changed from one showing liquid flow at a relatively low temperature to one with a well-developed rubbery plateau, at least up to ~190°C. Only one major T_g was observed in the log E curves, that of polystyrene. The glass transition of the ionic regions was also observed as a shoulder on the loss tangent peak owing to the glass transition of polystyrene. However, this shoulder appeared on the low temperature side of the peak, suggesting that extensive plasticization by water had taken place. It should be stressed that block ionomers do not exhibit the mechanical properties of a cluster phase (in the EHM sense) as observed in the random ionomers, because the phase-separated regions in the block ionomers are quite large and are separated by considerable distances, much greater than the persistence length of the polymer chain. Therefore, the regions of reduced mobility do not overlap, though their existence was confirmed by Yano et al. (78) by dielectric measurements. Because the ionic regions are quite large and the regions of reduced mobility are essentially thin layers around the multiplets, the total volume fraction of material of reduced mobility is small.

Long et al. (79) investigated diblock ionomers made from styrene and isobutyl methacrylate, in which the esters were partly hydrolyzed and neutralized. It was found that neutralization with potassium extended the rubbery plateau extensively without changing the glass transition temperature of the styrene.

The mechanical properties of triblock ionomers based on lightly sulfonated polystyrene (29.8 wt % of styrene in the polymer) end blocks connected by an ethylene-*co*-butylene midblock were investigated by Weiss et al. (80). Both the nonionic and the ionized styrene segments showed an intermediate plateau at temperatures between the glass transition of styrene and that of ethylene-*co*-butylene. Ionization increased the level of the plateau, suggesting enhanced separation of the styrene microphase. A peak was found in the log E'' plots of the 8.7 and 11.9% zinc sulfonate samples at 195 and 204°C, respectively, i.e., above T_g. Because the polystyrene T_g is at 80–90°C, ΔT_g for this transition is ~120°C, which for these ion contents is approximately that of the cluster glass transition in other materials (Chapter 4). Therefore, it is reasonable to attribute this peak to a cluster glass transition of the sulfonated polystyrene regions.

The mechanical properties of a triblock with a polystyrene midblock and poly(sodium methacrylate) end blocks were investigated by Desjardins and Eisenberg (81). The dynamic mechanical properties (log E' versus temperature) are illustrated in Figure 7.18 for a sample containing polystyrene midblocks of 490 units end capped with poly(sodium methacrylate) of 8, 22, and 44 units. A pronounced rubbery plateau is seen for the 44-unit end block, which extends to almost 300°C. Clearly, the triblock ionomers show properties similar to those of a hard, rubbery material. A sample containing esterified end blocks—i.e., poly(*tertiary*-butyl methacrylate)—shows an almost identical primary glass transition (that of polystyrene) with a rapid drop of the modulus and a shorter plateau. Not surprisingly, the same behavior is observed for a diblock copolymer containing approximately 40 sodium methacrylate units and for the styrene homopolymer, which suggests the absence of network formation in the diblocks, as might be expected.

The effect of changes in the end block length on the mechanical properties is also illustrated in Figure 7.18. The materials containing end blocks of 8 and 22 sodium methacrylate units show relatively normal behavior, whereas the material containing an end block of 44 units has a high rubbery modulus, as noted. The reason for this large difference in the modulus is most likely owing to a difference in morphology of the polymers containing 8 and 22 units on the one hand and the sample containing 44 units on the other. A cylindrical morphology is not unreasonable for the 44-unit system in the light of the findings of Venkateshwaran et al. (76).

The difference between the behavior of diblocks and triblocks is consistent with similar studies on nonionic diblocks versus triblocks. Clearly, the triblocks are cross-linked, whereas the diblocks are reverse micelle-like structures, in which the ionic cores are surrounded by emanating polymer chains that are not connected to any other ionic aggregates. Thus their flow properties resemble those of starlike structures at high temperatures. Yoshikawa et al. (82) recently investigated the rheology of the diblocks. They found that the viscoelastic properties of diblock ionomers of polystyrene-*b*-poly(sodium methacrylate) resemble those of a star polymer. This behavior is reasonable, because the micelle-like structures persist to high temperature owing to the strong ionic interactions. Furthermore, an additional relaxation was found at low frequencies, owing to either thermal rotational diffusion of the micelles (which is limited by the retraction of polystyrene blocks) or filler behavior.

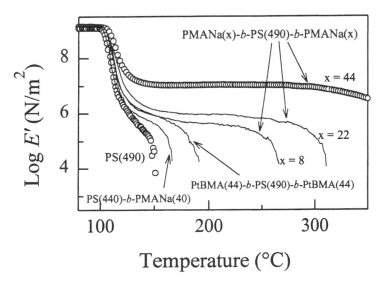

Figure 7.18. Log E' versus temperature for PMANa(x)-*b*-PS(490)-*b*-PMANa(x) triblocks, a nonionic triblock, a diblock, and PS(440). Modified from Desjardins and Eisenberg (81).

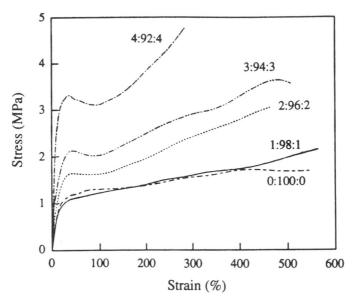

Figure 7.19. Stress–strain curve for triblock copolymers containing a nonionic midblock (EHMA) and potassium methacrylate salt end blocks. See the text for details. Modified from Venkateshwaran et al. (76).

The effect of the length of the ionic block on the stress–strain properties of a series of triblock copolymers containing a nonionic midblock (2-ethylhexyl methacrylate) (EHMA) and potassium methacrylate salt end blocks is shown in Figure 7.19 (76). The ionic block can be made from a t-butyl methacrylate block (TBMA) by hydrolysis and subsequent neutralization of the block. $0:100:0$ refers to the 2-ethylhexyl methacrylate homopolymer of molecular weight of ~50,000; $1:98:1$, $2:96:2$, $3:94:3$, and $4:92:4$ refer to the triblock containing potassium salt end blocks. The numbers describe the composition of the ester form sample by weight percent. For example, $1:98:1$ refers to the material containing two end blocks each of 1 wt % TBMA (hydrolyzed and then neutralized with KOH), connected to a segment containing 98 wt % EHMA. The modulus and the tensile strength of the material increase with increasing ionic block length.

7.2.2. Monochelics, Telechelics, and Stars

The next family of materials discussed is based on ion terminated systems. We distinguish three different types: the monochelics, in which an ion or ion pair is placed at only one end of a polymer chain; the telechelics, in which an ion or ion pair is placed at each end of a polymer chain; and stars in which each of three or more arms emanating from a center point is terminated by an ion or ion pair. The properties of these three groups of materials are sufficiently different to warrant separate discussions. In many ways, the monochelics and telechelics resemble the block ionomers, especially in their aggregation behavior, because the strong driving force for phase separation can make even ion pairs aggregate or separate from the matrix material.

7.2.2.1. Monochelics. The monochelics have been investigated only briefly. Their morphologies are generally spherical, with aggregation numbers typically on the order of 10 (83,84). Aggregation numbers of this order are also seen in the telechelics (85). Nonspherical morphologies have been reported by a number of authors (86), especially in a zwitterionic system (87). Barium sulfonate–terminated polystyrene chains exhibit a morphology that is typical of the monochelics; they have been recently studied by SAXS both in bulk and in solution (85,88). An equation for the distance between (spherical) multiplets d was proposed:

$$d = K_g \left(\frac{M_n}{fc}\right)^{1/3} \tag{7.1}$$

where f is the functionality, i.e., 1 or 2 (for monochelics or telechelics), c is the concentration of polymer in solution, and M_n is the number-average molecular weight. In the monochelics it is possible to use chains terminated with a cationic pendent ion and chains of another polymer terminated with an anionic pendent ion to make structures resembling block copolymers (89). An example is a blend of ω-carboxylic acid polystyrene and α,ω-diaminopolybutadiene (α,ω indicates chain end functionalization). This approach has been used to enhance miscibility in block copolymers (discussed in Chapter 9).

7.2.2.2. Telechelics. The telechelics have been the subject of extensive studies recently, primarily from Teyssié and Jérôme's group. A number of reviews have appeared on these materials (90, 91). As a result, the coverage here will be brief. Our primary purpose is to allow the reader to compare the morphologies and properties of telechelics with those of random ionomers. In general, the morphology of the telechelics resembles that of the monochelics and block copolymers. The variation of the distance between multiplets as a function of M_n/fc (eq. 7.1) was given above for both monochelics and telechelics of barium neutralized sulfonated polystyrene at different concentrations (85) (Fig. 7.20). The exponent 0.326 implies the existence of a liquidlike arrangement of stable spherical multiplets.

A review that deals more extensively with mechanical properties of the telechelics was published in 1989 (90). An excellent example of the relative importance of molecular weight and ion concentration in determining the mechanical properties is seen from the results of the dynamic mechanical studies (plotted as G versus ωa_T) of α,ω-dicarboxylic acid polyisoprene, both in its nonionic and in its ionic forms (92). The results for materials in the acid form are shown in Figure 7.21, where it is seen that increasing the molecular weight simply extends the curves to lower frequencies, equivalent to the development of a rubbery plateau. The neutralized polymers illustrate clearly the effect of both ion concentration and molecular weight (which ranges from 7,000 to 69,000). The longest rubbery plateau is exhibited by

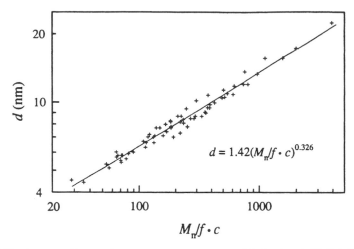

Figure 7.20. Average distance between aggregates versus M_n/fc. $d = 1.42 (M_n/fc)^{0.326}$. See the text for details. Modified from Vanhoorne et al. (85).

the 7,000 MW sample, indicating the importance of ionic cross-links. As the molecular weight increases to 20,000, the ion concentration decreases; and concomitantly there is an appreciable shift of the high modulus region to higher frequencies. However, the low modulus region shifts much more strongly toward the low frequency region. For the 36,000 MW sample, the shift has gone even farther to the high frequency side. As the molecular weight increases to 69,000, a shift back to lower frequency is observed, indicating that now the molecular weight effect dominates the ion concentration effect. Thus the 36,000 MW material represents the material in which the two effects are comparable and that is subject to the most rapid relaxation, both because of chain slippage or reptation and ion hopping.

Shift factors in the halato-telechelics tend to be of either the Williams–Landel–Ferry (WLF) type or the Arrhenius type, depending on the material and the ion concentration. For high molecular weight rubbery polyisoprenes, the shift factors are, in general, of the WLF type. For high ion contents (i.e., low MW), the shift factors tend to be of the Arrhenius type (93).

Broze et al. (94) investigated the cation dependence of the mechanical properties of α,ω-dicarboxylic acid polybutadienes for the MW 4600 sample (Fig. 7.22). The divalent Ba^{2+}, Mg^{2+}, and Ca^{2+} samples show high moduli, indicating the most effective cross-links. The drop of the modulus occurs at the lowest temperature for Zn^{2+}, indicating the highest rates of ion hopping. It was suggested that the smaller size and the higher electrostatic interaction of Be^{2+} carboxylates produce small and tight aggregates. Thus Be^{2+} carboxylates show the lowest modulus but the longest rubbery plateau among the divalent cations. Different packing within multiplets may also be involved for this ion. The dependence of the modulus on the cation is shown in Figure 7.23 (92). From this figure, the authors suggest that the distribution of relaxa-

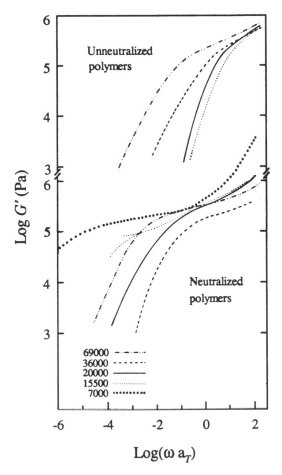

Figure 7.21. Partial master curve of log G' versus log ωa_T for unneutralized and magnesium-neutralized polyisoprene telechelics of various molecular weights. ----, 69,000; ---, 36,000; ——, 20,000; ···, 15,500; •••, 7,000. Modified from Jérôme and Broze (92).

tion times is related to the cation size, with the width increasing as the ionic radius decreases from Ca^{2+} to Mg^{2+}. The Be^{2+} samples seem to be subject of a double relaxation phenomenon. The flow activation energies for various ions as a function of ionic radius are shown in Figure 7.24 (94).

A comparison study of sulfonated and carboxylated telechelics based on polyisoprene was made by Venkateshwaran et al. (95). As expected from the behavior of the random copolymers, it was found that the sulfonated copolymers have a much slower decrease in modulus with time than do the carboxylates. For example, at 23°C, the stress for potassium sulfonate decreases (on a log plot) from -0.5 to

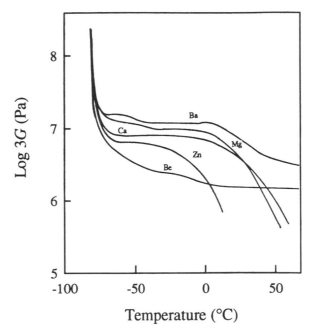

Figure 7.22. Log $3G$ versus temperature for polybutadiene telechelics neutralized with various cations. Modified from Broze et al. (94).

-0.7 MPa between 1 and 10 min, whereas that for the carboxylates decreases from -0.5 to -1.5 MPa over the same time scale (Fig. 7.25).

7.2.2.3. Stars

Three-arm star ionomers have been studied extensively for isobutylene based materials (96) synthesized originally by Kennedy and Storey. The structure of these ionomers is shown in scheme **2.2**.

The properties of these stars differ in some important ways from the properties of the other ionomers. To illustrate, there is an extra variable called the functionality, i.e., the number of arms emanating from the center for monochelics, telechelics, and three-arm stars (having one, two, and three branches, respectively). The properties of these three different systems differ appreciably. Figure 7.26 compares the penetration behavior as a function of temperature for monochelics (M-11-K) of 11,000 MW; telechelics (D-6.5-K) of 13,000 MW (i.e., 6,500 for each arm); and three-arm stars (T-8.3-K), in which each of the branches has a molecular weight of 8,300, all terminated with a potassium sulfonate group (97). The three-arm star with sulfonic acid terminal groups is also shown (T-8.3-SO$_3$H) and demonstrates the weak cross-linking by the sulfonic acid group compared to the potassium salt.

Several findings are apparent from Figure 7.26. First, the monochelics behave in a manner similar to nonionic polymers, i.e., they show a rapid drop in the modulus

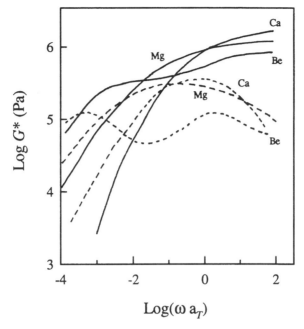

Figure 7.23. Log G^* (complex modulus) versus log ωa_T for carboxylated butadiene telechelics ($M_n = 4600$) neutralized with various cations. Solid lines: G', dotted lines: G''. T_{ref} varies from 295 to 303 K. Modified from Jérôme and Broze (92).

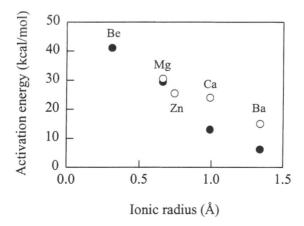

Figure 7.24. Activation energy versus ionic aggregate radius for carboxylate butadiene telechelics ($M_n = 4600$) neutralized with various cations in bulk (○) and in xylene solution (10%) (●). Modified from Broze et al. (94).

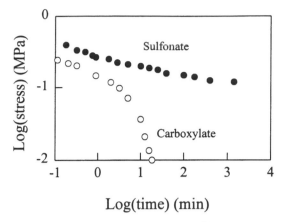

Figure 7.25. Stress–relaxation curves for a sulfonated (●) and carboxylated (○) polyiso-prene-based telechelic ionomer. Modified from Venkateshwaran et al. (95).

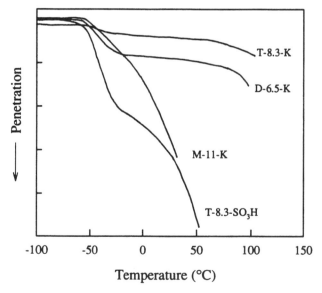

Figure 7.26. Penetration depth versus temperature for PIB stars (T), telechelics (D), and monochelics (M). The numbers refer to the molecular weight ($\times$ 1000) for each branch of the polymer; K indicates potassium-neutralized samples. The acid-terminated star is T-8.3-SO$_3$H. Modified from Bagrodia et al. (97).

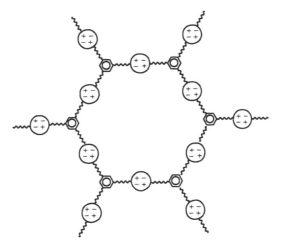

Figure 7.27. Aggregation in three-arm stars. Modified from Broze et al. (94).

because of the absence of network formation. It has been shown independently that sulfonate-terminated isobutylene chains have a multiplet structure consisting of two ion pairs (98). Thus one might expect the M-11-K ionomer to be, for all practical purposes, equivalent to a homopolymer chain of 22,000 MW isobutylene. The difunctional chain is expected to behave differently in that even if the multiplets are quarters (i.e., two ion pairs), chain extension should yield a polymer of theoretically infinite molecular weight. Such a polymer would be expected to have a large number of entanglements and to yield a physical network in which the physical cross-link density is dictated by the entanglement spacing. The material would flow only when the multiplets start to dissociate (by ion hopping) (98). These features are indeed observed experimentally: Flow begins in the vicinity of 100°C in this material. The trifunctional star T-8.3-K represents a different situation; The network is now cross-linked rather than merely entangled. The molecular weight between cross-links is expected to be twice the molecular weight per branch, because the network junctions are the centers of the star and not the ionic quartet (Fig. 7.27) (94).

Overneutralization has been studied for a number of materials in this family. Figure 7.28 shows an example of a penetometry study as a function of temperature of the completely neutralized and 100% overneutralized materials (97). It is seen that the completely calcium-neutralized sample (T-34-Ca) begins to show a decrease in the ionic modulus at approximately 90°C, whereas the 100% overneutralized material (T-34-Ca-100%) drops at about 200°C. The material has a MW of 11,000 per branch. Another example of the effect of the overneutralization is shown in Figure 7.29, which illustrates stress–strain behavior (99).

Overneutralization undoubtedly affects the properties (i.e., life times) of the multiplets, making them much more resistant to high temperatures. A possible explanation

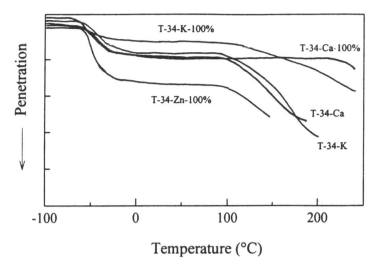

Figure 7.28. Penetration depth versus temperature for three-arm stars (34,000 MW) neutralized with calcium and potassium and 100% overneutralized with calcium, potassium and zinc. Modified from Bagrodia et al. (97).

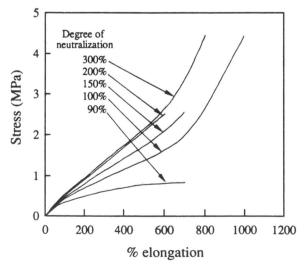

Figure 7.29. Stress–elongation curve for three-arm stars (14,000 MW) neutralized to various extents with potassium. Modified from Mohajer et al. (99).

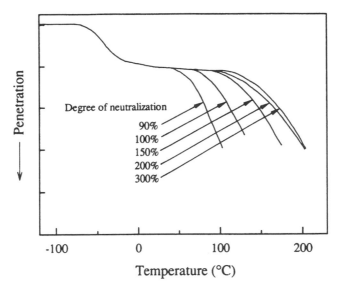

Figure 7.30. Penetration depth versus temperature for the same samples shown in Fig. 7.29. Modified from Bagrodia et al. (97).

may be that the extra neutralizing agent can reduce the effect of water in the multiplets (99) by decreasing the number of water molecules per ion pair in the multiplets. Another possibility is enlargement of multiplets by the incorporation of extra neutralizing agent, and the presence of small particles of a type found by Hara's group, which produce a filler effect (99,100). Enlargement of the multiplets is the more likely explanation, because now it is the multiplets that are responsible for holding the network together.

The degree of the penetration as a function of the degree of neutralization for the T-14-K salt is shown in Figure 7.30 (97). The effect of molecular weight on the flow region is illustrated in Figure 7.31 (100). It is clear that the higher the molecular weight, the longer the rubbery plateau in this system, because for high MW, flow begins only at low frequencies. In contrast, in the 8300 MW material, no rubbery plateau tail is seen at all. The stress–strain behavior is shown as a function of MW in Figure 7.32 (101) where both the decrease in the modulus and the improvement in elongation with increasing MW of the polymer are clearly seen.

7.2.3. Ionenes

The ionenes are one of the few families of materials in which precise spacing of ionic groups along the backbone is possible. The schematic structure of an aliphatic ionene is shown in Figure 2.5. The aliphatic sequences can be replaced by aromatic or mixed aromatic and aliphatic sequences, so that an enormous range of materials can be synthesized. High molecular weight analogs of these materials can also be

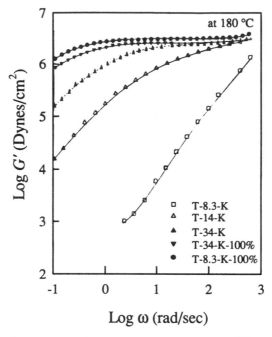

Figure 7.31. Log G' versus log ω for various three-arm stars (at 180°C) neutralized or 100% overneutralized with potassium. □, T-8.3-K; △, T-14-K; ▲, T-34-K; ▼, T-34K-100%; ●, T-8.3-K-100%; numbers indicate MW ($\times$ 1000). Modified from Bagrodia et al. (100).

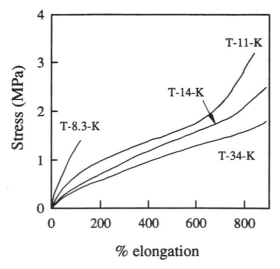

Figure 7.32. Stress–elongation curves for three-arm stars of various arm lengths. Numbers indicate MW ($\times$ 1000). Modified from Bagrodia et al. (101).

made, i.e., materials in which the spacings between the ammonium groups are oligomers or low molecular weight polymers.

Tsutsui (102) reviewed the early work on ionenes. The first syntheses of materials of this type were performed by Marvel's group (103) in the 1930s; Kern and Brenneisen (104) synthesized such materials in 1941. Rembaum et al. (105) reported the first recent synthesis of ionenes, which was the forerunner of a large series of papers from that group. Ionenes are most easily synthesized by using N,N,N',N'-tetramethyl-diaminoalkane with a dihaloalkane, which yields aliphatic systems. A series of Menschutkin reactions produces the polymer (106). The ionenes, especially those with a regular structure, are characterized by a high degree of crystallinity, which influences the bulk properties appreciably; thus the study of these systems is complicated.

In an early investigation of the glass transition of ionenes, Eisenberg et al. (107) worked on plasticized materials and extrapolated to zero plasticizer content to estimate the glass transition temperature. They found that the extrapolated value depended on the type of plasticizer used. Figure 7.33 shows the extrapolated glass transition of the 6,8 ionene as a function of the dielectric constant of the plasticizer. The sudden change in extrapolated T_g at a dielectric constant of approximately 80 suggests that a conformational transition may be involved. Glass transitions in the presence of crystallinity are obviously strongly influenced by that crystallinity. In the absence of crystallinity, a q/a effect operates, at least within narrow composition regions.

The dynamic mechanical properties of ionenes have been investigated by Tsutsui et al. (108). The dynamic mechanical loss spectra of 12,y ionenes are shown in

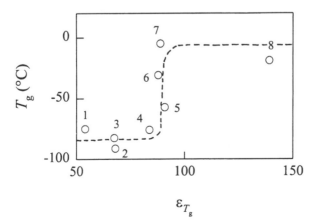

Figure 7.33. Glass transition temperature of 6,8 ionene (extrapolated to zero plasticizer content) using various plasticizers versus the dielectric constant of the plasticizer at T_g. *1*, methanol; *2*, ethylene glycol; *3*, glycerine; *4*, glycerine/water at 8/2; *5*, glycerine/water at 6/4; *6*, glycerine/water at 4/6; *7*, water; *8*, formamide. Modified from Eisenberg et al. (107).

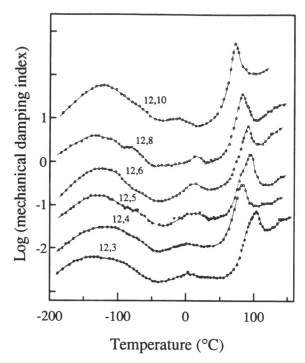

Figure 7.34. Mechanical damping versus temperature for 12,y ionenes. Modified from Tsul-sui et al. (108).

Figure 7.34. Three main peaks are identified: the α peak (75–100°C) is the result of micro-Brownian motions of the polymer units in the amorphous regions, the β peak (~0°C) is related to molecular motions in the ion-rich phase, and the γ peak (~ – 150°C) is attributed to local segmental motions of the methylene units.

An important paper in the field of ionenes was published by Klum et al. (109), who investigated structure–property relationships in a wide range of cross-linkable ionenes. Tertiary diamines were synthesized from diepoxides and secondary amines; alternatively, end capping of diols with tolylene diisocyanate, and subsequent reaction with N,N-dimethylethanolamine, also yielded these materials, as did terminations of living poly(tetrahydrofuran) with dimethylamine. Dihalides were made by ring opening reaction of diepoxides with ω-bromoacids. One example of these polymers is shown in Scheme 7.7. A detailed tabulation of structure–property relation-

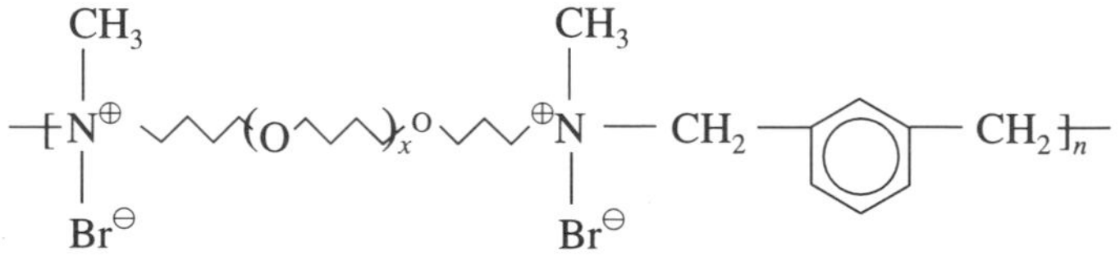

Scheme 7.7

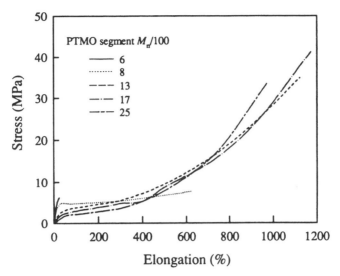

Figure 7.35. Stress–strain curves for PTMO–bipyridinium ionomers. Length of PTMO segment (× 100): ——, 6; ⋯, 8; ---, 13; –·–, 17; –––, 25. Modified from Feng et al. (112).

ships for these materials was given for many of structural parameters, including molecular weight, dihalide nucleofugacity, and amine nucleophilicity.

A recent study of ionenes came from the Wilkes group (110), who found that materials made from short polytetramethyloxide soft segments and bromooxylene hard segments had a rodlike morphology, identified by both x-ray scattering and transmission electron microscopy. In a subsequent study, Venkateshwaran et al. (111) investigated ionic domain plasticization in an ionene cationomer. The ionene was prepared by the reaction of dimethylamino-terminated poly(tetramethylene oxide) oligomers with various benzyl dihalide compounds, and the material was plasticized with zinc stearate. They showed that zinc stearate lowered the softening temperature by acting as a plasticizer. The softening temperature dropped from 180 to ~120°C, which is sufficiently low to permit melt processing. It was also suggested that zinc stearate, which crystallizes on cooling, enhances the mechanical properties in the solid state by acting as a reinforcing filler.

A subsequent study by Feng et al. (112) investigated bipyridinium-based ionene elastomers with tetramethylene oxide chains (Section 2.1.3.1). These materials exhibit high tensile properties and high elongation. The authors attributed the specific properties of this system to a high degree of microphase separation of the ionic domains together with strain-induced crystallization of the poly(tetramethylene oxide) soft segments. Typical stress–strain curves are shown in Figure 7.35. The microdomains in this system, by contrast to those described above, are spherical.

7.2.4. Polymer–Salt Mixtures

Because low molecular weight ethers are capable of dissolving salts to a considerable extent, it is not surprising that polymeric ethers are also able to form polymer–salt mixtures. Such materials were the subject of two publications in 1966 (113,114), in which the interactions between PEO and PPrO and salts such as lithium perchlorate (LiClO$_4$) and potassium iodide (KI) were recognized. Lundberg et al. (113) observed that mixtures of KI and PEO were subject to strong interactions, which manifest themselves as a considerable decrease in the crystallinity of PEO. The crystallinity dropped progressively with increasing salt content. While the polymer by itself is insoluble in methanol, the polymer salt complex is soluble, suggesting that a solution, which has many of the properties of polyelectrolyte solutions, is formed as a result of the strong interactions between the polymer and the salt.

Moacanin and Cuddihy (114) investigated mixtures of LiClO$_4$ and PPrO. They found a dramatic decrease in the specific volume of the PPrO with increasing LiClO$_4$ concentration, which again indicates strong interactions between the components. While the low molecular weight PPrO–LiClO$_4$ mixtures are single phase, demonstrated by the presence of a single glass transition temperature, the high molecular weight polymer mixtures show evidence of two-phase behavior: Two loss tangent peaks are observed for different ion contents, with the peak height changing as a function of salt concentration. The position of the low temperature peak is constant, but that of the high temperature peak changes slightly with composition. The low temperature peak is attributed to the glass transition of the noncrystalline polymer, while the high temperature peak is owing to the polymer–salt complex (Fig. 7.36).

Eisenberg et al. (115) reinvestigated the loss tangent as a function of temperature in low molecular weight systems (Fig. 7.37). Figure 7.38 shows the glass transition temperatures for several poly(propylene glycol) samples along with Moacanin and Cuddihy's results. It was observed that the horizontal shifting of the viscosity curves

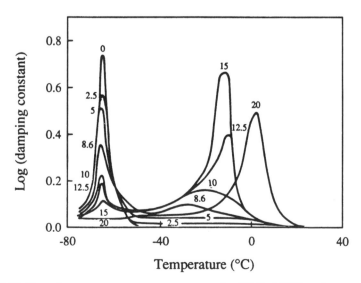

Figure 7.36. Damping constant versus temperature for PPrO–LiClO$_4$ mixtures. Numbers indicate LiClO$_4$ content (wt %). Modified from Moacanin and Chuddihy (114).

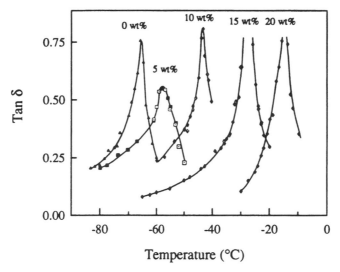

Figure 7.37. Tan δ versus temperature for PPrO(4000)–LiClO$_4$. Numbers indicate LiClO$_4$ contents. Modified from Eisenberg et al. (115).

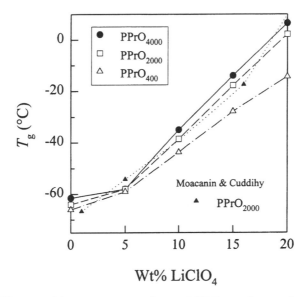

Figure 7.38. Glass transition temperatures of several PPrO samples as a function of LiClO$_4$ content. ●, PPrO$_{4000}$; □, PPrO$_{2000}$; △, PPrO$_{400}$; ▲, data from Moacanin and Cuddihy (114). Modified from Eisenberg et al. (115).

for different salt contents produced a master curve and that viscosity–temperature behavior could be described by the WLF equation. The WLF constants for the polymer–salt mixtures centered around 15 for C_1 and 30 for C_2, for 5–20 wt % of LiClO$_4$. The behavior of the PEO system was quite similar to that of the PPrO system.

Liu (116) investigated the details of the ion–dipole interaction by NMR in the KI–PEO system. Although the interaction with low molecular weight oligomers ($n \leq 6$) was found to be weak, its strength increased when n was >7. This suggests the existence of a cooperative effect. It was also found that the positively charged potassium ion in methanol interacts with the polymer. However, it was observed that in aqueous solution there was no interaction between PEO and KI, unlike in methanol, probably because of differences in the interaction strength between the two solvents and the solutes. A subsequent publication by Liu and Anderson (117) revealed that the spin-lattice relaxation time is independent of the polymer molecular weight, while the viscosity changes are dramatic. This observation suggests that the interactions between the polymer and the salt are short range.

Wetton et al. (118) subsequently investigated the interaction between zinc chloride and PPrO and showed that the divalent cation is also capable of interacting with the polymer to give similar ionic effects on the glass transition as those observed with monovalent cations. The details of the interaction between the cation and the polymer chain depend on the nature of the cation. For example, it was shown that with cobalt chloride, the coordination can be intermolecular as well as intramolecular, unlike zinc chloride in which the complexes are only intramolecular. This behavior correlates with the greater increase in T_g in the cobalt salt compared to the zinc salt. Hannon and Wissburn (119) studied polar phenoxy polymer–calcium thiocyanate complexes and showed that the range of the salt that can be used in these systems is quite broad. As in the other systems, increases in the glass transition temperature and the melt viscosity of the polymer were observed. A striking observation concerned an increased resistance of the material to stress cracking by polar organic liquids in this system. It was suggested that this effect may be related to the increase in the glass transition temperature or to changes in the solubility parameter, which manifest themselves in the solubility of the polymer–salt mixtures in liquids that are nonsolvents for the pure polymer.

Other polymer–salt mixtures were also studied. Thus inorganic nitrates were also found to be soluble in a range of polymers such as cellulose acetate, poly(vinyl alcohol), poly(vinyl acetate), poly(methyl methacrylate) and poly(methyl acrylate) (120). The strong interaction between the salts and the polymer was deduced not only from mechanical effects but also from large shifts in the infrared spectra of inorganic nitrate salts and polymer carbonyl and ester ether frequencies. The difference between various transition metal and alkaline earth metal ions was attributed to the increased degree of covalent character with ions such as zinc and copper, in contrast to calcium and cadmium, which tend to be ionic.

Another system that illustrates the range of materials with polymer–ion interactions was investigated by Reich and Michaeli (121), who showed that ions could be incorporated into polyacrylonitrile. In that study, the ions were also used as a means of determining the glass transition temperature by Mössbauer spectroscopy. Another interesting observation in the study was the decrease in the glass transition temperature with increasing water content, which accompanies the increase in ion content of the polymer.

The crystallinity of complexes of sodium iodide and sodium thiocyanate with PEO has also been investigated (122). A lamellar morphology was observed, with crystalline lamellae of 150–200 Å in thickness. It was also found that there is a lamellar thickening with increasing PEO molecular weight and with annealing. However, adding excess NaI prevents lamellar thickening in the NaI-PEO system. A band morphology was found in the annealed solution-deposited high molecular weight PEO–NaSCN samples, and shish kebab–like structures were seen in melt recrystallized NaSCN complexes. It was suggested that PEO–NaSCN has a more stable and rigid structure than does the corresponding PEO–NaI system. For the LiBF$_4$–PEO crystalline complexes, a double-helix model was used to interpret the wide-angle x-ray scattering pattern (123).

Polyether salt complexes have also been investigated in situations in which the polyether is present in side chains. One example of such a study is that of Xia and Smid (124), who investigated polymethacrylate containing pendent oligo-oxyethylene chains, in which the incoorporated salt was a sodium or lithium triflate.

Recently, extensive studies have been performed on the conductivity of polymer–salt complexes (125–128). The main reason for the interest in these systems lies in their potential use as solid polymer electrolytes, because the low T_g materials exhibit high conductivity in the complete absence of water. For example, Killis et al. (129) observed that the conductivity was high and that the conductance as a function of temperature is characterized by a WLF behavior dependence rather than an Arrhenius type.

Besner and Prud'homme (130) investigated the glass transition temperatures of LiSCN, KSCN, and CsSCN dissolved in an amorphous PEO. The glass transition temperatures for these three salt mixtures are given in Figure. 7.39 as a function of the salt content. It is clear that the glass transition temperatures at low ion content are independent on the cation. A leveling effect is seen for all of these ions. The leveling off occurs at the lowest temperature for Cs$^+$, at a slightly higher temperature for K$^+$, and at a still higher temperature for the Li$^+$ ion. The leveling was ascribed to a saturation effect. In the region in which the T_g was proportional to the salt content, independent of the type of cation, the increase of the glass transition is found to be 4°C/mol % of salt. The identity of the dT_g/dc value was ascribed to a nearly perfect balance of a coordination number effect and a binding energy effect, through which the binding energy is inversely proportional to the apparent coordination number of the cation. The sodium salt behaves somewhat differently from the other three cations. A subsequent series of studies from the same group (131–135) explored other aspects of polymer–salt mixtures. In one study (133) two glass transition temperatures were seen in the PEO–LiClO$_4$ system, which were ascribed to a phase-separation

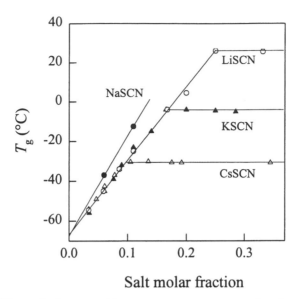

Figure 7.39. Plots of glass transition temperature versus molar fraction of salt in PEO–LiSCN, PEO–NaSCN, PEO–KSCN, and PEO–CsSCN mixtures. Modified from Besner and Prud'homme (130).

phenomenon analogous to that seen in block copolymers. The relationship between phase separation and the electrical conductivity was also investigated (134,135).

7.3. REFERENCES

1. Brown, H. P. *Rubber Chem. Technol.* **1957,** *30*, 1347–1386.
2. Cooper, W. *J. Polym. Sci.* **1958,** *28*, 195–206.
3. Tobolsky, A. V.; Lyons, P. F.; Hata, N. *Macromolecules* **1968,** *1*, 515–519.
4. Otocka, E. P.; Eirich, F. R. *J. Polym. Sci. A2* **1968,** *6*, 921–932.
5. Otocka, E. P.; Eirich, F. R. *J. Polym. Sci. A2* **1968,** *6*, 933–946.
6. Anfimov, B.; Ferracini, E.; Ferrero, A.; Riva, F. *Euro. Polym. J.* **1973,** *9*, 1021–1028.
7. Pineri, M.; Meyer, C.; Levelut, A. M.; Lambert, M. *J. Polym. Sci. Polym. Phys. Ed.* **1974,** *12*, 115–130.
8. Agarwal, P. K.; Makowski, H. S.; Lundberg, R. D. *Macromolecules* **1980,** *13*, 1679–1687.
9. Makowski, H. S.; Lundberg, R. D.; Westerman, L.; Bock, J. In *Ions in Polymers*; Eisenberg, A., Ed.; Advances in Chemistry Series 187; American Chemical Society: Washington, DC, 1980; Chapter 1.
10. Dieterich, D.; Keberle, W.; Witt, H. *Angew. Chem. Int. Ed. Engl.* **1970,** *9*, 40–50.
11. Yang, S.; Xiao, H. X.; Higley, D. P.; Kresta, J.; Frisch, K. C.; Farnham, W. B.; Hung, M. H. *J. Mater. Sci. Pure Appl. Chem.* **1993,** *A30*, 241–252.

12. Hwang, K. K. S.; Yang, C.-Z.; Cooper, S. L. *Polym. Eng. Sci.* **1981,** *21,* 1027–1036.

13. Yang, C.-Z.; Hwang, K. K. S.; Cooper, S. L. *Makromol. Chem.* **1983,** *184,* 651–668.

14. Miller, J. A.; Hwang, K. K. S.; Cooper, S. L. *J. Macromol. Sci. Phys.* **1983,** *B22,* 321–341.

15. Hwang, K. K. S.; Speckhard, T. A.; Cooper, S. L. *J. Macromol. Sci. Phys.* **1984,** *B23,* 153–174.

16. Speckhard, T. A.; Hwang, K. K. S.; Yang, C. Z.; Laupan, W. R.; Cooper, S. L. *J. Macromol. Sci. Phys.* **1984,** *B23,* 175–199.

17. Hsu, S. L.; Xiao, H. X.; Szmant, H. H.; Frisch, K. C. *J. Appl. Polym. Sci.* **1984,** *29,* 2467–2479.

18. Al-Salah, H. A.; Frisch, K. C.; Xiao, H. X.; McLean, J. A. Jr. *J. Polym. Sci. A Polym. Chem.* **1987,** *25,* 2127–2137.

19. Al-Salah, H. A.; Xiao, H. X.; McLean, J. A. Jr., Frisch, K. C. *J. Polym. Sci. A Polym. Chem.* **1988,** *26,* 1609–1620.

20. Zielinski, R.; Rutkowska, M. *J. Appl. Polym. Sci.* **1986,** *31,* 1111–1118.

21. Rutkowska, M. *J. Appl. Polym. Sci.* **1986,** *31,* 1469–1482.

22. Visser, S. A.; Cooper, S. L. *Polymer* **1992,** *33,* 3790–3796.

23. Ding, Y. S.; Register, R. A.; Yang, C.-Z.; Cooper, S. L. *Polymer* **1989,** *30,* 1204–1212.

24. Lee, J. C.; Kim, B. K. *J. Polym. Sci. A Polym. Chem.* **1994,** *32,* 1983–1989.

25. Xiao, H.; Xiao, H. X.; Frisch, K. C.; Malwitz, N. *J. Appl. Polym. Sci.* **1994,** *54,* 1643–1650.

26. Xiao, H.; Xiao, H. X.; Frisch, K. C.; Malwitz, N. *J. Mater. Sci. Pure Appl. Chem.* **1995,** *32,* 169–177.

27. Visser, S. A.; Cooper, S. L. *Macromolecules* **1991,** *24,* 2576–2583.

28. Visser, S. A.; Cooper, S. L. *Polymer* **1992,** *33,* 920–929.

29. Lee, D.-C.; Register, R. A.; Yang, C.-Z.; Cooper, S. L. *Macromolecules* **1988,** *21,* 998–1004.

30. MacKnight W. J.; Taggart; W. R., Stein, R. S., *J. Polym. Sci. Symp.* **1974,** *45,* 113–128.

31. Lee, D.-C.; Register, R. A.; Yang, C.-Z.; Cooper, S. L. *Macromolecules* **1988,** *21,* 1005–1008.

32. Ding, Y. S.; Register, R. A.; Yang, C.-Z.; Cooper, S. L. *Polymer* **1989,** *30,* 1213–1220.

33. Ding, Y. S.; Register, R. A.; Yang, C.-Z.; Cooper, S. L. *Polymer* **1989,** *30,* 1221–1226.

34. Visser, S. A.; Cooper, S. L. *Macromolecules* **1991,** *24,* 2584–2593.

35. Register, R. A. Ph.D. Thesis, University of Wisconsin, 1989.

36. Dreyfus, B.; Gebel, G.; Aldebert, P.; Pineri, M.; Escoubes, M.; Thomas, M. *J. Phys. France* **1990,** *51,* 1341.

37. Visser, S. A.; Cooper, S. L. *Macromolecules* **1992,** *25,* 2230–2236.

38. Matsuura, H.; Eisenberg, A. *J. Polym. Sci. Polym. Phys. Ed.* **1976,** *14,* 1201–1209.

39. Eisenberg, A.; Matsuura, H.; Tsutsui, T. *J. Polym. Sci. Polym. Phys. Ed.* **1980,** *18,* 479–492.

40. Eisenberg, A.; Navratil, M. *J. Polym. Sci. Polym. Lett.* **1972,** *10,* 537–542.

41. Hird, B.; Eisenberg, A. *J. Polym. Sci. B. Polym. Phys.* **1990,** *28,* 1665–1675.

42. Kim, J.-S.; Jackman, R. J.; Eisenberg, A. *Macromolecules* **1994,** *27,* 2789–2803.

43. Duchesne, D. Ph.D. Thesis, McGill University, 1985.

44. Duchesne, D.; Eisenberg, A. *Can. J. Chem.* **1990,** *68*, 1228–1232.

45. Wollmann, D.; Williams, C. E.; Eisenberg, A. *Macromolecules* **1992,** *25*, 6775–6783.

46. Ma, X.; Sauer, J. A.; Hara, M. *Macromolecules* **1995,** *28*, 3953–3962.

47. Chang, M. C. O.; Thomas, D. A.; Sperling, L. H. *J. Appl. Polym. Sci.* **1987,** *34*, 409–422.

48. Brandrup, J.; Immergut, E. H. *Polymer Handbook*, 3rd ed.; Wiley: New York, 1979.

49. Ma, X.; Sauer, J. A.; Hara, M. *Macromolecules* **1995,** *28*, 5526–5534.

50. Gronowski, A. A.; Jiang, M.; Yeager, H. L.; Wu, G.; Eisenberg, A. *J. Membr. Sci.* **1993,** *82*, 83–97.

51. Graiver, D.; Baer, E.; Litt, M.; Baney, R. H. *J. Polym. Sci. Polym. Chem. Ed.* **1979,** *17*, 3559–3572.

52. Graiver, D.; Litt, M.; Baer, E. *J. Polym. Sci. Polym. Chem. Ed.* **1979,** *17*, 3573–3587.

53. Graiver, D.; Litt, M.; Baer, E. *J. Polym. Sci. Polym. Chem. Ed.* **1979,** *17*, 3589–3605.

54. Graiver, D.; Litt, M.; Baer, E. *J. Polym. Sci. Polym. Chem. Ed.* **1979,** *17*, 3607–3624.

55. Graiver, D.; Litt, M.; Baer, E. *J. Polym. Sci. Polym. Chem. Ed.* **1979,** *17*, 3625–3636.

56. Salamone, J. C.; Volksen, W.; Olson, A. P.; Israel, S. C. *Polymer* **1978,** *19*, 1157–1162.

57. Schilz, D. N.; Peiffer, D. G.; Agarwal, P. K.; Larabee, J.; Kaladas, J. J.; Soni, L.; Handwerker, B.; Garner, R. T. *Polymer* **1986,** *27*, 1734–1742.

58. Liaw, D.-J.; Lee,. W.-F.; Whung, Y.-C.; Lin, M.-C. *J. Appl. Polym. Sci.* **1987,** *34*, 999–1011.

59. Monroy Soto, V. M.; Galin, J. C. *Polymer* **1984,** *25*, 121–128.

60. Zheng, Y.-L.; Galin, M.; Galin, J.-C. *Polymer* **1988,** *29*, 724–729.

61. Pujol-Fortin, M.-L.; Galin, J.-C. *Macromolecules* **1991,** *24*, 4523–4530.

62. Ehmann, M.; Galin, J.-C. *Polymer* **1992,** *33*, 859–865.

63. Ehrmann, M.; Mathis, A.; Meurer, B.; Scheer, M.; Galin, J.-C. *Macromolecules* **1992,** *25*, 2253–2261.

64. Ehrmann, M.; Galin, J.-C.; Meurer, B. *Macromolecules* **1993,** *26*, 988–993.

65. Eisenberg, A.; Hird, B.; Boore, R. B. *Macromolecules* **1990,** *23*, 4098–4108.

66. Hird, B.; Eisenberg, A. *Macromolecules* **1992,** *25*, 6466–6474.

67. Ehrmann, M.; Muller, R.; Galin, J.-C.; Bazuin, C. G. *Macromolecules* **1993,** *26*, 4910–4918.

68. Bazuin, C. G.; Zheng, Y. L.; Muller, R.; Galin, J. C. *Polymer* **1989,** *30*, 654–661.

69. Mathis, A.; Zheng, Y.-L.; Galin, J. C. *Polymer* **1991,** *32*, 3080–3085.

70. Gouin, J.-P.; Williams, C. E.; Eisenberg, A. *Macromolecules* **1989,** *22*, 4573–4578.

71. Gouin, J.-P.; Bossé, F.; Nguyen, D.; Williams, C. E.; Eisenberg, A. *Macromolecules* **1993,** *26*, 7250–7255.

72. Nguyen, D.; Zhong, X.-F.; Williams, C. E.; Eisenberg, A. *Macromolecules* **1994,** *27*, 5173–5181.

73. Lu, X.; Steckle, W. P.; Weiss, R. A. *Macromolecules* **1993,** *26*, 5876–5884.

74. Lu, X.; Steckle, W. P.; Weiss, R. A. *Macromolecules* **1993,** *26*, 6525–6530.

75. Lu, X.; Steckle, W. P.; Hsiao, B.; Weiss, R. A. *Macromolecules* **1995,** *28*, 2831–2839.

76. Venkateshwaran, L. N.; York, G. A.; DePorter, C. D.; McGrath, J. E.; Wilkes, G. L. *Polymer* **1992,** *33*, 2277–2286.

77. Gauthier, S.; Eisenberg, A. *Macromolecules* **1987,** *20,* 760–767.

78. Yano, S.; Tadano, K.; Jérôme, R. *Macromolecules* **1991,** *24,* 6439–6442.

79. Long, T. E.; Allen, R. D.; McGrath, J. E. In *Chemical Reactions on Polymers*; Benham, J. L.; Kinstle, J. F., Eds.; ACS Symposium Series 364; American Chemical Society: Washington, DC, 1988; Chapter 19.

80. Weiss, R. A.; Sen, A.; Pottick, L. A.; Willis, C. L. *Polymer* **1991,** *32,* 2785–2792.

81. Desjardins, A.; Eisenberg, A. *Plast. Rubber Compos. Process. Appl.* **1992,** *18,* 161–168.

82. Yoshikawa, K.; Desjardins, A.; Dealy, J. M.; Eisenberg, A. *Macromolecules* **1996,** *29,* 1235–1243.

83. Möller, M.; Omeis, J.; Möhleisen, E. In *Reversible Polymeric Gels and Related Systems*; Russo, P. S., Ed., ACS Symposium Series 350; American Chemical Society: Washington, DC, 1987; Chapter 7.

84. Zhong, X. F.; Eisenberg, A. *Macromolecules* **1994,** *27,* 1751–1758.

85. Vanhoorne, P.; Van den Bossche, G.; Fontaine, F.; Sobry, R.; Jérôme, R.; Stamm, M. *Macromolecules* **1994,** *27,* 838–843.

86. Jalal, N.; Duplessix, R. *J. Phys. France* **1988,** *49,* 1775–1783.

87. Brédas, J. L.; Chance, R. R.; Silbey, R. *Macromolecules* **1988,** *21,* 1633–1639.

88. Vanhoorne, P.; Maus, C.; Van den Bossche, G.; Fontaine, F.; Sobry, R.; Jérôme, R.; Stamm, M. *J. Phys. IV* **1993,** *3,* 63–66.

89. Horrion, J.; Jérôme, R.; Teyssie, P. *J. Polym. Sci. C Polym. Lett.* **1986,** *24,* 69–76.

90. Jérôme, R. In *Telechelic Polymers: Synthesis and Applications*; Goethals, E. J., Ed.; CRC: Boca Raton, FL, 1989; Chapter 11.

91. Vanhoorne, P.; Jérôme, R. In *Ionomers: Characterization, Theory, and Applications*; Schlick, S., Ed.; CRC: Boca Raton, FL, 1996; Chapter 9.

92. Jérôme, R.; Broze, G. *Rubber Chem. Technol.* **1985,** *58,* 223–242.

93. Horrion, J.; Jérôme, R.; Teyssié, P.; Marco, C.; Williams, C. E. *Polymer* **1988,** *29,* 1203–1210.

94. Broze, G.; Jérôme, R.; Teyssié, P.; Marco, C. *J. Polym. Sci. Polym. Phys. Ed.* **1983,** *21,* 2205–2217.

95. Venkateshwaran, L. N.; Tant, M. R.; Wilkes, G. L.; Charlier, P.; Jérôme, R.; *Macromolecules* **1992,** *25,* 3996–4001.

96. Mohajer, Y.; Tyagi, D.; Wilkes, G. L.; Storey, R. F.; Kennedy, J. P. *Polym. Bull.* **1982,** *8,* 47–54.

97. Bagrodia, S. R.; Wilkes, G. L.; Kennedy, J. P. *J. Appl. Polym. Sci.* **1985,** *30,* 2179–2193.

98. Hara, M.; Eisenberg, A.; Storey, R.; Kenndedy, J. P. In *Coulombic Interactions in Macromolecular Systems*; Eisenberg, A.; Bailey, F. E., Eds., ACS Symposium Series 302; American Chemical Society: Washington, DC, 1986; Chapter 14.

99. Mohajer, Y.; Bagrodia, S.; Wilkes, G. L.; Storey, R. F.; Kennedy, J. P. *J. Appl. Poly. Sci.* **1984,** *29,* 1943–1950.

100. Bagrodia, S.; Pisipati, R.; Wilkes, G. L.; Storey, R. F.; Kennedy, J. P. *J. Appl. Polym. Sci.* **1984,** *29,* 3065–3073.

101. Bagrodia, S.; Mohajer, Y.; Wilkes, G. L.; Storey, R. F.; Kennedy, J. P. *Polym. Bull.* **1983,** *9,* 174–180.

102. Tsutsui, T. In *Development in Ionic Polymers 2*; Wilson, A. D., Ed.; Elsevier: New York, 1980; Chapter 4.

103. Littmann, E. R.; Marvel, C. S. *J. Am. Chem. Soc.* **1930,** *52,* 287–294.

104. Kern, W.; Brenneisen, E. *J. Prakt. Chem.* **1941,** *159,* 193–218.

105. Rembaum, A.; Baumgartner, W.; Eisenberg, A. *J. Polym. Sci. B Polym. Lett.* **1968,** *6,* 159–171.

106. Menschutkin, N. *Z. Physik. Chem* **1890,** *5,* 589–600.

107. Eisenberg, A.; Matsuura, H.; Yokoyama, T. *Polym. J.* **1971,** *2,* 117–123.

108. Tsutsui, T.; Tanaka, R.; Tanaka, T. *J. Polym. Sci. Polym. Phys. Ed.* **1976,** *14,* 2259–2271.

109. Klun, T. P.; Wendling, L. A.; van Bogart, J. W. C.; Robbins, A. F. *J. Polym. Sci. A Polym. Chem.* **1987,** *25,* 87–109.

110. Feng, D.; Wilkes, G. L.; Leir, C. M.; Stark, J. E. *J. Macromol. Sci. Chem.* **1989,** *A26,* 1151–1181.

111. Venkateshwaran, L. N.; Leir, C. E.; Wilkes, G. L. *J. Appl. Polym. Sci.* **1991,** *43,* 951–966.

112. Feng, D.; Wilkes, G. L.; Lee, B.; McGrath, J. E. *Polymer* **1992,** *33,* 526–535.

113. Lundberg, R. D.; Bailey, F. E.; Callard, R. W. *J. Polym. Sci. A1* **1966,** *4,* 1563–1577.

114. Moacanin, J.; Cuddihy, E. F. *J. Polym. Sci. C* **1966,** 313–322.

115. Eisenberg, A.; Ovans, K.; Yoon, H. N. In *Ions in Polymers;* Eisenberg, A., Ed.; Advances in Chemistry Series 187; American Chemical Society: Washington, DC, 1980; Chapter 17.

116. Liu, K.-J. *Macromolecules* **1968,** *1,* 308–311.

117. Liu, K.-J.; Anderson, J. E. *Macromolecules* **1969,** *2,* 235–237.

118. Wetton, R. E.; James, D. B.; Whiting, W. *J. Polym. Sci. Polym., Lett. Ed.* **1976,** *14,* 577–583.

119. Hannon, M. J.; Wissbrun, K. F. *J. Polym. Sci. Polym. Phys. Ed.* **1975,** *13,* 113.

120. Wissbrun, K. F.; Hannon, M. J. *J. Polym. Sci. Polym. Phys. Ed.* **1975,** *13,* 223–241.

121. Reich, S.; Michaeli, I. *J. Chem. Phys.* **1972,** *56,* 2350–2353.

122. Lee, C. C.; Wright, P. V. *Polymer* **1982,** *23,* 681–689.

123. Payne, D. R., Wright, P. V. *Polymer* **1982,** *23,* 690–693.

124. Xia, D. W.; Smid, J. *J. Polym. Sci. Polym. Lett. Ed.* **1984,** *22,* 617–621.

125. Watanabe, M.; Sanui, K.; Ogata, N.; Inoue, F.; Kobayashi, T.; Ohtaki, Z. *Polym. J.* **1984,** *16,* 711–716.

126. Watanabe, M.; Sanui, K.; Ogata, N.; Inoue, F.; Kobayashi, T.; Ohtaki, Z. *Polym. J.* **1985,** *17,* 549–555.

127. Watanabe, M.; Sanui, K.; Ogata, N.; Kobayashi, T.; Ohtaki, Z. *J. Appl. Phys.* **1985,** *57,* 123–128.

128. Papke, B. L.; Ratner, M. A.; Shriver, D. F. *J. Electrochem. Soc.* **1982,** *129,* 1694–1701.

129. Killis, A.; Le Nest, J.-F.; Cheradame, H. *Makromol. Chem. Rapid Commun.* **1980,** *1,* 595–598.

130. Besner, S.; Prud'homme, J. *Macromolecules* **1989,** *22,* 3029–3037.

131. Dumont, M.; Boils, D.; Harvey, P. E.; Prud'homme, J. *Macromolecules* **1991,** *24,* 1791–1799.

132. Besner, S.; Vallée, A.; Bouchard, G.; Prud'homme, J. *Macromolecules* **1992,** *25,* 6480–6488.

133. Vachon, C.; Vasco, M.; Perrier, M.; Prud'homme, J. *Macromolecules* **1993,** *26,* 4023–4031.

134. Lascaud, S.; Perrier, M.; Vallée, A.; Besner, S.; Prud'homme, J.; Armand, M. *Macromolecules* **1994,** *27,* 7469–7477.

135. Vachon, C.; Labrèche, C.; Vallée, A.; Besner, S.; Dumont, M.; Prud'homme J. *Macromolecules* **1995,** *28,* 5585–5594.

CHAPTER 8

PLASTICIZATION

Just as in nonionic polymers, plasticization can modify the properties of ionomers dramatically. However, in the ionomers, plasticization provides a wider scope for property modification because of the presence of both hydrophilic and hydrophobic regions. Thus it was recognized early in the study of ionomers that polar plasticizers interact with the ion-dense regions (i.e., the ionic aggregates), whereas plasticizers of low polarity interact primarily with the hydrophobic matrix (1). Amphiphiles can also be used as plasticizers and provide additional scope for property modification.

An early example of multiple plasticization in ionomers is the work of Lundberg et al. (1), who showed that sulfonated polystyrene containing a relatively low concentration of functional groups (1–5 mol %) could be plasticized in the ionic regions by incorporating a material such as glycerol into the ionic aggregates. Such a procedure lowers the bulk viscosity of the material at elevated temperatures because it disrupts the ionic interactions in the multiplets. By contrast, the use of a plasticizer of low dielectric constant has a much smaller effect on the melt viscosity, because the ionic aggregates remain intact.

A large number of papers and a review (2) have been published on the plasticization of ionomers. Here, we take *plasticization* to mean both the incorporation of low molecular weight materials into the polymer (external plasticization) and the attachment of low glass transition segments to the polymer (internal plasticization). The discussion is divided into four parts. The first part treats the incorporation of low molecular weight species of high polarity. The second is devoted to a description of the effects of low molecular weight species of low polarity, including a description of the effect of oligomers, because they lower the glass transition temperature T_g of the matrix. However, it should be noted that the incorporation of high molecular weight homopolymers, based on the same repeat unit as the matrix, has important effects, discussed in Chapter 9 because no appreciable T_g lowering is involved.

The effects of amphiphiles are discussed in the third section. Finally, internal plasticization, achieved by the attachment of long alkyl chains to the polymer backbone or by using neutralizing agents that contain long alkyl chains, is discussed in the last part of the chapter.

Although the words *high* and *low* in connection with polarity have subjective meanings, in the context of the present discussion, materials of high polarity should be taken to mean materials that interact primarily with the ionic groups. In this sense, water ($\epsilon = 80.1$), glycerol ($\epsilon = 42.5$), and some of alcohols with short alkyl chains (ranging from $\epsilon = 32.6$ for methanol to $\epsilon = 17.8$ for 1-butanol) are examples of materials of high polarity; the hydrocarbons and materials such as dioctyl phthalate ($\epsilon = 6.4$) are examples of materials of low polarity. It is not easy to define clearly the difference between blending and plasticization, as will be shown in the next chapter, because some blends show an effect on the properties of the high T_g component akin to plasticization. In conformity with earlier usage, *plasticization* is taken to mean interaction as a result of incorporation of materials of relatively low molecular weight or relatively short segment length (external or internal plasticization), and *blending* is taken to mean mixing with relatively high molecular weight species. This definition presents a problem when dealing with plasticization or blending using oligomers, because a wide range of molecular weights can be involved. In those borderline cases, the definitions employed by the original authors will generally be followed.

8.1. SMALL MOLECULE PLASTICIZERS OF HIGH POLARITY

8.1.1. Mechanical Properties

Tobolsky et al. (3) performed an early study of the effect of polar plasticizers on the stress–relaxation of a carboxylic rubber (a terpolymer of butadiene–acrylonitrile–methacrylic acid). They found that treatment with acetic acid lowered the rubbery plateau modulus of a sulfur-cured Zn^{2+} salt sample by a factor of three, and the T_g changed only slightly.

Navratil and Eisenberg (4) showed that incorporation of 25 wt % of dimethyl sulfoxide (DMSO) into poly(styrene-*co*-sodium methacrylate) [P(S-*co*-MANa)] shifted the position of the stress–relaxation pseudo–master curve to a much shorter time scale. At a modulus of 10^7 N/m^2, the value of time was shifted by nine orders of magnitude relative to the unplasticized system. However, the shift was considerably smaller at a higher modulus. Thus the plot becomes more similar to that of the nonionic system. These effects might occur because DMSO reduces the ionic interactions in the multiplets and reduces clustering, thus allowing the ionomer to show properties resembling those of the nonionic polymer with only minor changes in the matrix T_g.

In an extensive subsequent study, Lundberg et al. (1) investigated the viscosity and the T_g of sulfonated polystyrene [P(S-*co*-SS)] ionomers as a function of plasti-

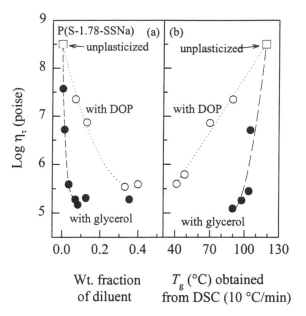

Figure 8.1. a, Melt viscosity at 220°C versus weight fraction of plasticizer for P(S-1.78-SSNa) plasticized with DOP (○) and glycerol (●) and unplasticized (□). **b**, Melt viscosity versus T_g (from DSC at 10°C/min) for the same systems. Modified from Lundberg et al. (1).

cizer content. Figure 8.1a shows the melt viscosity at 220°C as a function of weight fraction of diluent for a P(S-1.78-SSNa) sample. It is seen that the incorporation of even small amounts of glycerol lowers the viscosity drastically. The effect of dioctyl phthalate (DOP), a plasticizer of much lower polarity, will be discussed later. Figure 8.1b shows a correlation between the melt viscosity at 220°C and the T_g of plasticized ionomers containing various amounts of plasticizers. Again, the figure shows that the incorporation of even a small amount of glycerol has a dramatic effect on the melt viscosity without greatly affecting the T_g of the polymer, in marked contrast to the behavior of DOP. These are crucial features of high dielectric constant plasticizers in ionomers. Materials of high dielectric constant, such as glycerol, interact primarily with the ionic regions. Thus the T_g of the surrounding matrix regions is not affected appreciably. Naturally, however, the T_g of the cluster regions is affected in a major way (shown below).

Dynamic mechanical properties of styrene ionomers plasticized with either glycerol or diethylbenzene (DEB) were investigated by Bazuin and Eisenberg (5). A plot of G' as a function of temperature is shown in Figure 8.2. It is clear that glycerol lowers the matrix glass transition temperature slightly and removes altogether the inflection point related to the glass transition of the cluster regions, which suggests that clustering has effectively been eliminated. It should be stressed that in many systems plasticized with nonvolatile plasticizers of high dielectric constant, the multiplets remain intact, although the strength of the interaction of the ionic groups in the multiplets is weakened considerably. In the glycerol plasticized system, a drop in the modulus is also seen at temperatures between -100 and $0°C$. This low temperature drop is most likely owing to the glass transition temperature of phase-separated glycerol-rich regions.

In a subsequent study, Fitzgerald and Weiss (6) showed that at ~ 10 wt % glycerol in a P(S-1.65-SSMn) sample, the log E'' peak attributed to the cluster T_g still remains, although it is depressed appreciably, whereas the peak attributed to the matrix T_g is depressed only slightly. In the Na^+ salt case, however, for a comparable ion content, the log E'' peak owing to the cluster T_g disappears (7,8). Evidence for the continued existence of multiplets in the Mn^{2+} samples plasticized with 10 wt % glycerol comes from the work of Fitzgerald and Weiss (6), who showed by electron spin resonance (ESR) that ionic aggregation is still present despite the high glycerol content. The glass transition temperature for P(S-2.59-SSNa) was measured as a function of plasticizer content by Weiss et al. (8) (Fig. 8.3.). The difference in behavior between DOP (low polarity) and glycerol (high polarity) is clearly seen. Glycerol depresses the matrix T_g slightly, up to about 3 wt % of plasticizer, and then the effect levels off. By contrast, DOP continues to decrease the T_g. It is interesting to note that the modulus below the T_g is considerably lower in the glycerol plasticized system. This, as noted above, is undoubtedly owing to the presence of a low temperature relaxation.

Water can also serve as a polar plasticizer, which results in the progressive decrease of the well-defined cluster tan δ peak (9,10). Naturally, an ionomer can absorb a lot of water because of the presence of ionic aggregates in the material. The uptake of water probably results in a drop in the cluster glass transition of the ionomer, and thus water-swollen systems can also be considered as plasticized systems. Be-

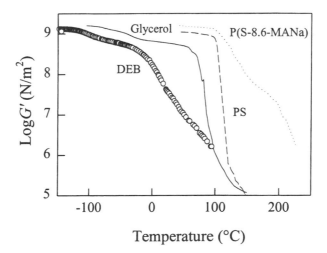

Figure 8.2. Log G' versus temperature for PS, P(S-8.6-MANa), and the ionomer plasticized with glycerol (~25 wt %) and with DEB (~25 wt %). Modified from Bazuin and Eisenberg (5).

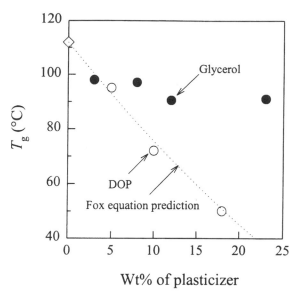

Figure 8.3. Matrix T_g versus plasticizer for P(S-2.59-SSNa) plasticized with DOP (○) and glycerol (●). *Dotted line*, Fox equation prediction. Modified from Weiss et al. (8).

cause the mechanical properties of water-swollen materials were discussed earlier (Sections 5.8.6, 6.1.7, and 6.2.7) when we examined water uptake of various ionomers, they are not discussed further here.

The discussion so far has centered on styrene ionomers. A recent study by dynamic mechanical thermal analysis of plasticization of methyl methacrylate based ionomers using glycerol as a plasticizer was performed by Ma et al. (11). In that system, glycerol can probably partition into both the ionic regions and the matrix and thus plasticizes both. The effect is most likely the result of the higher dielectric constant of poly(methyl methacrylate) compared to that of polystyrene, i.e., 3.0 versus 2.5. The study also showed that a 12.4 mol % ionomer, on plasticization with ~9 and ~19 wt % glycerol, exhibits a strong increase in the overall integrated area under the loss tangent peak. Both matrix and cluster peaks are present, but the intensities are considerably higher; the matrix peak height is higher than that of the cluster peak, in contrast to unplasticized systems, which confirms the partition of glycerol into both the multiplets and the backbone regions.

8.1.2. Morphology

The morphology of sulfonated polystyrene containing either Zn^{2+} or Mn^{2+} salts (6,8) was investigated as a function of plasticizer content. For the P(S-3.85-SSZn), it was found that, at 1 or 2.3 wt % methanol, the small-angle x-ray scattering (SAXS) peak was intact; at 6.3 wt % of methanol the peak shifts to a somewhat lower angle, (i.e., larger length scale), accompanied by an appreciable drop in the intensity (8). For the Mn^{2+} sulfonate, the peak position (as a function of the methanol content) shifts to lower angles up to 10 wt % methanol, at which point the peak essentially merges with the small-angle upturn. These results are probably related to swelling of the multiplets and a consequent drop in the electron density. Reorganization of the microstructure is probably also involved.

In the case of glycerol, by contrast, even at relatively low glycerol contents (1 wt %), the peak shifts immediately to lower angles and shows a dramatic increase in intensity (6,8). The sharpness of the peak increases with increasing glycerol content. The increase in the intensity probably results from a reorganization of the multiplets, leading to an increase in the volume fraction of phase-separated aggregates. The difference in behavior for the two plasticizers is undoubtedly related to the different methods of interaction between the ionic groups and the liquid, although the detailed mechanistic differences are not clear at this point. It is conceivable that glycerol may progressively phase separate at higher temperatures into regions devoid of ionic groups. It was also found that with increasing temperature, the SAXS peak position of the glycerol plasticized sample moves to higher angles, accompanied by a decrease in the intensity, the effect paralleling the behavior seen with decreasing plasticizer content.

The study of the effect of water on ionomer morphology started about three decades ago. Wilson et al. (12) first observed that the SAXS peak becomes sharper and more intense when a small amount of water is added to the ionomer, whereas water saturation reduces the peak intensity. Marx et al. (13) studied the effect of water on the SAXS profiles of poly(butadiene-*co*-sodium methacrylate) ionomers and found that the SAXS peak shifts to a smaller angle, and its intensity increases as the water content increases. A more extensive SAXS study of poly(ethylene-*co*-methacrylate) [P(E-*co*-MA)] ionomers containing 3.8 mol % of ions was performed by MacKnight et al. (14). They found that in the dry state the peak maximum corresponds to distances in the range of 23 to 35 Å, depending on the type of cation; whereas on saturation, the water completely destroys the ionomer peak, and only a broad small-angle upturn is observed. Kutsumizu et al. (15) also studied the effect of water sorption on the SAXS profiles of a P(E-*co*-MANa) ionomer containing 5.4 mol % of ions. They found that with increasing water content, the SAXS peak moves to smaller angles (i.e., larger spacing) and its intensity decreases. They suggested that the shift of the peak is the result of the swelling of the ionic aggregates, with both the number of the ionic aggregates and the number of ion pairs per aggregate remaining unchanged by hydration. It was also suggested that the decrease in the peak intensity is owing to a decrease in the electron density contrast between the ionic aggregates and the surrounding matrix regions, resulting from hydration of the ionic regions.

Yarusso and Cooper (16) studied the effect of water on the SAXS profile of zinc-sulfonated polystyrene ionomers. They found that the ionomer peak shifts to lower angle, and its intensity increases as water is added. They suggested that a morphological rearrangement occurs upon swelling and that the rearrangement results in a change both in the fraction of ions that form aggregates and in the number of aggregates in the material, unlike P(E-*co*-MANa). At low water contents, the water is absorbed by the ionic groups that are present in the matrix. At higher water contents, the hydrated species tend to aggregate. Thus the intensity of the SAXS peak increases. Fitzgerald (17), however, found that with increasing water content the SAXS peak intensity decreases, and the peak position shifts to lower angles. He suggested that in the case of Yarusso and Cooper's study (16), the samples might be supersaturated, because the workers used a pressure cooker to diffuse the water into the ionomers. Thus the ionomer peak might be owing to water clusters. It was also suggested that under such conditions the absorbed water changes the ionic aggregation by allowing ion pairs that are separate in the dry matrix to form quartets or higher aggregates on exposure to water. Because the size of the aggregates would increase with increasing water content, the electron density difference would decrease, resulting in a deceased peak intensity.

The effect of other polar plasticizers on the SAXS profile was also studied by Marx et al. (13). They found that the polar solvents (e.g., methacrylic acid, acetic acid, and formic acid) decrease the SAXS peak intensity but do not change the peak position.

8.2. PLASTICIZERS OF LOW POLARITY

8.2.1. Small Molecule Plasticizers

8.2.1.1. Mechanical Properties. In an early study of the plasticization of P(S-*co*-MANa) with DOP, Navratil and Eisenberg (4) observed that the stress–relaxation curves tended to broaden on introduction of the plasticizer, in contrast to the behavior with DMSO, which narrowed the distribution of relaxation times. The broadening was also observed by Weiss et al. (8) on sodium-sulfonated polystyrene containing 1.82 mol % of functional groups in dynamic mechanical master curves.

In an extensive study of plasticization, Lundberg et al. (1) investigated the drop in melt viscosity at 220°C for a sulfonated styrene ionomer containing 1.78 mol % of sulfonic acid plasticized with dioctyl phthalate (Fig. 8.1). It is noteworthy that the drop in the glass transition temperature of ~ 75°C is accompanied by a drop in the melt viscosity of less than three orders of magnitude. The reason for this behavior becomes apparent when one considers the dynamic mechanical properties.

Bazuin and Eisenberg (5) investigated plasticization of two polystyrene ionomers containing either 2 mol % sodium methacrylate or 5.1 mol % sodium sulfonate by DEB, a plasticizer of low volatility that structurally resembles polystyrene. In the P(S-*co*-MANa) sample, it was shown that with increasing plasticizer content, the modulus–temperature curves in the primary glass transition region become progressively less steep. Furthermore, it was shown that both the cluster tan δ peak and the matrix tan δ peak attributed to the glass transition of those regions move to lower temperatures with increasing plasticizer content. In the sample containing methacrylate, the separation between the two glass transitions values ΔT_g increases with increasing plasticizer content. Thus at 0 wt % plasticizer, ΔT_g is 40°C; and at 83

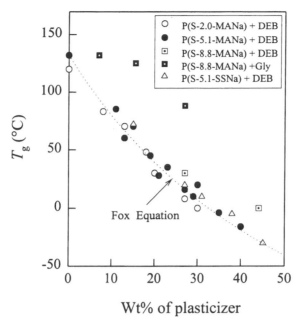

Figure 8.4. Matrix T_g versus plasticizer for various styrene ionomer samples. $\bigcirc$, P(S-2.0-MANa) + DEB; $\bullet$, P(S-5.1-MANa) + DEB; $\boxdot$, P(S-8.8-MANa) + DEB; $\blacksquare$, P(S-8.8-MANa) +Gly; $\triangle$, P(S-5.1-SSNa) + DEB, *dotted line*, Fox equation prediction. Modified from Bazuin and Eisenberg (5).

wt % plasticizer ΔT_g is 67°C. Figure 8.4 shows the glass transition temperature as a function of the weight percent of plasticizer. In the sulfonate system, the modulus–temperature plots show similar slopes in the transition region over a fairly wide range of plasticizer contents; and again, both tan δ peaks move to lower temperature with increasing plasticizer concentration. This behavior is not surprising within the framework of the Eisenberg–Hird–Moore (EHM) model (18), because both the cluster and matrix regions are materials of high styrene content, and the plasticizer should partition more or less equally between them. Therefore, the effect of plasticizers of low polarity on both glass transitions should be similar. It should be noted, however, that a parallel drop in T_g with plasticizer content, although common, is not always observed.

The essentially even distribution of plasticizer between the matrix and the cluster also explains why the viscosity remains relatively high, because the multiplets remain intact. However, if one only measures the glass transition temperature, one needs to take into account not only the drop in the matrix glass transition temperature with plasticizer content, but also that of the cluster glass transition. Lundberg et al.'s (1) measurements were all performed at 220°C, close to the cluster T_g of the plasticized polymer for a wide region of plasticizer contents. As the plasticizer content increases, the matrix T_g drops to 45°C (at 40% plasticizer), while the cluster T_g, although

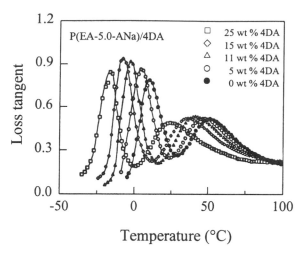

Figure 8.5. Tan δ versus temperature for P(EA-5.0-ANa) with several 4DA contents. $\square$, 25 wt % 4DA; $\diamondsuit$, 15 wt %; $\triangle$, 11 wt %; $\bigcirc$, 5 wt %; $\bullet$, 0 wt %. Modified from Tong and Bazuin (19).

decreasing, is still high. Therefore, even though the measuring temperature is above the T_g of the clusters, the cluster glass transition is much closer to the measuring temperature than to the matrix T_g (which was the T_g measured by Lundberg et al.). Therefore the viscosity is still high.

An interesting study of plasticization was performed by Tong and Bazuin (19) on ethyl acrylate ionomer plasticized with 4-decylaniline (4DA). Although 4DA is a plasticizer that contains a highly polar group, it behaves like a nonpolar plasticizer as it is distributed evenly between the cluster and matrix regions. Thus both T_g values are depressed nearly in parallel with increasing plasticizer content (Fig. 8.5), which was explained by suggesting that a strong hydrogen bond is formed between the aniline group and the carbonyl group of the ethyl acrylate. This interaction allows the plasticizer to be partitioned evenly between the two regions, rather than having the aniline group reside preferentially in the multiplet, as one might expect, and thus plasticizes only the cluster region. The phenomenon of preferential plasticization will be discussed more extensively below, in connection with amphiphilic plasticizers.

Another interesting phenomenon was observed in the same study. At an ion content of 10 mol %, the authors found that the relative heights of the cluster and matrix tan δ peaks changed with the addition of plasticizer, along with the usual shift in the peak positions. Specifically, as the plasticizer content increases, the intensity of the cluster peak decreases, while that of the matrix peak increases. Apparently, clustered regions are converted into unclustered regions with increasing plasticizer content.

For some systems, the opposite behavior is observed, i.e., clustering is enhanced on plasticization. Duchesne and Eisenberg (20) noted that poly(butyl acrylate), which has a glass transition temperature of $-54°C$, behaves as a normal ionomer (i.e., it

shows both a cluster and a matrix T_g) with the ionic group N-methyl-4-vinylpyridin-
ium iodide present. By contrast, if the same ionic group is incorporated into ethyl
acrylate, which has a glass transition temperature 30°C higher (i.e., -24°C) (20), the
mechanical properties show no evidence of cluster formation. However, clustering is
induced in the ethyl acrylate–based ionomer containing 10.5 mol % N-methyl-4-
vinylpyridinium iodide on addition of 10 wt % of dimethyl malonate, which mimics
the backbone structurally. The explanation for this effect was provided in a subsequent study of styrene N-alkyl-4-vinylpyridinium iodide copolymers (21). In a SAXS
investigation, it was found that the dry ionomer showed no evidence of the presence
of multiplets, because no characteristic SAXS peak was seen. However, when DEB
was added as a plasticizer to this material, in quantities as low as 5% (based on
styrene units), clustering was induced. The clustering was proven both by dynamic
mechanical studies, which show a two-peak behavior in the plasticized material (Fig.
8.6), and by SAXS studies, which show a typical ionomer peak (Fig. 8.7). In contrast
to the vinylpyridinium systems, carboxylated ionomers based on ethyl acrylate show
normal ionomer behavior in that two tan δ peaks are seen (19).

In unplasticized styrene ionomers containing vinylpyridine, no multiplets exist,
presumably because the electrostatic interactions are too weak at the glass transition
(>100°C) to form multiplets on cooling. However, once the T_g is lowered and
additional free volume is introduced or the mobility is increased by the presence of
plasticizer, multiplets can form; and the proximity of the multiplets (seen from the
SAXS peak) is apparently sufficient to result in behavior typical of a clustered
system. In fact, the 5, 7, 9, and 15 % (based on styrene units) plasticized ionomers
appear to be highly clustered, as can be seen from the ratio of the two tan δ peak
heights; the cluster peak is considerably more intense than the matrix peak (21).

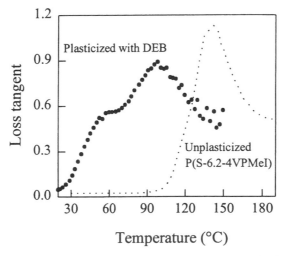

Figure 8.6. Tan δ versus temperature for unplasticized P(S-6.2-4VPMeI) (•) and the same
sample plasticized with 10 wt % DEB (●). Modified from Wollmann et al. (21).

The same effect is apparently operative in ethyl acrylates. However, it is important to include the dielectric constant of the polymer when considering electrostatic interactions. Even though the T_g of the styrene system is considerably higher than that of ethyl acrylate, styrene has a lower dielectric constant. This factor probably acts as an additional driving force in multiplet formation upon plasticization. Thus the styrene-co-4-vinylpyridine, which is unclustered, can become a clustered material by the addition of a relatively small amounts of the plasticizer.

It is interesting to relate this effect to the methyl methacrylate system. Ma et al. (22) have shown that methyl methacrylate-co-sodium methacrylate ionomers [P(MMA-co-MANa)] are clustered. In contrast, the Cs⁺ salt, P(MMA-co-MACs), which has a much larger cation, does not appear to be clustered (23). In poly(methyl methacrylate) (PMMA), the T_g is high (~100°C) and the dielectric constant is larger than that of styrene (3.0 versus 2.5), which suggests that differences in the electrostatic interactions are a major factor in determining whether the system is clustered or not. The polymer is clustered for the sodium carboxylate salt and unclustered for the cesium carboxylate. Again, it is possible that plasticization of the P(MMA-co-MACs) system might induce clustering by a plasticizer of low dielectric constant.

8.2.1.2. Morphology. For a P(S-2.65-SSMn) sample plasticized with DOP, the SAXS peak intensity decreases, but the shape and position of the peak does not change as the plasticizer content increases (6) It was suggested, therefore, that the introduction of DOP does not affect the ionic aggregates, but causes preferential swelling of the hydrocarbon phase. The decrease in the peak intensity is owing to the decrease in the volume fraction of ionic aggregates as well as an increase in the electron density of the hydrocarbon phase, causing a decrease in the contrast between the ionic aggregates and the hydrocarbon phase.

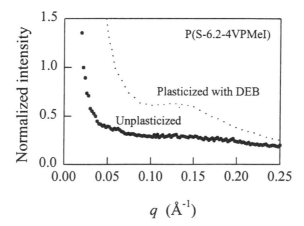

Figure 8.7. SAXS profile plotted as intensity versus q for unplasticized P(S-6.2-4VPMeI) (●) and the same sample plasticized with 10 wt % DEB (•). Modified from Wollmann et al. (21).

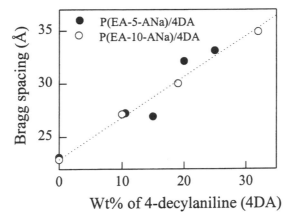

Figure 8.8. Bragg spacing versus 4DA for P(EA-5-ANa) (●) and P(EA-10-ANa) (○). Modified from Tong and Bazuin (19).

In a SAXS study of poly(ethyl acrylate) ionomers plasticized with 4DA, it was found that the spacing increases linearly with increasing weight percent of plasticizer (Fig. 8.8) (19). It was suggested that 4DA swells the intermultiplet regions, and thus the spacing increases. It was also found that the intensity of the peak increases with plasticizer content, which is probably the result of the low electron density of the 4DA and thus enhanced contrast between the multiplets and the surrounding material. The authors suggested that the differences between their SAXS results and those of the P(S-*co*-SSMn)–DOP system studied by Fitzgerald and Weiss (6) may be owing to one or more of the following possibilities: in the P(S-*co*-SSMn)–DOP case (a) the plasticizer content is lower, (b) the ion content is lower, and (c) the specific interactions between the matrix and the plasticizers are weaker than in the plasticized ethyl acrylate system.

A plasticization study involving dibutylphthalate was performed on the *n*-butyl acrylate-(sulfopropyl)ammonium betaine copolymers (10). Because of their affinity for the polymer backbone, dibutylphthalate plasticizes both the matrix and cluster regions. The SAXS patterns of the plasticized system show a progressive decrease in the peak position with increasing plasticizer content and eventual disappearance at 80 wt % of plasticizer. This suggests that initially the distance between multiplets increases, but eventually the plasticizer causes the disappearance of the SAXS peak.

8.2.2. Oligomeric Plasticizers

Oligomers can, of course, also act as plasticizers, because their T_g is considerably lower than that of the high molecular weight homopolymer (24,25). The melt rheology of styrene ionomers plasticized with styrene oligomers at various compositions showed that the ionic microphase-separated regions are affected by plasticization but are not destroyed (24). Failure of time–temperature superposition was observed at high plasticizer contents. Villeneuve and Bazuin (25) showed that styrene oligo-mers can act as normal plasticizers of low polarity, by depressing the glass transition temperatures of both the cluster and the matrix phases. However, because the depres-sion of the matrix T_g is somewhat more pronounced (i.e., the two glass transition temperatures do not decrease in parallel), it was suggested that the oligomer initially resides preferentially in the regions of high mobility, i.e., the matrix, because the matrix phase is more ion free than is the cluster phase and can, therefore, accommo-date somewhat more oligomer. The effect is not large; e.g., the depression is ~0.6°C/wt % of oligomeric styrene for the cluster, and ~1.1°C/wt % for the matrix.

8.3. AMPHIPHILIC PLASTICIZERS

A mechanism of plasticization that differs appreciably from either of the two types described above involves amphiphilic plasticizers. These systems are of particular interest because, as the name implies, amphiphilic plasticizers can influence directly both the ionic and the nonionic regions. Low dielectric constant plasticizers are distributed into both the cluster and the matrix regions, depressing the glass transi-tions of both. Polar plasticizers interact primarily with the multiplets, reducing their cohesiveness and depressing the cluster glass transition. In the case of amphiphilic plasticization, both processes can take place simultaneously. The materials are usu-ally classical amphiphiles, e.g., zinc stearate, which contain segments of high polarity and segments of low polarity. However, in some cases, complications can arise, as we show below.

8.3.1. Surfactants

The effect of the addition of zinc stearate on the mechanical properties of zinc sulfonated ethylene–propylene terpolymer (EPDM) has been studied extensively (26). It was found that at 100–150°C, zinc stearate starts to melt, weakens the strong ionic association, and causes the sample to behave like a high molecular weight amorphous material. The T_g and the rubbery modulus decrease with increasing zinc stearate content. Thus it was suggested that the zinc stearate has a dual effect. On the one hand, it acts as a filler and raises the T_g of the material slightly. Presumably, this behavior is caused by the action of solid crystalline zinc stearate. More important, zinc stearate also acts as an ionic domain plasticizer, probably by the incorporation of some ionic head groups of the zinc stearate into the sulfonate multiplets, which lowers the onset temperature of ion hopping in the multiplets. The presence of ionic zinc stearate groups in the multiplet thus serves to increase the bond interchange kinetics at a constant temperature, because the carboxylate groups act as a plasticizer for the sulfonate multiplets by lowering the average electrostatic interaction in the multiplets relative to that in the pure sulfonates. Thus the material can flow at much lower temperature than in the absence of zinc stearate.

Kim et al. (27) investigated the plasticization of sulfonated polystyrene by sodium dodecylbenzenesulfonate (SDBS). The sulfonate group in the SDBS has exactly

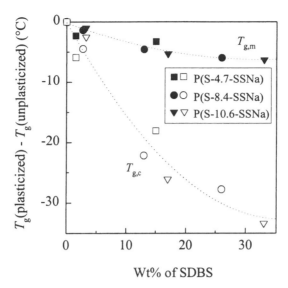

Figure 8.9. Glass transition lowering in P(S-*co*-SSNa) of various ion contents of the matrix $T_{g,m}$ (*solid symbols*) and cluster $T_{g,c}$ (*open symbols*) values by sodium dodecylbenzenesulfonate. *Squares,* P(S-4.7-SSNa); *circles,* P(S-8.4-SSNa); *triangles,* P(S-10.6-SSNa). Modified from Kim et al. (27).

the same ionic head group as the primary multiplet; therefore, weakening of the electrostatic interaction by plasticization is not expected. The sulfonate head group does, however, become incorporated, so the dodecylbenzene segment is kept close to the multiplet. Consequently, the dynamics of the polymer chains in the immediate vicinity of the multiplet are accelerated relative to those in the unplasticized system. The regions that are far from the multiplets are not affected, except by the presence of (probably rare) micelles of the surfactant materials. Thus, because the dynamics of only the immediate vicinity of the multiplet are affected (i.e., the regions of reduced mobility), one expects that only the cluster glass transition should be reduced substantially and that there should be no appreciable effect on the glass transition of the matrix regions, as was, indeed, observed. Figure 8.9 shows plots of the T_g depression versus the amount of SDBS. It is seen that the T_g of the matrix region changes by only ~6°C, whereas the T_g of the cluster region is depressed by ~33°C for 32 wt % of the plasticizer.

A most interesting comparison of three different types of plasticizer—sodium benzenesulfonate (SBS), dodecylbenzene (DB), and SDBS—in sulfonated polystyrene was performed by Orler et al. (28), who used plasticizers as shown in Scheme **8.1**. A comparison of these three plasticized systems illustrates the different effects

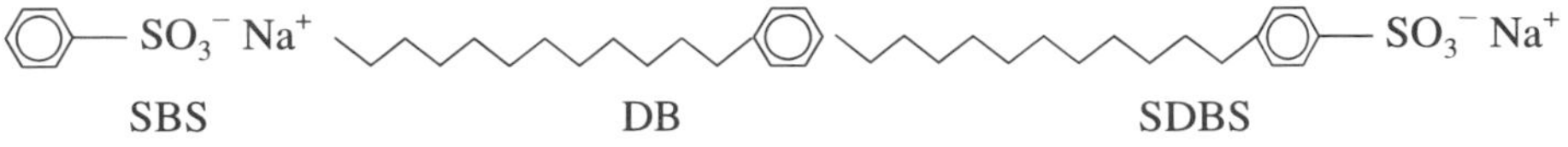

$$\text{SO}_3^- \text{ Na}^+ \qquad\qquad\qquad\qquad\qquad\qquad\qquad \text{SO}_3^- \text{ Na}^+$$

SBS   DB   SDBS

Scheme 8.1

of the plasticization of random ionomers. SBS depresses the cluster T_g slightly without affecting the matrix T_g. Clearly, some plasticization of the regions surrounding the multiplets occurs here. Sodium sulfonate groups become incorporated into the surface of the multiplet; and the phenyl rings induce some additional mobility in the immediate vicinity of the multiplet in a similar manner, but with smaller changes than in systems plasticized with SDBS. The DB, a nonionic species, depresses the glass transition of both the matrix and cluster phase essentially in parallel, as do other nonpolar plasticizers. Finally, the SDBS depresses the T_g of the cluster region drastically without affecting the matrix T_g, as found by Kim et al. (27).

Tong and Bazuin (29) observed an unusual effect when crystalline molecules containing functional groups, similar to those in the ionomer, were used as the plasticizer, e.g., hexadecanoic acid or sodium hexadecanoate in poly(ethyl acrylate-co-sodium acrylate). Sodium hexadecanoate is crystalline and remains phase separated when it is mixed with the ionomer. Thus, rather than acting as an ionic domain plasticizer, it is merely a filler particle; and until temperatures rise above the cluster T_g, it raises the modulus instead of affecting the glass transition. In contrast, the action of hexadecanoic acid resembles more that of a classical ionic domain plasticizer: It reduces the cluster glass transition by accelerating the ion hopping process through the incorporation of the -COOH groups into the multiplet. Some of the surfactant may become neutralized via ion exchange with the ionic groups on the ionomers.

Another study of amphiphilic plasticizers was performed by Natansohn et al. (30), which involved the phase structure of mixtures of poly(ethyl acrylate) ionomers plasticized with sodium hexadecanoate. They found that the poly(ethyl acrylate) ionomer–sodium hexadecanoate mixture has a morphology that involves bi-layers of carboxylate ions at the interface. This trend was observed in nonionic poly(ethyl acrylate)–sodium palmitate mixtures; but in the nonionic case, the ester groups of poly(ethyl acrylate) are preferentially located at the interface. The suggested morphology is shown in Figure 8.10. The study confirms the findings of Tong and Bazuin (29).

8.3.2. Oligomeric Amphiphiles

8.3.2.1. Mechanical Properties. Plante et al. (31) studied plasticization by oligomeric styrene amphiphiles. The polymer was p-carboxylated polystyrene containing 7 mol % of ionic groups, and the plasticizer was a styrene oligomer ($M_w = 800$) containing one terminal p-carboxylate unit. Two cations, Ba^{2+} and Cs^+, were used to neutralize the polymers. Figure 8.11 shows the tan δ peak positions reflecting both the cluster and matrix glass transitions for the barium ionomer plasticized with barium oligomer, the cesium ionomer plasticized with cesium oligomer, and nonionic polystyrene plasticized with the cesium oligomer. It is clear that for the cluster glass transition, plasticization by the Cs^+ carboxylate-terminated chain is much more effective than that by the Ba^{2+} carboxylate-terminated chain. In the Ba^{2+} salt, the cluster glass transition drops by ~1°C/wt % of oligomer; in the Cs^+ salt, the drop is ~6°C/wt % of oligomer and disappears at 5 wt % oligomer content. The matrix

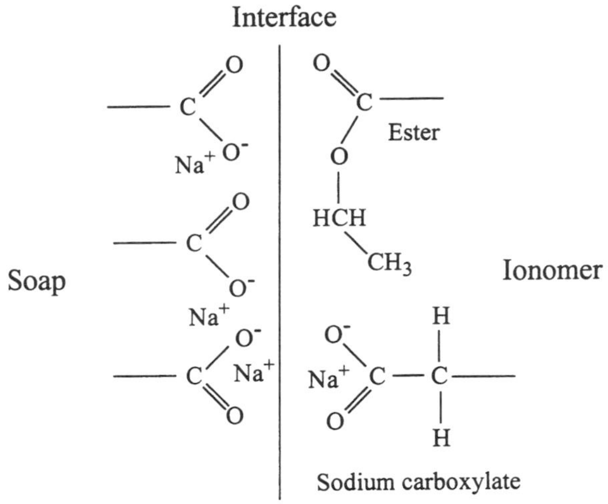

Interface

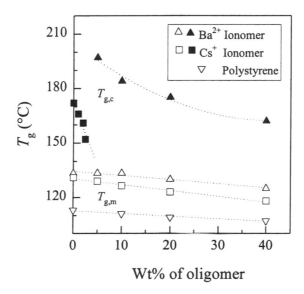

Figure 8.10. The interface between the ionomer and the soap domains. Modified from Natansohn et al. (30).

Figure 8.11. Glass transition temperature (from tan δ peak at 1 Hz) versus oligomer content for P(S-7-SC) neutralized with Ba^{2+} (*triangles*) and Cs^{+} (*squares*). *Solid symbols*, $T_{g,c}$; *open symbols*, $T_{g,m}$; ∇, homopolymer T_g values. Modified from Plante et al. (31).

glass transitions are depressed nearly in parallel. The different strengths of the ionic interactions between Cs^+ and Ba^{2+} are also reflected in differences in the morphology (discussed below).

In a subsequent study (32) involving monofunctional and bifunctional oligomers of a higher molecular weight, two polymers with different ion contents were chosen to explore plasticization effects in cluster-dominated and matrix-dominated systems. Monofunctional oligomers were found to plasticize both phases. In the matrix-dominated systems, the bifunctional oligomer has an antiplasticization effect owing to the improved dispersion of the ionic aggregates in the matrix and the immobilization of the oligomer at both ends in multiplets. In cluster-dominated systems, the oligomer phase-separates because of its much lower ion content.

8.3.2.2. Morphology.
In the Cs^+ ionomer, the SAXS peak, which appears at $q = \sim 0.21$ Å^{-1}, decreases in intensity as the oligomer salt is added (31). The peak is present at 2.5 wt % of oligomer, but disappears at 5 wt %. A peak reappears >10 wt % of oligomer, which increases in intensity and moves to smaller angles up to 20 wt %, after which no further change in the SAXS profile is observed upon further oligomer addition. This behavior is in marked contrast to that of the Ba^{2+} salt, for which the peak moves to lower angle with increasing oligomer content and decreases only slightly in intensity, even in the presence of 40 wt % of the oligomeric salt.

The primary reason for the difference is the different strength of interaction between the Cs^+ carboxylate units and the Ba^{2+} carboxylate units. Ba^{2+}, being divalent, clearly binds to the multiplet much more strongly than does Cs^+. A consequence of this difference is an increase in the T_g of the cluster regions in the Ba^{2+} carboxylate compared to the Cs^+ carboxylate. Also, because the interaction in the Cs^+ carboxylate is weaker than in the Ba^{2+} carboxylate, the addition of a Cs^+ oligomer to the immediate region surrounding the multiplet may well be more effective in disrupting the multiplet, by introducing additional free volume in its immediate vicinity, than is the Ba^{2+} carboxylate. It should be remembered that one divalent cation must interact with two anions, which means that the hopping unit, if it is to be uncharged, is a triplet ($-COO^-$ Ba^{2+} $^-OOC-$). The Ba^{2+} cation and one carboxylate ion, as a pair, however, have one positive charge. All these factors in the Ba^{2+} salt manifest themselves in a much stronger interaction, which translates into a much weaker glass transition depression with plasticizer content.

It is also of interest to compare the effectiveness of plasticization of the carboxylate-terminated oligomer with that of SDBS. The Ba^{2+} carboxylate-terminated oligomer and the SDBS amphiphile plasticize the multiplets and lower the cluster T_g values equally, showing a drop of ~30°C at 30 wt % of plasticizer. However, the Cs^+ carboxylate-terminated oligomer behaves differently; it is much less effective than the Ba^{2+} oligomer or SDBS, again underlining the inherent weakness of the Cs^+ carboxylate multiplet.

One interesting morphological feature of this study was the detection of a SAXS peak that disappears and reappears with increasing concentration of the Cs^+ carboxylate oligomer. The spacing (35 Å) for the higher oligomer concentrations was suggested to be the result of the diameter of a spherical shell containing a micelle of

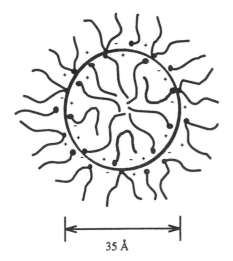

Figure 8.12. Proposed model of an aggregate in P(S-7-SC) plasticized with monochelic oligomer of 800 MW. Modified from Plante et al. (31).

35 Å

Cs^+ carboxylate oligomers on the inside and normal ionomer segment with Cs^+ carboxylate units (as well as some Cs^+ carboxylate oligomer) on the outside (Fig. 8.12). This morphology is novel in ionomers plasticized by this type of material.

8.3.3. Amines

The alkylamines are another type of amphiphilic plasticizer, and several studies on them have been published. The glass transition temperature of copolymers containing acid groups in many cases drops on neutralization with amines (especially those amines containing moderately long alkyl chains), and they are described here. In one study, Weiss et al. (33) showed that sulfonated polystyrene in the acid form can be plasticized effectively by long-chain alkylamines. The resulting materials can be considered either as combs, as simple neutralized systems, or as ionomers plasticized by amphiphiles. A wide range of amines was investigated in the study, from ammonia through methylamine, dimethylamine, and butylamine to many longer chain amines, including dimethylstearylamine and dimethyl C_{20-22} amine. The sulfonic acid content of the polymer was in the range of 1 to 21 mol %. It was found that the variation of the glass transition temperature with sulfonate concentration behaves in a normal way for the free acid, the zinc salt, and the ammonium salt, i.e., T_g increases at ~3°C/mol % of sulfonate groups. When plasticized by monoalkyl-substituted ammonium salts $^+NRH_3$ containing methyl and butyl groups, it was found that the T_g values increase at ~2.9°C/mol %. In contrast, the glass transition remains basically unchanged with varying ion content for the laurylammonium salt (C_{12}); but for the stearylammonium (C_{18}) and C_{20-22} salts, the glass transition actually drops with increasing ion content. The behavior of dialkyl- and trialkyl-substituted ammonium salts is generally similar; the salts containing alkyl chains shorter than the butyl group show an appreciable increase in the T_g (2.7°C/mol

%), whereas the butyl-substituted salt shows a much weaker dependence. Major drops in T_g are observed for the longer chain salts.

In a subsequent study, Smith and Eisenberg (34) showed that the T_g of 8.1 mol % sulfonated polystyrene decreases at $-2.3°C$ per carbon atom in the alkyl chain of the amine plasticizer. It was also shown that the T_g of this material can be increased if rigid amines are used. For example, 1-aminopyrene raises the T_g from 122 to 130°C, 3-aminofluoranthene to 134°C, and 1-adamantanamine to 137°C. The extent in the glass transition temperature increase is related to the size and the rigidity of the hydrocarbon portion of the amines.

A follow-up study for carboxylated poly(methyl methacrylate) ionomers was published by Smith and Goulet (35), who found that proton transfer occurred from the acid to the amines, giving H-bonded ion pairs that behaved like grafts. Linear amines were found to depress the T_g, whereas the rigid adamantanamine had little effect on T_g.

8.4. INTERNAL PLASTICIZATION

8.4.1. Mechanical Properties

Internal plasticization is plasticization via attachment of a side chain consisting of a highly flexible unit of low intrinsic glass transition to the polymer backbone, which results in a depression of the glass transition of the polymer as a whole. This method is not as effective as plasticization by small molecule plasticizers, but it has the advantage that removal of the plasticizer becomes impossible, unless chemical bonds are broken. A number of studies have been performed on the internal plasticization of ionomers. One of these studies was discussed in Section 4.4.4, in connection with the T_g of ionomeric comb polymers prepared by quaternizing styrene-co-4-vinylpyridine random copolymers with iodoalkanes of different chain lengths (36). Equation 4.6 was given for the glass transition depression of samples quaternized with iodoalkanes exceeding four carbon atoms in length. The formation of comb ionomers is a general method for plasticization, if the comb length is sufficiently long and the T_g of the backbone is high enough to be depressed by the comb material. Naturally, the lower the T_g of the backbone, the less effective is any particular comb at depressing the glass transition.

Ionomeric combs (i.e., combs in which the ionic group is located at the end of a side chain) were discussed in Sections 3.1 and 4.4 in connection with the glass transition and morphology (37,38). This effect is another type of internal plasticization, and its effectiveness can clearly be seen by comparing glass transitions as a function of ionic comb length at equal ion contents.

In another internal plasticization study, P(S-co-MANa) containing 7 mol % methacrylate was alkylated by attachment of 1-decene at the *para* position of the phenyl ring (39). It was found that the T_g of both the matrix regions and the cluster regions shifts to lower temperature with increasing degree of alkylation. The matrix T_g also decreases with increasing concentration of side chains, although the cluster glass

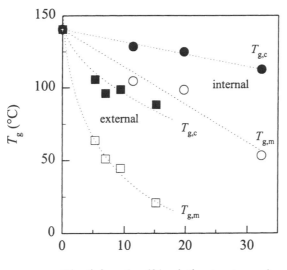

Figure 8.13. Effect of internal (*circles*) and external (*squares*) plasticization on matrix (*open symbols*) and cluster (*closed symbols*) T_g values versus plasticizer content for P(S-*co*-4VP) copolymers. Modified from Wollmann et al. (21).

transition initially increases and then drops. This increase at low degrees of plasticization is quite reproducible and has been attributed to improved multiplet formation in the presence of a low percentage of plasticizer. The improvement is attributed to a packing effect and is also reflected in the increase in the apparent cross-link density, as determined from the modulus owing to the ionic cross-links.

A further study of internal plasticization of styrene-*co*-*N*-methyl-4-vinylpyridinium iodide ionomer by alkylation at the *para* position of the phenyl ring was mentioned in connection with morphology and glass transition (Sections 3.3.5 and 4.4.5), where it was shown that cluster behavior could be induced by plasticization (21). The values of the glass transition temperatures of both internal (by alkylation) and external (by addition of DEB) plasticized systems are compared in Figure 8.13 for the 6 mol % samples as a function of the degree of plasticization. It is seen that the single glass transition temperature for the unplasticized material, which reflects the matrix glass transition, splits, indicating the formation of both cluster and matrix regions. It is also seen that internal plasticization is less effective than external plasticization at inducing cluster formation in unclustered materials. The reason for the greater effectiveness of external plasticizers is that the external plasticizer, being completely unattached, introduces more free volume into the polymer chain per unit mass than does an internal plasticizer of similar structure, which is attached at one end.

8.4.2. Morphology

The morphology of side chain ionomers as a function of both ion concentration and side chain length was discussed in Chapter 3. It is worth reiterating that an increase in the side chain length causes an increase in the aggregation number in the multiplet and a narrowing of the SAXS peak, reflecting increased uniformity in intermultiplet distances (37).

The morphology of the alkylated poly(styrene-*co*-sodium methacrylate) materials was also investigated by the SAXS (40) (Section 3.1.3.6). It was found that at moderate levels of alkylation, the definition of the ionomer peak improves, suggesting that internal plasticization improves ion aggregation. This was also seen in the study by Wollmann et al. (21) of internal plasticization in ionomers containing vinylpyridinium groups. However, >100% alkylation, the SAXS peak disappears, which suggests that the large volume of added material in the vicinity of the multiplet makes multiplet formation impossible. For intermediate levels of alkylation, the sizes of the multiplets and the spacings between them seem to be unchanged with the degree of alkylation.

8.5. SUMMARY

To conclude, we briefly summarize the four plasticization mechanisms. Nonpolar small molecule plasticizers distribute themselves more or less evenly throughout the matrix. They thus plasticize both the matrix regions and the regions of reduced mobility, therefore lowering the T_g of both, frequently in parallel. Polar molecules, by contrast, generally interact mainly with the multiplets. They thus soften the multiplet and lower the cluster T_g by increasing the mobility of the reduced mobility regions via accelerated ion hopping, resulting from multiplet plasticization.

Amphiphilic plasticizers can influence both the regions of reduced mobility surrounding the multiplet and the multiplet itself. For example, if a sulfonated polymer is plasticized with carboxylate amphiphiles such as stearic acid, the carboxylate group in the sulfonate multiplet increases the mobility within the multiplet, in addition to plasticizing the regions of reduced mobility around the multiplet as a result of the presence of the alkyl chain. Thus an appreciable decrease in the T_g of the cluster region must be expected. By contrast, an if a sulfonate polymer is plasticized with an amphiphile such as sodium dodecylbenzene sulfonate, the mobility of the multiplet would not be expected to increase but regions of the reduced mobility immediately surrounding the multiplet would be plasticized. Thus, again, a decrease in the T_g of the cluster region is to be expected, without an appreciable effect on the glass transition of the matrix.

Finally, internal plasticizers, while generally not as effective per unit weight as small molecule systems, cannot be removed from the material except by breaking primary bonds. At low concentrations, they can raise the cluster T_g by facilitating multiplet formation. At higher levels, multiplet formation is still facilitated, as can be seen from a sharpening of the SAXS peak, but the cluster T_g drops. At high levels, multiplet formation is disrupted for steric reasons.

8.6. REFERENCES

1. Lundberg, R. D.; Makowski, H. S. In *Ions in Polymers*; Eisenberg, A., Ed.; Advances in Chemistry Series 187; American Chemical Society: Washington, DC, 1980; Chapter 5.

2. Bazuin, C. G. In *Multiphase Polymers: Blends and Ionomers*; Utracki, L. A.; Weiss, R. A., Eds.; ACS Symposium Series 395; American Chemical Society: Washington, DC, 1989; Chapter 21.

3. Tobolsky, A. V.; Lyons, P. F.; Hata, N. *Macromolecules* **1968**, *1*, 515–519.

4. Navratil, M.; Eisenberg, A. *Macromolecules* **1974**, *7*, 84–89.

5. Bazuin, C. G.; Eisenberg, A. *J. Polym. Sci B Polym. Phys.* **1986**, *24*, 1137–1153.

6. Fitzgerald, J. J., Weiss, R. A. *J. Polym. Sci. B Polym. Phys.* **1990**, *28*, 1719–1736.

7. Fitzgerald, J. J.; Kim, D.; Weiss, R. A. *J. Polym. Sci. C Polym. Lett.* **1986**, *24*, 263–268.

8. Weiss, R. A.; Fitzgerald, J. J.; Kim, D. *Macromolecules* **1991**, *24*, 1064–1070.

9. Kim, J.-S.; Eisenberg, A. *J. Polym. Sci. B Polym. Phys.* **1995**, *33*, 197–209.

10. Galin, M.; Mathis, A.; Galin, J.-C. *Macromolecules* **1993**, *26*, 4919–4927.

11. Ma, X.; Sauer, J. A.; Hara, M. *Polymer* **1997**, *38* 4429–4431.

12. Wilson, F. C., Longworth, R., Vaughan, D. J. *Polym. Prepr. (Am. Chem. Soc., Div. Polym. Chem.)* 1968 9(1), 505–512.

13. Mark, C. L.; Caulfield, D. F.; Cooper, S. L. *Macromolecules* **1973**, *6*, 344–353.

14. MacKnight, W. J.; Taggert, W. P.; Stein, R. S. *J. Polym. Sci. Polym. Symp.* **1974**, *45*, 113–128.

15. Kutsumizu, S.; Nagao, N.; Tadano, K.; Tachino, H.; Hirasawa, E.; Yano, S. *Macromolecules* **1992**, *25*, 6829–6835.

16. Yarusso, D. J.; Cooper, S. L. *Polymer* **1983**, *26*, 371–378.

17. Fitzgerald, J. J., Ph.D. Dissertation, University of Connecticut at Storrs, 1986.

18. Eisenberg, A.; Hird, B.; Moore, R. B. *Macromolecules* **1990**, *23*, 4098–4108.

19. Tong, X.; Bazuin, C. G. *J. Polym. Sci. B Polym. Phys.* **1992**, *30*, 389–399.

20. Duchesne, D.; Eisenberg, A. *Can. J. Chem.* **1990**, *68*, 1228–1232.

21. Wollmann, D.; Williams, C. E.; Eisenberg, A. *Macromolecules* **1992**, *25*, 6775–6783.

22. Ma, X.; Sauer, J. A.; Hara, M. *Macromolecules* **1995**, *28*, 3953–3962.

23. Gronowski, A. A.; Jiang, M.; Yeager, H. L.; Wu, G.; Eisenberg, A. *J. Membr. Sci.* **1993**, *82*, 83–97.

24. Bazuin, C. G.; Eisenberg, A.; Kamal, M. *J. Polym. Sci. B Polym. Phys.* **1986**, *24*, 1155–1169.

25. Villeneuve, S.; Bazuin, C. G. *Polymer* **1991**, *32*, 2811–2814.

26. Agarwal, P. K.; Makowski, H. S.; Lundberg, R. D. *Macromolecules* **1980**, *13*, 1679–1687.

27. Kim, J.-S.; Roberts, S. B.; Eisenberg, A.; Moore, R. B. *Macromolecules* **1993**, *26*, 5256–5258.

28. Orler, E. B.; Calhoun, B. H.; Moore, R. B. *Macromolecules* **1996**, *29*, 5965–5971.

29. Tong, X.; Bazuin, C. G. *Chem. Mater.* **1992**, *4*, 370–377.

30. Natansohn, A.; Bazuin, C. G.; Tong, X. *Can. J. Chem.* **1992**, *70*, 2900–2905.

31. Plante, M.; Bazuin, C. G.; Jérôme, R. *Macromolecules* **1995**, *28*, 1567–1574.

32. Plante, M.; Bazuin, C. G.; Jérôme, R. *Macromolecules* **1995,** *28*, 5240–5247.

33. Weiss, R. A.; Agarwal, P. K.; Lundberg, R. D. *J. Appl. Polym. Sci.* **1984,** *29*, 2719–2734.

34. Smith, P.; Eisenberg, A. *J. Polym. Sci. B Polym. Phys.* **1988,** *26*, 569–580.

35. Smith, P.; Goulet, L. *J. Polym. Sci. P B Polym. Phys.* **1993,** *31*, 327–338.

36. Wollmann, D.; Gauthier, S.; Eisenberg, A. *Polym. Eng. Sci.* **1986,** *26*, 1451–1456.

37. Moore, R. B.; Bittencourt, D.; Gauthier, M.; Williams, C. E.; Eisenberg, A. *Macromolecules* **1991,** *24*, 1376–1382.

38. Gauthier, M.; Eisenberg, A. *Macromolecules* **1990,** *23*, 2066–2074.

39. Gauthier, M.; Eisenberg, A. *Macromolecules* **1989,** *22*, 3751–3755.

40. Moore, R. B.; Gauthier, M.; Williams, C. E.; Eisenberg, A. *Macromolecules* **1992,** *25* 5769–5773.

CHAPTER 9

IONOMER BLENDS

In view of the significant improvement in material properties that can be achieved in polymer blends over their individual components, it is not surprising that extensive efforts have been made to study miscibility in polymers (1–4). These studies have been performed not only in industrial laboratories but also in a number of academic laboratories, with the aim of improving properties and understanding the fundamental relationships among polymer miscibility, the properties of the individual materials, and the properties of blends.

Many polymer pairs are immiscible. Consider the thermodynamics of mixing; the entropy of mixing of polymers becomes small as the molecular weight increases because of chain connectivity. Thus even a mildly unfavorable enthalpy of mixing, which would not be enough to hinder mixing in small molecules because of the strong entropic driving force, is enough to render most polymer pairs immiscible. A number of miscible polymer pairs do exist (1–4), but most frequently these rely on specific interactions between repeat units or strong incompatibility between two segments of the same repeat unit. The latter favors miscibility with a homopolymer with which the two segments interact more favorably than they do with each other. In general, however, specific additives or modifications are needed to induce miscibility of polymer pairs.

A number of strategies have evolved that allow one to exercise some degree of control over compatibility or miscibility, including the addition of interfacial agents, which act as emulsifiers (e.g., block copolymers that contain segments of the two components); the copolymerization of the two components to a small extent in the polymers to be mixed; the application of techniques such as reactive processing (5) or high stress shearing (6) and co-cross-linking (7); and the formation of interpenetrating networks (8). In addition, chemical strategies, such as the incorporation of interacting groups, have been developed. These include hydrogen bonds (H-bonds)

(9–19), acid–base interactions (20), donor–acceptor complexes (21), dipole–dipole interactions (22–25), and charge transfer agents (26–29). Since the early 1980s, the use of ionic interactions in the control of miscibility has received considerable attention, and this chapter is devoted to a discussion of these techniques. An extreme example of this driving force is the formation of polyelectrolyte complexes in which a polyanion is mixed with a polycation to give stoichiometric materials that act like homogeneous polymer complexes (20,30–36).

9.1. TECHNIQUES FOR CHARACTERIZING MISCIBILITY

Before proceeding with the discussion of the detailed methods of miscibility enhancement using ionic interactions, it is useful to describe briefly the most popular techniques used to study polymer miscibility. Scattering measurements form a major family of techniques. Light scattering (LS) is perhaps the easiest. Cloudiness of a material is easy to detect, even with the unaided eye, and reveals the presence of scattering centers of different refractive indices, which have dimensions of approximately the wavelength of light (1,000–10,000 Å). Thus crystallinity (as spherulites) or the presence of large enough phase-separated regions in blends gives rise to cloudiness in samples. Small-angle x-ray scattering (SAXS) or small-angle neutron scattering (SANS) measurements rely on much shorter wavelengths and give information on heterogeneities in the electron density or the neutron scattering length density, respectively, which are on the order of 10–1,000 Å.

Mechanical and dielectric techniques make up another major area. These techniques are sensitive to different length scales; in the case of dynamic mechanical studies, the minimum dimensions are typically on the order of 50–100 Å. Thus, if phase separation has occurred on a length scale considerably smaller than 50 Å, dynamic mechanical techniques may not detect a separate glass transition. In contrast, if the regions are considerably larger and the phase-separated materials have glass transitions that are sufficiently far apart, two different glass transitions will be detected. The spatial dimensions necessary to resolve glass transitions in two different phases by differential scanning calorimetry (DSC) are probably still larger, but the technique is useful because of its speed of measurement and its automation.

Solid-state nuclear magnetic resonance (NMR) techniques have been used extensively, especially the spin-lattice relaxation time T_1 and spin-lattice relaxation performed in the rotating frame $T_{1\rho}$. Generally, T_1 measurements are sensitive to dimensions on the order of 200 Å, and $T_{1\rho}$ of ~ 20 Å. On occasion, infrared (IR) spectroscopy can be used to study miscibility. For example, if the IR band originating from a particular functional group is sensitive to its environment, then IR techniques may be able to distinguish whether the particular functional group is in the environment of its own material or in the material with which the polymer containing that group has been blended. Furthermore, optical spectroscopic techniques can be used to study phenomena such as proton transfer and charge transfer in blends. Naturally, electron microscopy can also be used, after appropriate sectioning and staining, if necessary. Spatial resolution in electron microscopy investigations is limited by the resolving power of the instruments and can be as low as a few angstroms in favorable cases. Fluorescence techniques, involving nonradiative energy transfer between a donor on one polymer chain and an acceptor on the other material, have also been used extensively. The mixed sample is irradiated at a wavelength that is absorbed by the donor and monitored at a wavelength emitted by the acceptor. If nonradiative energy transfer has occurred between the donor and the acceptor, then it can be concluded that the two species must be closer than some threshold distance to each other and that mixing has occurred. By contrast, if emission from the acceptor is not observed, but only emission characteristic of donor, then the two species are too far apart for nonradiative energy transfer to have occurred, and mixing is excluded.

In view of the above discussion, it should be borne in mind when the blend literature is reviewed that if a particular blend is considered miscible, it is so only within the limits of the technique used. Thus a material can appear to be homogeneous by the study of transparency, which shows that there are no heterogeneities at a level on the order of the wavelength of light; however, dynamic mechanical techniques can show two glass transition temperatures, indicating that on the 100 Å level, the material is phase separated. Similarly, a material can appear to be homogeneous by dynamic mechanical tests, but these tests cannot distinguish heterogeneity on the order of 20 Å, which can, however, be explored by $T_{1\rho}$ NMR measurements.

We must also carefully define the words *miscibility* and *compatibility*. Miscibility is a thermodynamic concept and implies homogeneity at the molecular level in the usual thermodynamic sense, whereas compatibility indicates improvement in properties of the blend. Thus blends can be compatible without necessarily being miscible. It is the control of the properties that is important from a technical point of view, and not necessarily the achievement of homogeneity at all length scales.

The first part of this chapter presents a menu of the different types of ionic interactions that can be used to control the miscibility of otherwise immiscible polymer pairs. In the next two sections, the morphology of blends and concepts of crosslinking and plasticization will be discussed briefly. The final sections will be devoted to more detailed discussions of the use of the various interactions, including ion–ion, ion pair–ion pair, ion–dipole, and ion–coordination. In each section, the general method will be discussed, and examples will be given of its application to real systems. The concluding section discusses a number of special topics, including the kinetics of mixing of dissimilar polymers, which will lead to a discussion of the equilibrium properties of ionomer blends. The existence of true thermodynamic miscibility in dissimilar polymer pairs, in which miscibility has been improved by use of ionic interactions, will also be discussed. The question will be posed whether one can talk about true thermodynamic miscibility in ion–ion systems and, by extrapolation, miscible polysalt complexes.

9.2. MENU OF INTERACTIONS

Several different types of ionic interactions can be used to control miscibility; a menu is provided in Figure 9.1. Before proceeding with the descriptions of these

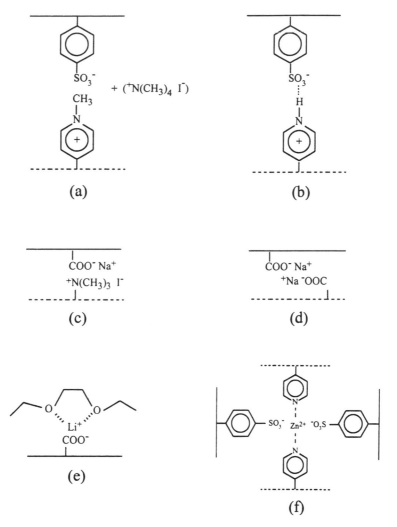

Figure 9.1. Menu of ionic interactions leading to miscibility enhancement: (**a**) ion–ion, (**b**) hydrogen bond–assisted ion–ion, (**c**) complementary ion pair–ion pair, (**d**) identical ion pair–ion pair, (**e**) ion–dipole, and (**f**) ion–coordination.

interactions, it should be recalled that it is not always the strongest interactions that provide the best results. The strongest interactions may lead to a homogeneous material at a relatively low ion content. However, they have the disadvantage of producing high melt viscosities, which may make processing of the materials difficult. Thus, in selecting the type of interaction for a particular application, it is useful to bear in mind that the weakest interaction necessary to accomplish the goal is often preferable to the strongest. Of course, sometimes strong interactions are needed; and therefore, because ionic interactions do provide a range of interaction strengths, these types of interaction are particularly useful in miscibility control.

Figure 9.1a illustrates a simple ion–ion interaction, which arises between such groups as N-methylpyridinium and benzenesulfonate. Of course, these groups must be placed on different chains to provide enhanced miscibility. The simplest way to prepare materials with this type of interaction is by mixing a sulfonated copolymer neutralized with a large cation, such as a tetraalkyl ammonium ion, and a second polymer containing pendent pyridinium ions with large counteranions, such as bromide or iodide. When the copolymers containing these various species are mixed in a solvent of an intermediate dielectric constant, such as dimethyl sulfoxide (DMSO), the micro-ion, tetraalkyl ammonium iodide in this case, remains in the solvent, and the polymer blend, because of the presence of strong interactions, precipitates as a gel.

Another example of ion–ion interactions, which does not involve the elimination of micro-ions, results from proton transfer. For example, if a polymer containing vinylpyridine is mixed with another polymer containing a pendent sulfonic acid—e.g., P(S-co-SSA)—protons can transfer upon mixing from the sulfonic acid to the vinylpyridine. This generates a polycationic chain from the chain that contains the pyridine and a polyanionic chain from the chain that contains the sulfonic acid. Thus ion–ion interactions can be achieved without eliminating micro-ions, perhaps with the added benefit of hydrogen bonding between the pyridinium cation and the sulfonate anion. The proton transfer assisted ion–ion interaction is illustrated in Figure 9.1b.

Ion pair–ion pair interactions provide another type of useful mechanism of miscibility enhancement. One can, for example, use a sodium carboxylate ion pair on one polymer chain to interact with a trialkyl ammonium halide group on the other polymer (Fig. 9.1c). Micro-ion elimination would lead to ion–ion interactions of the first type discussed above. However, without the elimination of micro-ions (e.g., as a result of preparation of the material in bulk), the microcounterions could remain with the pendent ions and lead to ion pair–ion pair interactions. It is also possible, however, that the micro-ions would crystallize (in the form of microcrystals). There has been little systematic exploration of blends of this type. Naturally, identical ion pairs, e.g., sodium carboxylate, can be used on both chains (Fig. 9.1d).

Ion–dipole interactions are illustrated in Figure 9.1e. The cation from lithium carboxylate—e.g., poly(styrene-co-Li methacrylate)—can interact with a polar polymer, such as poly(ethylene oxide), to convert it into a polycation, while the pendent carboxylate ions form a polyanionic chain. The strength of interaction is determined not so much by the lithium carboxylate interaction but by the ion–dipole interactions between the Li^+ cation and the ethylene oxide repeat unit. This type of interaction has been explored in a wide range of polymers and will be discussed below (see also Section 7.2.4).

Another important type of interaction is ion–coordination. In this case, a polymer such as poly(ethyl acrylate) that contains pendent vinylpyridine units [P(EA-co-4VP)] can interact with another polymer such as styrene that contains zinc sulfonate groups (Fig. 9.1f). As noted in connection with homoblends (Section 3.3.5), it is likely that the zinc cation exists in a tetracoordinated environment, involving two pyridine groups and two sulfonate groups. In the case of homoblends, the polymer chains carrying both sulfonates and pyridines are identical, e.g., polystyrene. However, in the case of normal blends, these two species would be placed on different polymer chains. It should be stressed that the interaction between zinc and pyridine involves coordinate covalent bonding rather than simple anion–cation interactions. Again, the interaction is strong and results in effective miscibility enhancement. It has been explored for a number of different cases (discussed below). Ion–coordination interactions are most useful for transition metal ions, whereas the simple ion–dipole interactions can involve ions such as alkali and alkaline earth metal cations. The interested reader is referred to the treatments in inorganic chemistry texts for a more detailed exploration of coordinate covalent bonding (37,38).

9.3. MORPHOLOGY

As one can imagine, the addition of a polymer containing groups such as vinylpyridine to a ionomer such as polystyrene zinc sulfonate causes disruption of multiplets. Douglas et al. (39) pointed out that a mechanical cluster peak is present in both metal coordination blends and proton transfer blends, even when no ionomer SAXS peak is seen. The reason for this second tan δ peak was discussed in Section 4.5.3, in connection with glass transitions. It was pointed out that behavior analogous to that of a normal multiplet is expected for a single zinc ion with a tetracoordinated shell consisting of two pyridinium and two benzenesulfonate moieties or for two ion pairs that form a four-centered aggregate in proton transfer blends. The chains attached to, and surrounding, such species have reduced mobility; and if these species are placed at appropriate distances from each other, the restricted mobility regions can overlap. Therefore, cluster behavior is certainly to be expected. Naturally, no SAXS peak is seen in such systems because, in some cases, no cations of high electron density are present (e.g., in proton transfer blends) and, in the other cases, single, randomly distributed zinc ions do not give rise to a SAXS peak. Register et al. (40) also noted the absence of the ionomer SAXS peak in the blend of copper-neutralized carboxy-terminated polybutadiene and P(S-3.42-4VP).

In other ion–coordination blends, e.g., a zinc polystyrene sulfonate [P(S-*co*-SSZn)]–polyamide (PA) blend, in which the phenomenon has been studied in detail, the SAXS ionomer peak is progressively disrupted with increasing weight percent of the PA. At a sufficiently high PA content, no SAXS peak is observed at all (41). These blends are discussed in more detail in Section 9.8.

9.4. CROSS-LINKING AND PLASTICIZATION

Before proceeding with the detailed discussion of the specific interactions used in miscibility enhancement, it is useful to review several topics of special relevance to ionomer blends. The morphology of the homoblends was discussed in Chapter

4, where it was shown that two benzenesulfonate–vinylpyridinium ion pairs yield the equivalent of an eight-functional cross-link that should behave like a multiplet of four ion pairs. The distances between these types of "multiplets" at ion concentrations in the range of 5–10 mol % are on the order of 20 Å in the styrene ionomers, i.e., well within the two persistence lengths postulated in the Eisenberg–Hird–Moore (EHM) model (42), which are required for the overlap of regions of restricted mobility. Thus cluster behavior is expected, and Douglas et al.'s (39) observation of a cluster peak in the homoblends and in the heteroblends is completely understandable.

Another concept that needs to be addressed is the cross-linking by ion–ion interactions. Otocka and Eirich (43), in their work on homoblends, showed that in butadiene homoblends, a vinylpyridinium carboxylate ion pair produces a cross-linking effect, because a rubbery plateau associated with ionic interactions (i.e., an ionic plateau) is present, the modulus of which increases with increasing ion content. In contrast, in the proton transfer assisted styrene homoblends, an ionic plateau is not found (39,44,45). To understand this behavior, bear in mind two experimental findings. One is the finding by Douglas et al. (39) that cluster and matrix glass transitions are close to each other and that, in the minimum between the two loss tangent peaks, the Young's modulus suggests a functionality of about eight for the ionic cross-links. It is on the basis of this finding that the two-ion-pair morphology mentioned in Section 3.3.5 was proposed.

Another factor is the finding by Smith and Eisenberg (45) that at elevated temperatures (i.e., close to the cluster T_g of the blends), no ionic plateau is seen. These factors are completely self-consistent. Recall that at the cluster glass transition, ion hopping occurs at a relatively high frequency (46). Below the cluster glass transition, i.e., at a point at which multiplets are still intact, the functionality of eight is not unreasonable. However, close to the cluster glass transition, i.e., in the region studied by Smith and Eisenberg (45), no cross-links should be observed on the time scale of the experiment (10s). It is useful to keep these factors in mind when one is reading the ionomer blend literature to be reviewed below.

In discussing the heteroblends, the concept of plasticization also needs to be borne in mind. If, for example, the styrene-*co*-styrene sulfonic acid copolymers [P(S-*co*-SSA)] are mixed with P(EA-*co*-4VP) copolymers, a T_g intermediate between those of styrene and ethyl acrylate is seen, at least when a single T_g is observed (see below) (39). This phenomenon represents a form of plasticization, because the T_g of polystyrene is depressed. This type of plasticization, however, is different from plasticization by small molecules or by internal plasticizers discussed in the previous chapter. The cluster T_g of a material such as a zinc–sulfonate ionomer is progressively disrupted with increasing functional group content in the blends. For example, for the ion–coordination blend case, at 5 mol % of functional group content in the heteroblends, the cluster peak (~235°C) owing to the zinc–sulfonate multiplet disappears, but a new cluster peak appears at a considerably lower temperature (~ 135°C). In the heteroblends and the homoblends, the depression of the cluster T_g can certainly be regarded as a form of plasticization. However, when one considers that the T_g of the matrix in the homoblend case is identical, the word *plasticization* is unsatisfactory. One must understand the phenomenon from a morphological point of view. The introduction of the pyridine group into the zinc–sulfonate cluster results in disruption of the classical ionomer cluster and the formation of a cluster of a completely different morphology. Thus it is the change of cluster morphology, not classical plasticization, that lowers the T_g of the cluster in the blend relative to that in the unblended ionomer.

9.5. ION–ION BLENDS

It has been known for a long time that the mixing of two polyelectrolytes of opposite charge gives rise to coprecipitation of polyelectrolyte complexes (20,30–36). These are essentially ladderlike chains containing equimolar amounts of each polyelectrolyte. The polyelectrolyte complexes—or polysalt complexes, as they are occasionally called—are necessary for subsequent discussion of ion–ion interactions in polymer blends.

Another component is the formation of homoblends (Section 4.5.3). Several studies of homoblends have been carried out; the earliest one was by Otocka and Eirich (43), who looked at copolymers of butadiene-*co*-(2-methyl-5-vinyl)pyridinium methyl iodide mixed with butadiene-*co*-Li methacrylate copolymers. A later study was performed by Smith (44), who investigated polystyrene homoblends in which a P(S-*co*-SSA) copolymer was mixed with P(S-*co*-4VP). This blend forms a true proton transfer system. Both of these series of investigations show the presence of strong interchain interaction and thus suggest that the mixing of one homopolymer that contains a relatively small number of proton donors or anionic groups with another homopolymer that contains proton acceptors or cationic groups should lead to enhancement of the miscibility.

The first study of true ion–ion blends was that of Smith and Eisenberg (47), who examined the blends of P(S-*co*-SSA) with P(EA-*co*-4VP). The two homopolymers—polystyrene (PS) and (PEA)—naturally give two completely independent loss tangent peaks: at 2 mol % of functional group content the peaks approach each other, and at 5 mol % only a single peak is observed. The experimental technique used was a torsion pendulum, which is a resonance instrument in which the resonance frequency changes extensively as the material undergoes its glass transition. Thus the cluster peak was not separated from the overlapping matrix peak. However, it was noted that the transition region in the G' curve was broadened relative to that of homopolymer. The loss tangent curves are shown in Figure 9.2 (48). It is seen clearly that at ~ 2 mol % of functional group content (the individual concentrations were 1.7 and 1.8 mol %, and the blend is designated 1.7/1.8), two peaks are observed; the lower temperature peak is attributed to the ethyl acrylate–rich phase, and the high temperature peak is attributed to the styrene. In the 4.1/3.8 mol % sample, a small ethyl acrylate peak is still seen at a somewhat higher temperature. However, the main peak has shifted to ~50°C, indicating blend formation. In the 5.1/5.8 and 9.8/10.5 mol % systems, the matrix peak has shifted to considerably higher temperatures, as expected. The cluster peaks are not seen in this study because of the methods of measurement (see above).

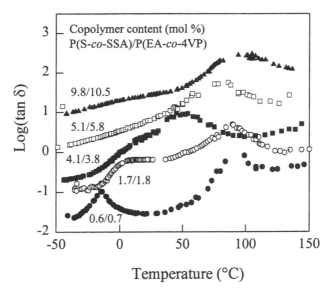

Figure 9.2. Log tan δ versus temperature for blends of P(S-*co*-SSA) with P(EA-*co*-4VP) of various functional group contents. ▲, 9.8/10.5 mol %; □, 5.1/5.8 mol %; ■, 4.1/3.8 mol %; ○, 1.7/1.8 mol %; ●, 0.6/0.7 mol %. Modified from Smith et al. (48).

A spectroscopic investigation of the energetics of this proton transfer reaction suggested that it is subject to an equilibrium at temperatures sufficiently far above T_g (i.e., T_g + 40°C) and that the strength of the interaction is ~25 kJ/mol of interacting groups (45,46). The degree of proton transfer was 80–90% for most of the ion contents studied. Bakeev and MacKnight (49) found even higher efficiencies, using fluorescence probe techniques. The existence of proton transfer and consequently of strong ionic interactions was confirmed by that technique for pyridine concentrations of 4.5 and 7.8 mol% in the blends of P(EA-*co*-4VP) with P(S-4.5-SSA).

Bazuin and Eisenberg (50) investigated the melt rheology of styrene–ethyl acrylate blends, in which miscibility results from proton transfer. Their results, expressed as master curves of the dynamic storage moduli as a function of reduced frequency, are shown in Figure 9.3. The dramatic increase in viscosity between 0 and 2 mol % and further increases with increasing ion content between 2 and 10 mol %, clearly illustrate the effects of the strong interaction. Increasingly strong interactions also suggest the increasing difficulty of melt processing these materials with increasing ion content. Far above the T_g, time–temperature superposition is applicable, and the shift factors are of the Williams–Landel–Ferry (WLF) type. For flow to occur in these blends, a number of ionic interactions along a chain segment must be loosened simultaneously, which makes flow difficult, even if individual associations have only a short lifetime.

The morphology of blends in this context is of great interest. As pointed out above, one can speak of clustering in the ion–ion blends studied by Douglas et al.

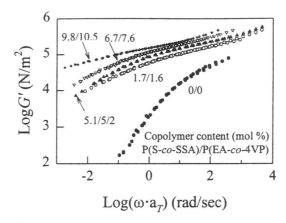

Figure 9.3. Log G' vs log ωa_T for blends of P(S-X-SSA) with P(EA-Y-4VP) of various functional group contents. ◆, 9.8/10.5 mol %; ▽, 6.7/7.6 mol %; ▲, 5.1/5.2 mol %; ○, 1.7/1.6 mol %; ●, 0/0 mol %. Modified from Bazuin and Eisenberg (50).

(51) in exactly the same sense as in the EHM model. It is also of interest to inquire whether there are domains of the original materials, in this particular case polystyrene or poly(ethyl acrylate), and if so, how large the domains are as a function of ion content. At 0 and 2 mol % functional groups, the dynamic mechanical curves clearly indicate the existence of phase-separated regions. The 5 mol % loss tangent curve indicates the presence of some ethyl acrylate–rich regions, but the interpretation of this type of curve in terms of the existence of phase-separated regions in the mixed material is difficult and is still an open question. On one hand, Douglas et al. (51) suggested that phase-separated regions should exist, even at relatively high ion contents. However, polysalt complexes are homogeneous materials, so that at some point between the 2 mol % region and the salt complex region, homogeneity must set in. The exact onset is difficult to determine and depends on one's definition of homogeneity.

As the discussion of solution properties by NMR (52) will show (below), there is strong evidence of homogeneity at 8 mol %, at least under the conditions of one experiment. The evidence for homogeneity comes from the observation of an aromatic shielding effect in the study of methyl methacrylate-*co*-4-vinylpyridine blended with the P(S-*co*-SSA). The study was performed in DMSO. DMSO is a proton transfer solvent (i.e., a proton can transfer in solution from the styrene sulfonic acid to the vinylpyridine). If the chains were not close, proton transfer could lead to a polyelectrolyte effect, which would induce intrachain repulsion, resulting in a stiffening of the backbone, as was suggested by Douglas et al. (51), which might cause an alignment of the rest of the chains. However, once the blend is formed in solution, the ions are generally paired; therefore, a polyelectrolyte effect is no longer to be expected. Under these circumstances, the existence of the aromatic shielding effect by the vast majority of the rings—which affects the methyl groups in the poly(methyl methacrylate) (PMMA)—means that the chains must be close together, even relatively far from the location of the functional groups. The kinetic aspects of this process are discussed more extensively later in this section. In general, more work is required to explore the size of the phase-separated regions in blends subject to miscibility enhancement via specific interactions. One such attempt is discussed below in connection with ion–dipole blends involving PA/P(S-*co*-SSA).

The use of proton transfer blends along the lines suggested by Smith and Eisenberg (47) has been applied to a number of other systems. In an early study, Zhou and Eisenberg (53) showed that sulfonated *cis*-1,4-polyisoprene (which became partly cyclized in the course of sulfonation) could be mixed with P(S-*co*-4VP) copolymer to yield a blend involving the same type of interactions as were found in the styrene–ethyl acrylate case. Similar effects were seen on the optical transparency of the samples, dynamic mechanical properties, and others. A later study involved the miscibility enhancement in a poly(tetrafluoroethylene) (PTFE) copolymer and poly(ethyl acrylate) blends, the PTFE copolymer sample being one of the perfluorosulfonates, Nafion (DuPont) in this specific case (54). In that study, a composition-dependent glass transition was observed.

Still later examples involve miscibility enhancement in a segmented polyurethane containing methyldiethanolamine (MDEA) as a part of the backbone in the hard segment, mixed with sulfonated polystyrene (55–58). It was shown that the miscibility of the styrene sulfonic acid with the hard segment was accompanied by an expulsion of the soft segment from the initially homogeneous polyurethane. Thus, as the styrene content increases, the mechanical effects of soft segment phase separation becomes more pronounced. However, the interaction between the styrene and the hard segments was very much in evidence both during the course of preparation, i.e., gel formation, and through the measurement of the glass transitions of a mixture of a polyurethane containing only hard segments with a sulfonated polystyrene. Initially, it was thought that in that blend the proton from the sulfonic acid was transferred to the amine on the MDEA segment in the hard segment of the polyurethane. Natansohn et al. (59), however, in a subsequent study using NMR, showed that the proton transfer could take place to all of the possible sites, i.e., the tertiary nitrogen of the MDEA chain extender in the hard segment, the urethane, the allophanate, and the other secondary structures.

Tan et al. (60) studied rigid-rod molecular composites prepared via ion–ion interactions. They found that when a polyelectrolyte, i.e., poly(2-acrylamido-2-methyl-propanesulfonic acid) is mixed with a rigid-rod molecule, i.e., poly(*p*-phenylene bisbenzothiazole), no phase separation is observed by scanning electron microscopy on the 500 Å scale. A more recent study using the same type of approach was performed by Eisenbach et al. (61). In the ionomer blends of rigid poly(diacetylenes) containing functional side groups (proton acceptors) mixed with sulfonated polystyrene, miscibility was confirmed by a single but relatively broad tan δ peak at ~110°C. These studies (60,61) are particularly noteworthy because rigid chains were successfully mixed with flexible chains to yield a blend subject to strong interactions. It is essential to keep in mind when preparing ion–ion blends by proton transfer that one needs to have donor-acceptor pairs that are strong enough to yield actual proton transfer, and not merely hydrogen bonding. Thus, for many materials, a carboxylic acid on one chain and a pyridine ring on the other would not give proton transfer but would result in hydrogen bonding. These hydrogen bonds, in many cases, are sufficient to induce useful properties in the blends, including, in some cases, strong miscibility enhancement; such systems have been studied extensively. To have a true ion–ion blend, however, proton transfer is essential.

As was mentioned, it is possible to obtain ion–ion blends without proton transfer, e.g., a P(S-*co*-SSA) copolymer neutralized with a large cation (quaternary ammonium or alkyl ammonium) and mixed with ethyl acrylate-*co*-N-methyl-4-vinylpyridinium iodide [P(EA-*co*-4VPMeI], which also has a large counterion (62) can produce an ion–ion blend. Upon mixing, either in benzene–methanol or in DMSO, the micro ions are effectively eliminated, probably because they remain in solution; and the resulting gelled polymer blend can be isolated and investigated. The same degree of miscibility enhancement is observed in these systems as in the proton-transfer blends. If one uses a single glass transition as the criterion of miscibility (i.e., on the length scale of 50–100 Å), one observes miscibility at 5% of functional groups and above. Below this point (e.g., at 0 and 3 mol %) two glass transitions are observed (Fig. 9.4). The presence of micro-ions in the supernatant solution can be confirmed by various analytical techniques, such as NMR. It should be pointed out that a similar type of ion–ion blend can be achieved by mixing one copolymer containing sulfonic acid with another containing 4-vinylpyridinium methyl iodide (62). Upon mixing, hydrogen iodide is formed, which can be eliminated as a gas under reduced pressure, especially in nonaqueous environments.

The nature of the ionic groups is important for miscibility enhancement in ion–ion blends. For example, if methacrylic acid is substituted for styrene sulfonic acid, then the miscibility enhancement is much less effective (63). Thus similarly prepared blends with styrene-*co*-methacrylic acid [P(S-*co*-MAA)] instead of P(S-*co*-SSA)

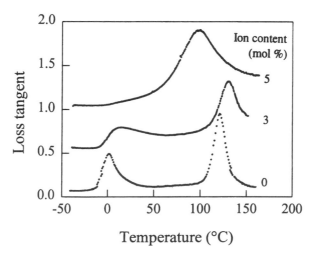

Figure 9.4. Tan δ versus temperature for P(S-*co*-SS⁻)–P(EA-*co*-4VPMe⁺) ion–ion blends of three functional group contents. Modified from Zhang and Eisenberg (62).

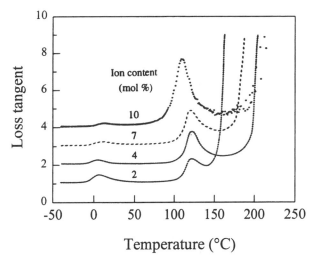

Figure 9.5. Tan δ versus temperature for P(S-*co*-MA$^-$)–P(EA-*co*-4VPMe$^+$) ion–ion blends of several functional group confents. Modified from Zhang and Eisenberg (63).

have two T_g values, even at a functional group content of 10 mol % (Fig. 9.5). This result confirms that one cannot regard pendent anions or cations on polymer chains as simple point charges. If they were to behave as point charges, then the difference in the behavior of the blends containing sulfonic acid or carboxylic acid would be much smaller than that obscrved here. Clearly other effects, such as interaction strengths and packing, are of importance.

Most of the blending procedures for the proton transfer blends involve the mixing of solutions; it was thus of interest to investigate the kinetics of the blending process. One means of doing this kind of investigation is to use the phenomenon of aromatic shielding in NMR in blends of methyl methacrylate containing 4-vinylpyridine with P(S-*co*-SSA). As a result of aromatic shielding, the position of a peak in the NMR spectrum of a group (e.g., CH$_3$) that is located within ~4 Å of an aromatic ring is shifted relative to that of the same group not in the vicinity of the aromatic ring. Thus kinetics of proton transfer can be followed by monitoring the relative heights of the peaks attributed to the shielded and unshielded groups. Proton transfer, as discussed above, leads to the formation of ionic chains. The ionic groups drive miscibility to the point at which aromatic shielding of the benzene ring of the styrene is observed on the methoxy groups of the poly(methyl methacrylate) chain (52). One study (64) confirmed that aromatic shielding was a consequence of the presence of electrostatic interactions, because it could be suppressed by the addition of small molecule salts that can satisfy the coordination sphere of the ions on the polymers.

In a subsequent study (65), the kinetics of the aromatic shielding process were investigated. A representation of the process is given in Figure 9.6. In the early stages, the process is second order. In the first step, proton transfer occurs from the sulfonic acid group to the vinylpyridine moiety, which produces the first ion pair

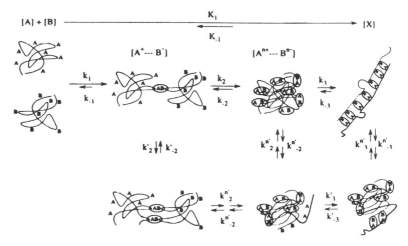

Figure 9.6. Suggested mechanism for the coil overlap process between P(S-*co*-SSA) (*thin line*) and P(MMA-*co*-4VP) (*thick line*). *A*, 4VP; *B*, SSA; *ovals*, ion pair. The lengths of the *arrows* represent, qualitatively, the relative rates of reactions. Modified from Bossé and Eisenberg (65).

between the two dissimilar chains. Once proton transfer has occurred, the chains are relatively close to one another, and the vast majority of the remaining protons on the two chains are transferred in a cascadelike process. This step leads to the formation of an interpolymer complex in which the ions are randomly paired. The third step involves the spatial reorganization of the complex to yield the final product, in which dissimilar chains are relatively well aligned in a manner resembling a ladder. The formation of this ladderlike complex leads to a gradual decrease in the original methoxy proton signal (the unshielded signal) and to the appearance of the shielded methoxy signal. The first two steps of the process (i.e., the initial contact and proton transfer leading to the random complex) are relatively rapid, whereas the spatial reorganization occurs more slowly. Overall, the kinetics could be modeled by an expression involving two opposing first-order reactions.

9.6. ION PAIR–ION PAIR BLENDS

In the earliest study of miscibility enhancement using ion pair–ion pair interactions, Vollmert and Schoene (66) investigated the transparency of samples of poly(styrene-*co*-metal acrylate) with poly(butyl acrylate-*co*-metal acrylate) and found that as the ion content increased, the transparency of the samples increased. However, complete miscibility was not achieved at 9 wt % acid content. Even at the highest ion contents studied (~20 mol % acid), the sample still retained some turbidity. In a subsequent study, Rutkowska and Eisenberg (67) mixed a polyurethane containing a quaternary ammonium salt in the hard segments with polystyrene containing sodium methacrylates, i.e., asymmetric ion pairs (unlike Vollmert and Schoene's study, which used a symmetric system, in that both polymers contained pendent carboxylates). The study showed that miscibility enhancement does take place, because the upper T_g (of the hard phase that had mixed with the polystyrene) depends on composition. Thus the 100 wt % P(S-9.6-MANa) has a T_g of ~150°C (i.e., the matrix T_g of the ionomer); upon mixing with the urethane, the T_g drops continuously to 35°C (i.e., the T_g of ionic polyurethane). The continuity of the T_g shows that these two systems are miscible, because of ion pair–ion pair interactions. In addition to this glass transition, one also sees a glass transition at -40°C owing to the soft segments of the polyurethane. This T_g is independent of composition.

In a later study, Zhang and Eisenberg (63) investigated blends of P(S-co-SSNa) and P(EA-co-4VPMeI) and compared the results with ion–ion interactions in the same parent polymer. A typical comparative plot is shown in Figure 9.7 Curve A, which is the loss tangent for P(S-co-SS$^-$)–P(EA-co-4VPMe$^+$), i.e., the ion–ion blend at ~7 mol % of functional groups, shows clear evidence of strong miscibility, in that only single loss tangent peak is present (at ~80°C). Curve B shows P(S-co-SSNa)–P(EA-co-4VPMeI) (~7 mol % of ions), in which the counterions have been left in the blend. This curve shows clear evidence of two-phase behavior, because two peaks are seen. As a matter of fact, these ion pair–ion pair interactions are not strong enough to enhance miscibility even in a 7 mol % blend. This is illustrated in Figure 9.8, which shows only a weak dependence of T_g on composition, because both T_g peaks retain their identity and do not move toward each other, eventually coalescing to give one peak at an intermediate temperature.

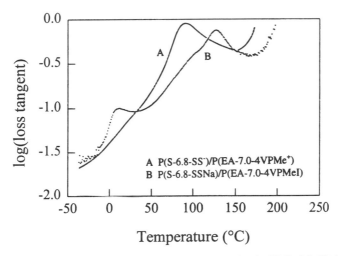

Figure 9.7. Log tan δ versus temperature for the ion–ion blend of P(S-6.8-SS$^-$)–P(EA-7.0-4VPMe$^+$) (*curve A*) and the ion pair–ion pair blend of P(S-6.8-SSNa)–P(EA-7.0-4VPMeI) (*curve B*). Modified from Zhang and Eisenberg (63).

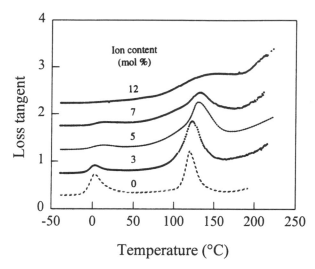

Figure 9.8. Tan δ versus temperature for ion pair–ion pair blends of P(S-*co*-SSNa)–P(EA-*co*-4VPMeI) of various functional group contents. Modified from Zhang and Eisenberg (63).

9.7. ION–DIPOLE BLENDS

The strong interaction between ions and polar groups was recognized long ago. One demonstration of this is the solubility of a range of salts in ethers. In an early application of this concept to polymers, it was shown that lithium perchlorate, $LiClO_4$ can be mixed with poly(ethylene oxide) (PEO) or poly(propylene oxide) (PPrO) over a wide concentration range and that an increase in the T_g is a result of the mixing (68,69). The miscibility was attributed to the strong interaction between the Li^+ ions and lone electron pairs of oxygen atom in the polyether.

The first example of the use of these interactions in miscibility enhancement was provided by Hara and Eisenberg (70), who showed that a styrene ionomer could be mixed with PEO or PPrO, whereas homopolystyrene is not miscible with these polymers. Again, miscibility was ascribed to the strong interactions between the cations and the oxygen atoms of the poly(alkylene oxide), which converts the neutral material into a polycationic chain. Miscibility enhancement was suggested by the presence of a composition-dependent T_g and by the transparency of the samples. It was also shown that an increase in the ion content of the ionomer improves miscibility.

In a subsequent study (71), cloud point curves were obtained for blends of poly(alkylene oxide) with polystyrene ionomers of different ion contents and with polystyrene ionomers of fixed ion content but with different cations. It was shown that the higher the ion content, the higher the temperature at which phase separation occurred, which is typical of mixed systems subject to specific interactions, which exhibit lower critical solution temperature (LCST) behavior. Furthermore, it was shown that the lithium ionomer requires the highest temperature for phase separation, with sodium and potassium giving progressively lower temperatures, as expected from the relative strengths of the interactions. A recent infrared study of the P(S-*co*-MANa)–PEO blends revealed the presence of specific interactions by a shift of the carboxylate ion and ether group stretching bands (72). In the P(S-*co*-MAA)–PEO system, hydrogen bonding can, of course, also be a strong driving force for miscibility, because the carboxylic acid can provide a strong bond between the styrene carboxylic acid copolymer and the alkylene oxide chain (15,71).

It should be noted that not all dipoles are equally effective in promoting miscibility. Thus in a study of blends of polar homopolymers with P(S-*co*-MALi) copolymers, it was found from peak shifts in loss tangent studies that poly(vinyl acetate) and poly(dimethylsiloxane) show no miscibility enhancement; polycaprolactone, poly(ethylene terephthalate), polyepichlorohydrin, poly(ethylene succinate) and polysulfide show some miscibility enhancement; and poly(vinyl chloride) and polyethyleneimine show favorable interactions leading to single peak behavior, suggesting homogeneity on the 50–100 Å level (48,73). For poly(vinyl acetate) and poly(dimethylsiloxane), the size of the substituent groups was invoked to explain the absence of miscibility. However, the low partial charge on the oxygen atoms in the poly(dimethylsiloxane) may also be of importance.

A well-explored family of blends is based on styrene copolymers and the polyamides. A number of investigations have been performed that used different mechanisms of miscibility enhancement. In an early series of investigations, hydrogen bonding between the PA and the acid containing styrene copolymer was studied (74–77). More recently, ionic interactions were used, both of the ion–dipole and the ion–coordination type. Although the division between ion–dipole and ion–coordination interactions is somewhat arbitrary in this particular case, it will, nonetheless, be made to keep the presentation within the framework of the four different types of interaction: ion–ion, ion pair–ion pair, ion–dipole, and ion–coordination. Thus part of the discussion of the styrene–PA and ethylene–PA blends will be presented in this section, if ions such as Li^+ are involved; and some of the studies will be described in the next section, primarily for cases involving Zn^{2+} cations.

An early investigation of hydrogen bonding–assisted miscibility enhancement was that of MacKnight et al. (74), involving polyamide-6 (PA-6) and ethylene-*co*-methacrylic acid copolymers. They found that the methacrylic acid content of the copolymer exerts a strong effect on the properties of the blends, especially in regard to the sizes of the domains of the dispersed ethylene-*co*-methacrylic acid phase in the PA matrix, which decrease as the acid content increases.

Fairley and Prud'homme (75) studied the mechanical properties of blends of polyethylene with PA-6 using poly(ethylene-*co*-methacrylic acid) (4–15 mol % of acid) as a compatibilizer. They found that changes in morphology were encountered with variation of the compatibilizer content. A subsequent investigation by Willis and Favis (76) involved poly(ethylene-*co*-methacrylate-*co*-isobutyl acrylate) terpolymer as a compatibilizer between polyolefins and polyamides; again effects on the particle size were found.

More recently, Kuphal et al. (77) observed phase separation in blends of styrene-*co*-acrylic acid copolymer (of 8 mol % of acid groups) with some of the aliphatic, partly crystalline polyamides, e.g., PA-6, PA-11, and PA-12. For blends of PA-6,6 and PA-6,9, the polyamides were partially miscible with styrene containing 8 mol % acrylic acid; at 20 mol % of acid groups, they were completely miscible owing to hydrogen bonding.

In more recent studies, it was shown that the sulfonic acid group, which is more acidic than the carboxylic acid group, has a much stronger effect on miscibility enhancement. Thus Molnár and Eisenberg (78) found that, if polystyrene functionalized with 2.2 mol % sulfonic acid groups was mixed with PA-6, there was strong miscibility enhancement, as seen by the dramatic decrease in particle size when the sulfonated polystyrene (instead of homopolystyrene) was mixed with the PA-6. Figure 9.9 shows scanning electron micrographs of the fracture surface of PA-6 blends with homopolystyrene and P(S-co-SSA). A cautionary comment is in order for systems involving sulfonic acid in the presence of PA. Sulfonic acid is a catalyst for the hydrolysis of PA. Therefore, if the mixed systems in the acid form are brought to a high temperature, chain scission can occur (79).

It should be mentioned that the PA-6 blends with sulfonated polystyrene are complicated systems. The degree of crystallinity as well as the sizes of phase-separated regions can change from system to system, and the extent of the clustering can change depending on both the pretreatment of the samples and the ion content, as well as the wt% of ionomer in the blends. Therefore, in this treatment, we confine our discussion only to miscibility enhancement and refer the reader to the original literature for detailed discussions of the various effects, many of which are quite important.

A partial phase diagram for the P(S-co-SSA)–PA system in *m*-cresol solution is

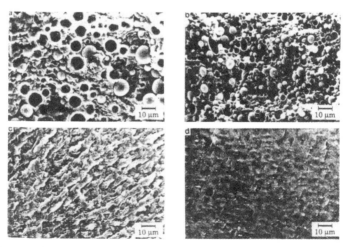

Figure 9.9. Scanning electron micrographs of fracture surfaces of PA-6 blends containing (**a**) 20 wt % PS, (**b**) 40 wt % PS, (**c**) 20 wt % P(S-2.2-SSA), and (**d**) 40 wt % P(S-5.4-SSA). Reprinted with permission from Molnár and Eisenberg (78).

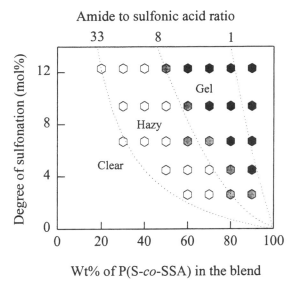

Figure 9.10. Partial phase diagram for 2% (w/v) solution blends of P(S-*co*-SSA) of various ion contents with PA-6 in *m*-cresol. *Dotted lines*, constant amidesulfonic acid ratios. Modified from Molnár and Eisenberg (79).

given in Figure 9.10 (79), which shows that, with an increase in the sulfonic acid content at a constant degree of sulfonation, the solutions go progressively from clear to hazy to gel-like. Eventually, of course, the solution must become clear again at a high relative percentage of P(S-*co*-SSA), because pure P(S-*co*-SSA) is soluble in *m*-cresol at that sulfonic acid content.

Another way of demonstrating the interaction is by studying the torque generated as a function of time observed when P(S-*co*-SSA) is added to PA-6 in a mixer (Fig. 9.11). The initial rapid increase is owing to the drop in temperature when the second component is added to the mixture. Thermal equilibrium is attained after approximately 2 min, and the rise in torque beyond that point is a function of the formation of strong interactions. NMR and other techniques have also been used to detect strong interactions between the P(S-*co*-SSA) and the PA (79).

A detailed infrared study of interactions between PA-6 and sulfonated polystyrene was also performed (80), and specific interactions were demonstrated. The study of the glass transition by DSC also revealed miscibility enhancement (80).

Miscibility enhancement between polystyrene ionomers and PA involves ion–dipole interactions. In one study, Molnár and Eisenberg (81) showed that the phenomena were quite complex, because various processes occur simultaneously: a decrease in crystallinity in PA as miscibility is increased, a decrease in the extent of clustering of the ionic regions as more ionic groups interact with PA-6, and an increase in the extent of formation of the miscible phase. The effect of various cations must also be considered; Li$^+$ has been shown to be much more effective in

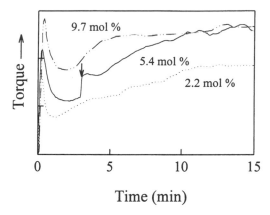

Figure 9.11. Torque intensity versus mixing time for a 40/60 blend of P(S-*co*-SSA) with PA-6. The P(S-*co*-SSA) copolymer was added to the pretreated PA-6 in the mixer. Numbers indicate sulfonic acid content; the *arrow* indicates the point at which a second batch of P(S-*co*-SSA) was added to bring the final concentration of the sulfonic acid to 40%. Modified from Molnár and Eisenberg (79).

the promotion of ion–dipole interactions between poly(styrene-*co*-sulfonate) and PA-6 than is sodium. Thus in the study of blends of 5.4 and 9.8 mol % P(S-*co*-SSLi) with PA-6, it was shown that, in the 50:50 blends, the 9.8 mol % Li$^+$ salt yields a single loss tangent peak, whereas the 9.8 mol % sodium salt shows two loss tangent peaks, indicating incomplete mixing. These phenomena are illustrated in Figure 9.12. In that study, a simple theory was proposed that suggests that the 5.4 mol % sulfonated ionomer shows miscibility only when the volume percent of the ionomer in the blends exceeds 60%. In the 9.8 mol % ionomer, the miscible region extends down to ~ 20%, and the 12.0 mol % ionomer should be completely miscible. By the same token, pure styrene should be completely immiscible. The details of the theory are beyond the scope of this presentation, and the reader is referred to the original literature (81).

In a subsequent NMR study of the same system, Gao et al. (82) explored the sizes of the phase-separated regions by using T_1 NMR measurements to probe dimensions on the order of 200 Å, glass transition measurements for the 50–100 Å range, and $T_{1\rho}$ to explore length scales on the order of 20 Å. The results of the study are shown in Figure 9.13. P(S-*co*-SSLi) containing 5.4 and 9.8 mol % ionic groups were mixed with PA-6, as were pure polystyrene and P(S-*co*-SSNa) as models of immiscible systems. Thus, in the lithium sulfonate case, it is shown that in the system containing 70 wt % of ionomer miscibility is present at the 20 Å level. In the 9.8 mol % Li$^+$ salt case, at 30 and 50 wt %, miscibility is seen in the 50 and 200 Å range. In the 5 mol % lithium sulfonate case, even at 30 and 50 wt %, miscibility is seen only at the 200 Å level. Naturally, in the sodium sulfonate and pure styrene homopolymer case, no miscibility is seen on any of the length scales probed by these techniques.

A follow-up study explored the correlation between miscibility (on the size scale accessible by dynamic mechanical measurements) and the degree of sulfonation of the P(S-*co*-SSLi) as well as the concentrations of amide groups in the PA (83). It is clear that both factors influence miscibility (Fig. 9.14). For example, for the 9.8 mol % lithium sulfonate ionomer, miscibility is encountered only at concentrations higher than ~20 mol % of amide groups (PA-6,10), whereas at 5 mol % sulfonation, miscibility does not occur below concentrations of ~35 mol % of amide groups (PA-4,6), as noted earlier. It is also possible to correlate the concentration of amide groups with the mole percent of lithium sulfonate in the polymer. Thus at the 9.8 mol % level, miscibility is found at ~20 mol % amide groups, which corresponds to a composition somewhat between PA-11 and PA-6,10. On the other hand, for the 5 mol % sulfonated system, miscibility is not seen until the mole percent of amide groups exceeds 30%, i.e., higher than PA-6.

A recent rheological study by Yoshikawa et al. (84) confirmed that the blends of P(S-9.8-SSLi)–PA-6 behave as a homogeneous elastic fluid over most compositions. In contrast, the blends with the sodium salt show behavior typical of an immiscible system.

Ng et al. (85) performed an extensive study of compatibilization of nylon 4 (PA-4) with P(S-*co*-SSLi). The degree of sulfonation was 9.7 mol %. Miscibility was observed over the entire range of compositions, which is not surprising in view of both the high degree of sulfonation and the high amide content. They also found that annealing of the samples increases the T_g, because of an increase in both the number and the size of crystallites.

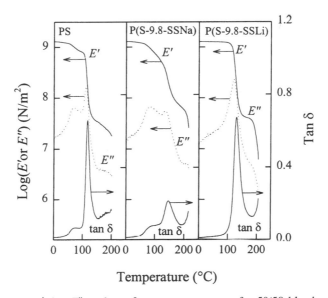

Figure 9.12. Log E', log E'', and tan δ versus temperature for 50/50 blends of PA-6 with (*left*) PS, (*middle*) with P(S-9.8-SSNa), and (*right*) with P(S-9.8-SSLi). Modified from Molnár and Eisenberg (81).

More recently, a study was made of the miscibility enhancement mechanism in sulfonated styrene–PA blends through a detailed infrared spectroscopic analysis (86). It was shown that the initial step is the interaction of the amide carbonyl with the Li^+ ion, accompanied by hydrogen bonding between the amide NH and the sulfonate oxygen, eventually leading to the formation of a Li^+ ion coordinated with approximately four carbonyl groups, while the sulfonates convert to the acid form by hydrogen bonding to the amide NH groups. The Na^+ ion has been shown to be ineffective, because its interaction is much weaker than that of the Li^+ ion; and it is unable to interact strongly enough with the amide group to leave its sulfonate environment.

Two brief studies were devoted to the exploration of ion–dipole interactions in the miscibility enhancement of polyurethanes with polystyrene (67,87). In the first (67), a polyoxytetramethylene diol of an average molecular weight 930 was incorporated into the polyurethane, which was subsequently mixed with styrene-*co*-Li methacrylate copolymers containing 10.4 mol % of ionic groups. From the invariance of the T_g with composition, it was deduced that miscibility was not achieved in that

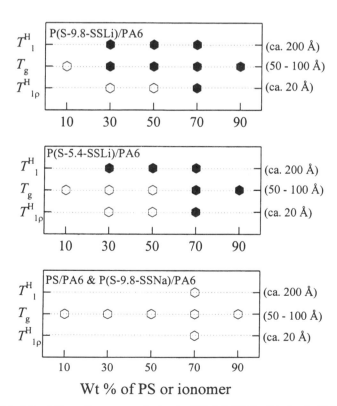

Figure 9.13. Miscibility diagrams of (*top*) P(S-9.8-SSLi)–PA-6, (*middle*) P(S-5.4-SSLi)–PA-6, and (*bottom*) with P(S-9.8-SSNa)–PA-6, and PS–PA-6. *Solid symbols*, one-phase systems. Modified from Gao et al. (82).

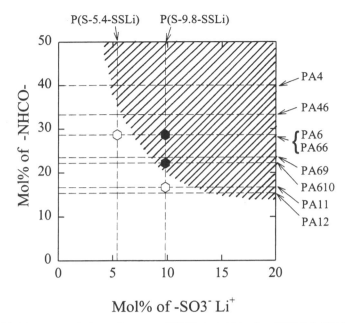

Figure 9.14. Fragment of phase diagram of 50/50 (w/w) blends of P(S-9.8-SSLi) with poly-amines. *Shaded area*, expected one-phase behavior of the blends. Modified from Molnár and Eisenberg (83).

system, probably because of the weak interactions between the oxytetramethylene diol segments and the ionomers. By contrast, when the polyether segment consisted of PPrO of molecular weight 1000 or 2000, miscibility with the P(S-10.5-MALi) was achieved (87). The difference in ion content is not believed to be important. However, the difference in the nature of the polar chain is probably highly significant and illustrates the importance of that aspect of ion–dipole interactions in miscibility enhancement.

In another application of ion–dipole interactions, Hara and Parker (88) mixed a sodium-sulfonated Kevlar ionomer [poly(*p*-phenylene terephthalamide propane sodium sulfonate)] with poly(4-vinylpyridine) and showed that the material was homogeneous. However, when the rigid chain molecule was mixed with the copolymer containing only 50 mol % of 4-vinylpyridine, strong evidence of phase separation was observed. In a subsequent study, it was shown that a considerable improvement in the mechanical properties can be achieved for molecular composites prepared in this manner (89).

9.8. ION-COORDINATION BLENDS

The formation of coordinate covalent bonds between transition metal ions and the nonbonded electron pairs on nitrogen atoms in amines or pyridines has been known for a long time. In polymers, the formation of complexes between poly(4-vinylpyridine) and transition metal ions was studied by Agnew (90). Lundberg and Makowski (91) showed that low molecular weight amines have strong interactions with the transition metal cations from metal sulfonate groups on polymers. In a subsequent study, Weiss (92) pointed out that zinc stearate is compatible with sulfonated ethylene-propylene-diene terpolymers (EPDM) in the acid form but not with the unfunctionalized material. These studies, along with the early studies of miscibility enhancement using simple ion–ion interactions (47), form the background for the use of metal coordination interactions in miscibility enhancement.

The first study involving the use of metal–coordination interactions in polymer miscibility enhancement was that of Peiffer et al. (93), who mixed zinc EPDM sulfonate containing approximately 0.65 mol % of functional groups with a P(S-co-4VP) copolymer containing approximately 8.5 mol % of 4VP. The melt viscosity was investigated as a function of blend composition. The authors found that 20 wt % of the P(S-co-4VP) copolymer (corresponding to approximately 1/1 ratio of sulfonate to 4VP) yielded the material with the highest melt viscosity, suggesting that the interactions are optimal at that point for the cations studied, i.e., Zn^{2+}, Cu^{2+}, and Co^{2+}. Zinc proved the most effective in giving polymers with useful properties, because both copper and cobalt salts yield much more strongly interacting complexes than do the other cations. As a matter of fact, in the case of cobalt, it was impossible to obtain homogeneous materials because of the strength of the interaction and, presumably, the resulting slow kinetics of mixing. Sodium and magnesium salts were found to be ineffective in promoting miscibility.

In a subsequent publication (94), the blend was compared with a mixture of homopolystyrene with zinc EPDM sulfonate. It was shown that the modulus of the zinc EPDM sulfonate with P(S-co-4VP) is much higher than that of the mixture with homopolystyrene and that the second loss tangent peak, owing to the glass transition of the styrene-rich phase, becomes a shoulder. The authors did not suggest that the mixtures were homogeneous but did propose strong miscibility enhancement. The sizes of phase-separated regions, however, are much smaller than the wavelength of light, because zinc EPDM sulfonate samples were shown to be quite transparent. An electron microscopy investigation confirmed the strong miscibility enhancement of the zinc-coordination blend.

The 1/1 stoichiometry is noteworthy. Although the authors did not point this out specifically, it is clear that the smallest electroneutral species must consist of one Zn^{2+} ion with two sulfonate anions. The 1/1 stoichiometry of functional groups also suggests that two pyridine groups are associated with this zinc-sulfonated species. Thus the coordination of the Zn^{2+} ion with four groups (i.e., two sulfonates and two vinylpyridines) is anticipated, and this was indeed found to be the case in the work of Sakurai et al. (95).

Functionally terminated chains can also be used to enhance miscibility by means of coordinate covalent interactions. Register et al. (40) used copper-neutralized carboxy-terminated polybutadiene mixed with P(S-co-4VP) copolymers. DSC results suggest that substantial mixing occurs in the copper–butadiene-rich phase but that complete miscibility is not achieved, because the P(S-co-4VP)-rich phase remains essentially unaffected. Electron microscopy shows an irregular bicontinuous morphology with ~100-nm-sized phase-separated regions. A SAXS study suggests that the ion aggregates present in the copper–butadiene ionomer are disrupted by blending, which is reasonable in view of the discussion in Section 9.3.

A recent study of coordination of metal ions with pyridines involved the addition of a material such as zinc acetate to polyurethanes containing pyridine groups (96). The results suggest that the addition of zinc acetate leads to the formation of cross-links, which can be explained only on the basis of coordination to the Zn^{2+} ion of at least two pyridine groups from different polymer chains.

An extensive study of ethylene-metal methacrylate copolymers mixed with P(S-co-4VP) copolymers was conducted by Peiffer (97), who showed that metal ions such as Na^+ only enhance miscibility weakly, in contrast to transition metal ions such as Zn^{2+}, which are strongly interacting. It was found in that study that the spherulite sizes remain essentially unchanged in nonassociating blends but are reduced to ~50% their original size in the associating blends. In nonassociating blends, the styrene (glassy) phase resides primarily in the interspherulite region. The tensile strength of the blends is substantially enhanced if there is miscibility enhancement, i.e., in the zinc blends, but not the sodium salt system. The nonassociating sodium blends are also characterized by poor interfacial adhesion at all compositions and clearly show failure in tensile properties such as tensile strength and elongation at break.

The ethyl acrylate–styrene blend system was also the subject of an extensive investigation in connection with ion–coordination blends as a miscibility enhancement mechanism. The materials involved were specifically EA-co-4VP copolymers mixed with the P(S-co-SSA) copolymers. The results of the study of these materials involving simple proton transfer blends were discussed earlier. Sakurai et al. (95) performed a detailed study of the bonding in this system involving coordinate covalent interactions between Zn^{2+} and vinylpyridine, with the Zn^{2+} originating in the P(S-co-SSZn). They suggested that, as was pointed out above, the Zn^{2+} is coordinated to two vinylpyridines with two benzenesulfonate groups providing electroneutrality. Thus four polymer chains can be thought to be cross-linked by this unit, which acts in a manner comparable to that of a four ion-pair multiplet, as was pointed out in Chapters 3 and 4.

A detailed dynamic mechanical study on the same system was published a year later by Douglas et al. (39). As mentioned in Section 3.3.5, a cluster peak was found. Because the experimental technique in that paper was dynamic mechanical thermal analysis in the shear mode, data above ~10^7 N/m^2 are not accessible. Therefore, similar samples were recently reinvestigated by our group. The results were discussed in connection with styrene homoblends (Chapter 5). It is significant that the dynamic mechanical properties of this blend system exhibit two loss tangent peaks. Even though the point was already made earlier, it is worth restating here that the two glass transition behavior results from complete reorganization of the multiplets to give the single cation ''multiplet.'' The concept of overlapping regions of restricted mobility explains the dynamic mechanical properties of this system, in exactly the same way as it does for materials containing normal multiplets.

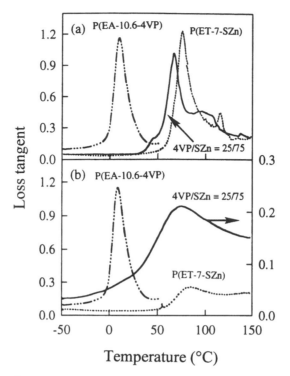

Figure 9.15. Tan δ versus temperature (1 Hz) for **(a)** quenched and **(b)** annealed samples of P(EA-10.6-4VP)–P(ET-7-SZn) (25/75 w/w) and its homopolymers. *Dotted and dashed line,* P(EA-10.6-4VP); *solid line,* 4VP/SZn; *dotted line,* P(ET-7-SZn). Modified from Ng et al. (98).

The P(EA-*co*-4VP) system was also mixed with zinc poly(ethylene terephthalate) sulfonate in a subsequent investigation by Ng et al. (98). A detailed FTIR analysis was performed on the blends. This system is more complicated than the styrene–ethyl acrylate system, because of the crystallinity of the poly(ethylene terephthalate) (PET). Thus the effect of annealing is much more important than in the noncrystallizable blends. Considerable miscibility enhancement was detected, but the system is not homogeneous. Loss tangent curves are shown in Figure 9.15 for quenched and annealed materials. It was shown that, during annealing, the tan δ peak position of the zinc sulfonate–PET-rich phase moves to higher temperature as a result of crystallization, in contrast to the behavior of the P(EA-*co*-4VP)-rich phase. This behavior is explained by phase reinforcement by the crystallization of the PET-rich phase and the increased phase purity of the P(EA-*co*-4VP)-rich phase.

A follow-up publication by Ng and MacKnight (99) was devoted to the crystalline phase of the PET in the blends. Most interesting, an increase in the rate of crystallization was observed, which was ascribed to enhanced nucleation efficiency originating from the specifically interacting sites. The melting behavior, studied by DSC, showed perturbation of crystal growth as judged from the broadening of the melting endotherms, lowering of the melting points, and overall reduction in the degree of bulk crystallinity. All were attributed to the presence of specific interactions.

Coordinate covalent interactions in miscibility enhancement were also explored in PA–polystyrene blends. In the first such study, Lu and Weiss (100) mixed PA-6 with P(S-10.1-SSMn) and found that the polymers were miscible over the entire composition range, as seen from the single composition-dependent T_g determined by DSC. The T_g was found to be higher than that calculated either from the Fox equation (101) or from the Couchman equation (102). It was suggested that hydrogen bonding, ion–dipole, and coordinate covalent interactions occur between the manganese sulfonate and the amide groups. The value of polymer–polymer miscibility parameter χ was calculated from melting point depression and was found to be large and negative, again reflecting strong interactions.

In a subsequent publication (41), the P(S-co-SSZn) ionomer was explored in an analogous manner. The formation of a complex between the metal sulfonate and the amide was invoked to explain the miscibility enhancement. The value of χ was determined to be -1.3, which again indicates strong intermolecular interactions. A SAXS study showed that the ionomer peak is progressively depressed with the addition of PA-6; at 80/20 P(S-10.1-SSZn)–PA-6, it disappears (Fig. 9.16).

An NMR study by Kwei et al. (103) of the same ion–coordination blend by the ^{13}C cross-polarization/magic angle spinning (CP/MAS) spectroscopic technique

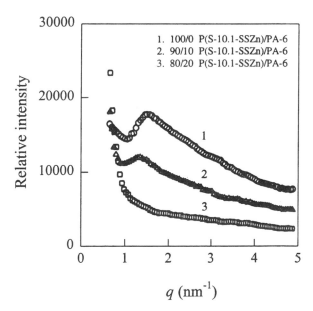

Figure 9.16. SAXS profile (plotted as scattering intensity versus q) for blends of P(S-10.1-SSZn) with PA-6. 1, 100/0 P(S-10.1-SSZn)–PA-6; 2, 90/10; 3, 80/20. Modified from Lu and Weiss (41).

revealed the existence of a complex between the zinc cation of the ionomer and the amide nitrogen atom. ^1H spin–lattice relaxation time (T_1^H and $T_{1\rho}^H$) measurements suggested that homogeneity persists down to a level of at least 20 Å.

In a recent study, Weiss and Lu (104) showed LCST behavior in a PA-6 mixture with sulfonated polystyrene neutralized with Li^+, Mg^{2+}, Zn^{2+} and Mn^{2+}. The value of the LCST depended on the ion content of the sulfonated polystyrene as well as on the type of ion (Fig. 9.17). The blends exhibited a much lower degree of hygroscopicity than did the parent polymers, which was attributed to the removal of hydrogen bonding sites for water by the formation of the complex, as had been suggested by Sakurai et al. (95) in an FTIR study of the blend of the P(S-co-SSZn) with P(EA-co-4VP).

In another study (105), it was shown that the phenylphenanthridine moiety in the PA backbone could act both as a proton acceptor and as a coordination site for zinc ions in the formation of either proton transfer blends or ion–coordination blends between the PA containing the phenylphenanthridine and the sulfonated polystyrene. The structure of the polyamides based on 3,8-diamino-6-phenylphenanthridine is shown in Scheme **9.1**.

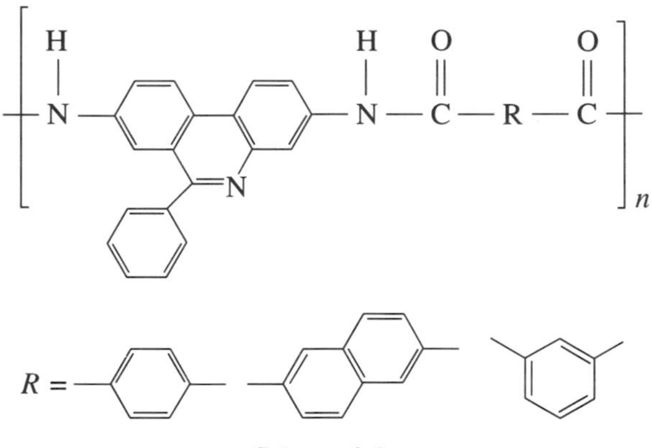

Scheme 9.1

Belfiore et al. (106) investigated transition metal coordination in poly(4-vinylpyridine) [P(4VP)] mixed with zinc acetate [$Zn(CH_3COO)_2$], laurate [$Zn(CH_3(CH_2)_{10}COO)_2$], stearate [$Zn(CH_3(CH_2)_{16}COO)_2$], and magnesium acetate. Their study describes mixtures of ionomers with small molecule amphiphiles and with polymers and was performed mainly by ^{13}C NMR spectroscopy, supported with glass transition and crystalline melting point studies of the zinc salts of the carboxylic acid. An example of the NMR results showing zinc stearate and zinc laurate mixed with P(4VP) in various proportions is shown in Figure 9.18. In the zinc stearate case, a peak at ~185 ppm characteristic of the pure material is seen. In the material containing 10 mol % of zinc stearate, the peak position is –181 ppm, indicating intermediate degrees of coordination to the pyridine ring. In contrast, the zinc laurate spectra for samples of intermediate composition (32–40 mol %) show two peaks at 181 and 185 ppm, indicating the presence of both coordinated zinc and crystalline zinc laurate.

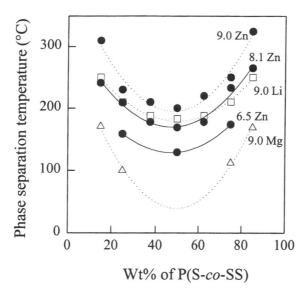

Figure 9.17. Liquid–liquid phase diagram in the LCST region for blends of P(S-*co*-SSZn) with PA-6 of the indicated sulfonate content. Modified from Weiss and Lu (104).

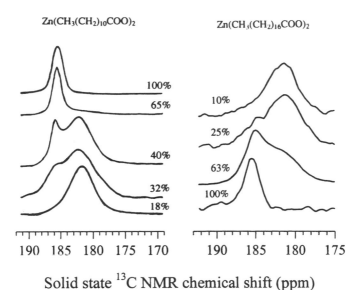

Figure 9.18. ^{13}C NMR spectra of the carboxylate chemical shift region for mixtures of (*left*) Zn (CH$_3$(CH$_2$)$_{10}$ COO)$_2$ and (*right*) Zn(CH$_3$(CH$_2$)$_{16}$COO)$_2$ with P(4VP). Numbers indicate the salt content. Modified from Belfiore et al. (106).

Infrared spectroscopy was used to prove that the lone electron pair of the pyridine nitrogen atom is involved in metal–ligand π bonding. This conclusion was deduced from a shift of the aromatic carbon–hydrogen stretching band to a higher frequency. There is a progressive drop in the T_g as the mole percent of zinc laureate increases, the glass transition reaching ~70°C at zinc laurate contents of ~25%. Beyond that point, the glass transition reaches a constant value and crystalline melting is observed, indicating phase separation of the zinc laurate. This study confirms that zinc does, indeed, bring together the carboxylate and vinylpyridine groups.

Stress–strain properties of the P(4VP)–zinc-neutralized ethylene ionomer mixtures are also improved when the nitrogen/zinc molar ratio is optimized (106). The synergistic mechanical properties of the blends of a partly zinc-neutralized (60%) ethylene-co-methacrylic acid copolymer (15 wt % of acid) [Surlyn 1706 (DuPont)] with P(4VP) were also investigated as a function of the 4VP concentration. It was found that the elastic modulus increases as stoichiometry is approached, reaching a maximum at ~85 mol % of nitrogen in P(4VP) to Zn^{2+} in the ethylene ionomer blends. In contrast to the behavior of the zinc salt, the magnesium salt does not show any coordination interaction, because the Mg^{2+} ion, which is identified as a hard acid (107), coordinates preferentially to the acetate ions and the waters of hydration rather than to the pyridine. This supposition was confirmed by NMR spectroscopy, by which it was shown that the ^{13}C signal of magnesium acetate does not change in the presence of 4VP.

9.9. MISCELLANEOUS BLENDS

In addition to the topics discussed above, a number of blends involving ionic polymers have been explored that do not fit conveniently into any of the defined categories. They are the subject of this section, which examines two different families of materials: one involves polymers in which miscibility enhancement is achieved via end group functionalization and the other involves the formation of blends between homopolymers and ionomers, under conditions in which ion–dipole interactions are not as dominant (Section 9.7).

9.9.1. Telechelic Blends

As a part of investigation of halatotelechelic polymers (Fig. 2.2c), Jérôme and Teyssié's group performed a number of studies of the miscibility of polymers containing functionalized end groups. In a typical case, a polymer functionalized at one (ω-) or both ends (α,ω-) with species such as a carboxylic acid or sulfonic acid is mixed with another polymer with which the first polymer is not normally miscible. However, miscibility can be enhanced if the second polymer is functionalized with an amino group or another proton acceptor in the terminal position. Both monofunctionally and difunctionally terminated polymers can be used.

In the first study of this type (108), polystyrene and polybutadiene, which are otherwise completely immiscible, were prepared in such a way that the polystyrene was end capped with carboxylic acids (T_g = 104°C) and the polybutadiene was end capped with dimethylamine groups (T_g = −76°C). A carboxy-terminated polystyrene mixed with carboxy-terminated polybutadiene was taken as the baseline system, because no miscibility enhancement was expected in that material. The molecular weights of the functional homopolymers were in the range of 4,000–6,000. Even in the interacting blends, two T_g values are observed—those of each homopolymer (at −78 and 62°C)—which, however, are perturbed because of the presence of the interactions. In the case of freshly prepared materials that had not been heated before, but were cast from a solvent, a third T_g was found, which disappeared after annealing. The intermediate (third) glass transition, occurring at 22°C, was attributed to an incompletely phase-separated material. Proton transfer was observed by an FTIR study for the carboxylic acid–dimethylamine blends, in contrast to the behavior of carboxylic acids in the presence of pyridine. This end functionalization leads to materials that closely resemble block copolymers.

In a subsequent study, Russell et al. (109) investigated the morphology of solution-cast mixtures of telechelic polymers in which, again, one was end capped with a tertiary amine, and the other with either a sulfonate or a carboxylate group. The sulfonate–*tert*-amine association was found to be more stable at elevated temperatures than its carboxylate analog. Most interesting, it was found that these materials show marked similarity to block copolymers in that they undergo a classical order–disorder transition at elevated temperatures. However, because of the presence of ionic interactions, the width of interface between the two phases remains sharp, even as the order–disorder transition region is approached. Furthermore, upper critical solution temperature (UCST) behavior was observed with increasing temperature.

These concepts were elaborated in a subsequent paper by Charlier et al. (110), who suggested that the lowest temperature of interest, i.e., the microphase separation temperature of the ionomers (MST), in some cases may be located even below the T_g of the mixture. Below the MST, the complementary functional groups not only form ion pairs, but these ion pairs are ordered in some way. Above the MST, ionic bridging is still present, but the ordered arrangement of the ionic groups relative to each other has disappeared. At a still higher temperature, called T_i, the authors suggest that the ionic interactions themselves are disrupted and that the functional groups of opposite polarity dissociate. A representative scheme is shown in Figure 9.19. The T_i is accompanied by phase separation of the material. The position of T_i can vary, depending on both the blend composition and the nature of the acid group (e.g., it is much higher in the sulfonate case than in the carboxylate case).

As might be expected, if the chain lengths of the functionally terminated polymers are short, the phase-separated regions are small, or the polymers are miscible, and only a single glass transition is observed; for higher molecular weight systems, two glass transitions can be seen. The similarity to block copolymers is seen from morphological studies, which suggest that under some circumstances, spherical, cylindrical, and lamellar morphologies are present, depending on the relative molecular weights of the components (109). The molecular weight (MW) dependence naturally also manifests itself in the strength of the interactions. In a mixture of polystyrene (MW = 30,000) with α,ω-dicarboxy-terminated polybutadiene (MW = 4,600),

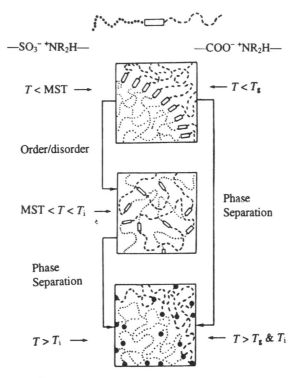

Figure 9.19. A model of phase behavior for blends of one polymer of sulfonate- or carboxylate-terminated telechelics with the other polymer of dialkyl ammonium–terminated chains. The chains are immiscible in the absence of ionic interactions. Modified from Charlier et al. (110).

large phase-separated regions were observed. However, much smaller regions were seen in α,ω-dicarboxy-terminated polystyrene (MW = 30,000) mixed with α,ω-dipiperazine-terminated polybutadiene (MW = 4,000) (107); the structure of piperazine group is shown in Scheme **9.2**.

$$— N \underset{}{\bigcirc} NH$$

Scheme 9.2

Still smaller regions were seen in α,ω-disulfonate–terminated polystyrene (MW = 30,000) mixed with the α,ω-dipiperazine-terminated polybutadiene (MW = 4,000). The molecular weight of the styrenic component also influences the size of phase-separated regions (111). Thus, when dimethylamino-terminated polyisoprene (MW = 20,000) is mixed with α,ω-dicarboxy-terminated poly(α-methylstyrene) [P(α-MS)]—as long as the molecular weight of the α,ω-dicarboxy-terminated P(α-MS) is relatively low (5,000–10,000)—the regions of phase separation are quite small. However, when the molecular weight of the α,ω-dicarboxy-terminated P(α-MS) reaches 20,000, large phase-separated regions are observed. This result suggests that, at 20,000 MW, the ionic interactions are no longer sufficient to overcome the enthalpic and entropic driving force for phase separation of the chain segments.

9.9.2. Blends of Ionomers with Their Nonionic Parent Polymers

This section deals with blends of ionomers with their nonionic parent polymers. First, blends of ethylene based ionomers with ethylene are discussed. Then we turn to blends of perfluorinated (PTFE-based) ionomers with poly(vinylidene fluoride), which, while not the parent polymer of the perfluorinated ionomer, is at least partly fluorinated. Finally, blends of styrene ionomers with styrene are covered.

Horrion and Agarwal (112) investigated the mechanical and thermal properties of blends of low-density polyethylene (LDPE) with ethylene–acrylic acid copolymers and their zinc salts. The range of ion contents was 0.75–6.2 mol %, and the degrees of the neutralization were 0 or 25%. The authors found that for both the 4.6 and the 6.2 mol % samples in the salt form, the polarity difference between the ionomer and the homopolymer was sufficient to prevent mixing. Crystallinity was maintained; however, while the degree of crystallinity decreased, the crystalline melting point did not change. The behavior was found to be independent of the degree of neutralization. Thus it can be concluded that the acrylic acid moieties are as efficient as zinc acrylate units in preventing crystallization. When the acrylic acid content was <5 mol %, partially compatible blends were formed. The physical properties of these partially compatible blends were found to be intermediate between those of the component polymers. Various cations were also investigated, but their effect on miscibility was not significant.

In a subsequent study, Cho et al. (113), investigated high-density polyethylene (HDPE) blends with ethylene ionomers. Again, immiscibility between the two components was observed. In a study of the tensile behavior, severe strain hardening was detected just above the yield point, which is correlated with a lower elongation at break and higher tensile strength compared to the homopolymer. This behavior was attributed to a network-like structure resulting from the presence of ionic aggregates. Furthermore the tensile properties showed severe negative deviation from linear additivity, which is another characteristic of incompatible blends. An analysis of the tear fracture surfaces of the blends suggested the presence of fibrillar structures for relatively low ionomer contents. However, at higher ionomer contents, less fibrillation on the fracture surface was seen, which again was believed to be related to the network-like structure of the ionomer.

Because EPDM is essentially a hydrocarbon-based polymer similar to polyethylene, except for the presence of the occasional methyl group and the diene unit, it is not surprising that EPDM and polyethylene, in which one of these components is ionic, have been mixed. One study involved mixtures of EPDM and ethylene ionomers (114). Most interesting, it was found that the zinc-neutralized ionomer showed better miscibility with EPDM than did the corresponding sodium-neutralized ionomer. Some similarity to thermoplastic interpenetrating polymer networks was claimed. The inverted situation was also investigated, i.e., the HDPE mixed with zinc EPDM sulfonate (115). At a 20 wt % content of the zinc EPDM sulfonate, the blend also showed some resemblance to an interpenetrating polymer network. However, when the zinc EPDM sulfonate content exceeded 20 wt %, the dispersed nature of the blend became clear.

The study of mixtures of perfluorosulfonated ionomers with poly(vinylidene fluoride) (PVDF) [-(CH$_2$CF$_2$)$_n$-] showed a small melting point depression in the DSC thermogram and considerable change in the dielectric α relaxation peak with blend composition (116). These results indicate that the Nafion (DuPont) PVDF blend may be partially miscible. However, above the crystalline melting temperature of the PVDF, the mixture becomes unstable and undergoes a thermally induced phase separation.

Blends of styrene and styrene-based ionomers have been the subject of extensive investigation for a number of years. Iwakura and Fujimura (117) reported the melt rheology of blends of styrene and styrene-co-methacrylic acid and its metal salt. More recently, Hara et al. (118,119) performed a series of investigations on styrene–sulfonated polystyrene blends. In the first of these studies (118), the morphology and deformation behavior were investigated. It was found that in blends of low ionomer concentration, phase separation of the ionomer was observed in the form of small dispersed particles. As the ion content increases, the average particle size increases; and because crazes develop in the matrix, the particles elongate and fibrillate. Good adhesion was observed between the ionic cross-linked particles and the matrix. Applied stress was shared between the two components, and the reinforcing effect of the ionic cross-linked particle on the matrix polymer was observed. In a subsequent study (119), it was found that finely dispersed ionomer particles exhibit a greater effective entanglement density in the matrix as a result of ionic cross-linking and that they provide reinforcement against early craze breakdown and fracture. Both tensile strength and fracture energy increase as the ionomer concentration in the blend increases; above ~10 wt % of ionomer, they become essentially independent of blend composition.

The tensile fracture properties were investigated in a follow-up publication (120). It was found that, over the entire composition range, a positive deviation from the rule of mixtures was seen in both tensile strength and tensile toughness, with an especially significant increase at low ionomer compositions (~10 wt %). The reason for this behavior was attributed to the fine dispersion of the rigid ionomer in the polystyrene matrix, coupled with good interfacial adhesion between the phases. An increase in ion content at constant blend ratio resulted in an increase in both the tensile strength and toughness of these two-phase blends. Because the ionomers are more fracture resistant than the parent polymer, the increase in ionomer content leads to an increase in fracture resistance of the blend. Some improvement in the mechanical properties was also found, because of the stress concentration effect of the ionomer.

In a recent study, Beck Tan et al. (121) investigated, by small-angle neutron scattering, the physical state of a blend of (deuterated) polystyrene with sulfonated (hydrogenated) polystyrene (1.67 mol % sulfonation) over a wide composition range. They found that the blends are phase separated and the polymer system is immiscible over the entire composition range from 5 to 95 wt % of deuterated polystyrene. The concentration of ionic groups in the ionomer is the lowest yet reported that causes immiscibility with the parent polymer.

9.9.3. Styrene Ionomer–Poly(Phenylene Oxide) Blends

A most illuminating series of studies involved the miscible pair of polystyrene and poly(phenylene oxide). In a study of miscibility of variously functionalized polystyrene and poly(2,6-dimethyl-1,4-phenylene oxide) (PPhO), it was shown that, in contrast to the miscible homopolymers, mixtures of sulfonated polystyrene with unfunctionalized PPhO or sulfonated PPhO with unfunctionalized polystyrene became immiscible if as little as 2–4 mol % of one of the components was functionalized (122). If both components were functionalized (sulfonated), then the miscibility of the two components was retained over a much broader range of functionalization (typically up to 10 mol %). In subsequent study (123), the blends of unfunctionalized poly(2,6-diphenyl-1,4-phenylene oxide) (PDPhPhO) and PS were found to be immiscible, which was attributed to the crystallinity of PDPhPhO. It was also found that in the case of sulfonated polystyrene mixed with sulfonated PDPhPhO, improved mechanical properties are retained, in spite of the phase separation, presumably because of the microphase separation of the ionic groups into aggregates containing ionic groups from both ionomers. This behavior is another example of the control of phase separation and mechanical properties provided by ionic interactions.

In a study of P(S-*co*-SSZn) mixed with PPhO homopolymer, Register and Bell (124) showed that up to a level of 7.7 mol % of sulfonation, the blends remain miscible as measured by thermal and mechanical criteria. It is noteworthy that miscibility is not accompanied by the destruction of the microdomain structure in the P(S-*co*-SSZn), even though it was observed that the PPhO coils are considerably larger than the spacing between ionic aggregates. A SAXS study showed that the average interaggregate spacing in the P(S-*co*-SSZn) is the same as in the 50/50 blend with PPhO. However, the extent of ionic phase separation is reduced on PPhO addition, suggesting that more ionic groups are dispersed in the matrix in the blend than in the ionic copolymer by itself.

A subsequent study (125), in which the miscibility was investigated for different ratios of ionomer to PPhO, showed that the ionomer can absorb the PPhO only up to a certain level, beyond which regions of pure phenylene oxide form. The saturation level decreases as the level of functionalization of the ionomer increases, but the type of ion is also important, with P(S-*co*-SSZn) showing the highest level of miscibility of all the ionomers that were investigated.

A detailed study of blends of PPhO and P(S-*co*-MANa) with different ion contents was performed by Bazuin et al. (126), who found that PPhO and P(S-*co*-MAA) were no longer miscible above 10 mol % of acid groups as a result of a copolymer effect, which reduces the interaction between PS and PPhO, because of the presence of the acid groups, which do not interact with PPhO. It was shown that the ionomer blends with PPhO are immiscible even at low ion contents (i.e., 2.4 mol %) and become increasingly so as the ion content is increased. The decrease in miscibility of polystyrene ionomer–PPhO blend relative to the homopolymer blend is attributed to a combination of factors: one is the increasing importance of the ionomer cluster phase, which excludes the homopolymer chains, and the other is the copolymer effect between the homopolymer and the unclustered part of the ionomer. Ion pairs that are not incorporated into multiplets decrease the favorable interactions between the phenylene oxide and styrene; this is encountered even more strongly for isolated multiplets. Thus the greater the degree of clustering, the lower the miscibility; By contrast, the lower the degree of incorporation of ion pairs into multiplets, the lower the perturbation of miscibility. The differences in miscibility between the various styrene ionomers and PPhO can be understood within that framework. It was also suggested that the sample preparation method and thermal history may contribute to the contrasting behavior.

9.10. Concluding Remarks

As the discussion has shown, a wide range of electrostatic interactions can be used to enhance miscibility in polymer blends, including ion–ion interactions (with and without proton transfer), ion pair–ion pair interactions, and ion–dipole interactions. Ion–coordination is another important variant. By considering this arsenal of possible interactions, one can choose the type and quantity of functional groups, which are best suited to miscibility enhancement. The fact that a range of interaction strengths can be used suggests that a large degree of control over ionomer properties can be exercised.

9.11. REFERENCES

1. *Polymer Blends*; Paul, D. R.; Newman, S, Eds.; Academic: New York, 1978.

2. Olabisi, O.; Robeson, L. M.; Shaw, M. J. *Polymer-Polymer Miscibility*; Academic: New York, 1979.

3. Utracki, L. A. *Polymer Alloys and Blends: Thermodynamics and Rheology*; Hanser: Munich, 1989.

4. *Multiphase Polymers: Blends and Ionomers*; Utracki, L. A.; Weiss, R. A., Eds.; ACS Symposium Series 395; American Chemical Society: Washington, DC, 1989.

5. Coran, A. Y.; Patel, R., Williams-Headd *Rubber Chem. Technol.* **1985,** *58*, 1014–1023.

6. Thornton, B. A.; Villasenor, R. G.; Maxwell, B. *J. Appl. Polym. Sci.* **1980,** *25*, 653–663.

7. Yoshimura, N., Fujimoto, K. *Rubber Chem. Technol.* **1969,** *42*, 1009–1013.

8. Sperling, L. H.; Taylor, D. W.; Kirkpatrick, M. L.; George, H. F.; Bardman, D. R. *J. Appl. Polym. Sci.* **1970,** *14*, 73–78.

9. Smith, K. L.; Winslow, A. E.; Petersen, D. E. *Ind. Eng. Chem.* **1959,** *51*, 1361–1364.

10. Djadoun, S.; Goldberg, R. N.; Morawetz, H. *Macromolecules* **1977,** *10*, 1015–1020.

11. Robeson, L. M.; McGrath, J. E. *Polym. Eng. Sci.* **1977,** *17*, 300–304.

12. Weeks, N. E.; Karasz, F. E.; MacKnight, W. J. *J. Appl. Phys.* **1977,** *48*, 4068–4071.

13. Ting, S. P.; Bulkin, B. J.; Pearce, E. M.; Kwei, T. K. *J. Polym. Sci. Polym. Chem. Ed.* **1981,** *19*, 1451–1473.

14. Fahrenholtz, S. F.; Kwei, T. K. *Macromolecules* **1981**, *14*, 1076–1079.

15. Pearce, E. M.; Kwei, T. K.; Min, B. Y. *J. Macromol. Sci. Chem.* **1984**, *A21*, 1181–1216.

16. Painter, P. C.; Park, Y.; Coleman, M. M. *Macromolecules* **1988**, *21*, 6672.

17. Weiss, R. A.; Shao, L.; Lundberg, R. D. *Macromolecules* **1992**, *25*, 6370–6372.

18. Taylor-Smith, R. E.; Register, R. A. *Macromolecules* **1993**, *26*, 2802–2809.

19. Taylor-Smith, R. E.; Register, R. A. *J. Polym. Sci. B Polym. Phys.* **1994**, *32*, 2105–2114.

20. Michaels, A. S. *Ind. Eng. Chem.* **1968**, *57*, 32–40.

21. Pugh, C.; Percec, V. *Macromolecules* **1986**, *19*, 65–71.

22. Prud'homme, R. E. *Polym. Eng. Sci.* **1982**, *22*, 90–95.

23. Allard, D.; Prud'homme, R. E. *J. Appl. Polym. Sci.* **1982**, *27*, 559–568.

24. Belorgey, G.; Aubin, M.; Prud'homme, R. E. *Polymer* **1982**, *23*, 1051–1056.

25. Djordjevic, M. B.; Porter, R. S. *Polym. Eng. Sci.* **1983**, *23*, 650–657.

26. Sulzberg, T.; Cotter, R. J. *J. Polym. Sci. A1* **1970**, *8*, 2747–2758.

27. Ohno, N.; Kumanotani, J. *Polym. J.* **1979**, *11*, 947–954.

28. Abe, K.; Haibara, S.; Itoh, Y.; Senoh, S. *Makromol. Chem.* **1985**, *186*, 1505–1512.

29. Alexandra, S.; Natansohn, A. *Macromolecules* **1990**, *23*, 5127–5132.

30. Michaels, A. S.; Miekka, R. G. *J. Phys. Chem.* **1961**, *65*, 1765–1773.

31. Kabanov, V. A. *Pure Appl. Chem. Macromol. Chem.* **1973**, *8*, 121.

32. Lysaght, M. J. In *Polyelectrolytes;* Fisch, K. C.; Klempner, D.; Patsis, A. V., Eds.; Technomic: Westport, CT, 1975; pp. 34–42.

33. Tsuchida, E.; Abe, K. *Adv. Polym. Sci.* **1982**, *45*, 1.

34. Kabanov, V. A.; Zezin, A. B. *Makromol. Chem. Suppl.* **1984**, *6*, 259–276.

35. Li, Y.; Xia, J.; Dubin, P. L. *Macromolecules* **1994**, *27*, 7049–7055.

36. Li, Y.; Dubin, P. L.; Havel, H. A.; Edwards, S. L. *Macromolecules* **1995**, *28*, 3098–3102.

37. Cotton, F. A.; Wilkinson, G. *Advanced Inorganic Chemistry,* 5th ed.; Wiley: New York, 1988.

38. Huheey, J. E.; Keiter, E. A.; Keiter, R. L. *Inorganic Chemistry: Principles of Structure and Reactivity,* 4th ed.; HarperCollins: New York, 1993.

39. Douglas, E. P.; Waddon, A. J.; MacKnight, W. J. *Macromolecules* **1994**, *27*, 4344–4352.

40. Register, R. A.; Sen, A.; Weiss, R. A.; Li, C.; Cooper, S. *J. Polym. Sci. B Polym. Phys.* **1989**, *27*, 1911–1925.

41. Lu, X.; Weiss, R. A. *Macromolecules* **1992**, *25*, 6185–6187.

42. Eisenberg, A.; Hird, B.; Moore, R. B. *Macromolecules* **1990**, *23*, 4098–4107.

43. Otocka, E. P.; Eirich, F. R. *J. Polym. Sci. A2* **1968**, *6*, 921–932.

44. Smith, P. Ph.D. Dissertation, McGill University, 1985.

45. Smith, P.; Eisenberg, A. *Macromolecules* **1994**, *27*, 545–552.

46. Hird, B.; Eisenberg, A. *Macromolecules* **1992**, *25*, 6466–6474.

47. Smith, P.; Eisenberg, A. *J. Polym. Sci. Polym. Lett. Ed.* **1983**, *21*, 223–230.

48. Smith, P.; Hara, M.; Eisenberg, A. In *Current Topics in Polymer Science,*, Vol. 2; Ottenbrite, R. M.; Utracki, L. A.; Inoue, S., Eds.; Hanser: Munich, 1987; Chapter 6.5.

49. Bakeev, K. N.; MacKnight, W. J. *Macromolecules* **1991**, *24*, 4575–4582.

50. Bazuin, C. G.; Eisenberg, A. *J. Polym. Sci. B. Polym. Phys.* **1986**, *24*, 1021–1037.

51. Douglas, E. P.; Sakurai, K.; MacKnight, W. J. *Macromolecules* **1991**, *24*, 6776–6781.
52. Natansohn, A.; Eisenberg, A. *Macromolecules* **1987**, *20*, 323–329.
53. Zhou, Z. -L.; Eisenberg, A. *J. Polym. Sci. Polym. Phys. Ed.* **1983**, *21*, 595–603.
54. Murali, R.; Eisenberg, A. *J. Polym. Sci. B Polym. Phys.* **1988**, *26*, 1385–1396.
55. Rutkowska, M.; Eisenberg, A. *Macromolecules* **1984**, *17*, 821–824.
56. Rutkowska, M.; Eisenberg, A. *J. Appl. Polym. Sci.* **1984**, *29*, 755–762.
57. Rutkowska, M.; Eisenberg, A. *Polimery* **1984**, *8*, 313–317.
58. Rutkowska, M.; Jastrzebska, M.; Kim, J. -S.; Eisenberg, A. *J. Appl. Polym. Sci.* **1993**, *48*, 521–527.
59. Natansohn, A.; Rutkowska, M.; Eisenberg, A. *Polymer* **1987**, *28*, 885–888.
60. Tan, L. -S.; Arnold, F. E.; Chuah, H. H. *Polymer* **1991**, *32*, 1376–1379.
61. Eisenbach, C. D.; Hofmann, J.; MacKnight, W. J. *Macromolecules* **1994**, *27*, 3162–3165.
62. Zhang, X.; Eisenberg, A. *Polym. Adv. Technol.* **1990**, *1*, 9–18.
63. Zhang, X.; Eisenberg, A. *J. Polym. Sci. B Polym. Phys.* **1990**, *28*, 1841–1851.
64. Bossé, F.; Eisenberg, A. *Macromolecules* **1994**, *27*, 2846–2852.
65. Bossé, F.; Eisenberg, A. *Macromolecules* **1994**, *27*, 2853–2863.
66. Vollmert, B.; Schoene, W. In *Colloidal and Morphological Behavior of Block and Graft Copolymers*; Molau, G. E., Ed.; Plenum: New York, **1977**, pp. 145–157.
67. Rutkowska, M.; Eisenberg, A. *J. Appl. Polym. Sci.* **1985**, *30*, 3317–3223.
68. Lundberg, R. D.; Bailey, F. E.; Callard, R. W. *J. Polym. Sci. A1* **1966**, *4*, 1563–1577.
69. Moacanin, J.; Cuddihy, E. F. *J. Polym. Sci. C* **1966**, *14*, 313–322.
70. Hara, M.; Eisenberg, A. *Macromolecules* **1984**, *17*, 1335–1340.
71. Hara, M.; Eisenberg, A. *Macromolecules* **1987**, *20*, 2160–2164.
72. Lim, J. -C.; Park, J. -K.; Song, H. -Y. *J. Polym. Sci. B Polym. Phys.* **1994**, *32*, 29–35.
73. Eisenberg, A.; Hara, M. *Polym. Eng. Sci.* **1984**, *24*, 1306–1311.
74. MacKnight, W. J.; Lenz, R. W.; Musto, P. V.; Somani, R. J. *Polym. Eng. Sci.* **1985**, *25*, 1124–1134.
75. Fairley, G.; Prud'homme, R. E. *Polym. Eng. Sci.* **1987**, *27*, 1495–1502.
76. Willis, J. M.; Favis, B. D. *Polym. Eng. Sci.* **1988**, *28*, 1416–1425.
77. Kuphal, J. A.; Sperling, L. H.; Robeson, L. M. *J. Appl. Polym. Sci.* **1991**, *42*, 1525–1535.
78. Molnár, A.; Eisenberg, A. *Polym. Commun.* **1991**, *32*, 370–373.
79. Molnár, A.; Eisenberg, A. *Polym. Eng. Sci.* **1992**, *32*, 1665–1677.
80. Sullivan, M. J.; Weiss, R. A. *Polym. Eng. Sci.* **1992**, *32*, 517–523.
81. Molnár, A.; Eisenberg, A. *Macromolecules* **1992**, *25*, 5774–5782.
82. Gao, Z.; Molnár, A., Morin, F. G.; Eisenberg, A. *Macromolecules* **1992**, *25*, 6460–6465.
83. Molnár, A.; Eisenberg, A. *Polymer* **1993**, *34*, 1918–1924.
84. Yoshikawa, K.; Molnár, A.; Eisenberg, A. *Polym. Eng. Sci.* **1994**, *34*, 1056–1064.
85. Ng, C.-W. A.; Bellinger, M. A.; MacKnight, W. J. *Macromolecules* **1994**, *27*, 6942–6947.
86. Rajagopalan, P.; Kim, J. -S.; Brack, H. P.; Lu, X.; Eisenberg, A.; Weiss, R. A.; Risen, W. M. Jr. *J. Polym. Sci. B Polym. Phys.* **1995**, *33*, 495–503.
87. Rutkowska, M.; Eisenberg, A. *J. Appl. Polym. Sci.* **1987**, *33*, 2833–2838.

88. Hara, M.; Parker, G. J. *Polymer* **1992**, *33*, 4650–4552.

89. Parker, G.; Hara, M. *Polymer* **1997**, *38*, 2701–2709.

90. Agnew, N. H. *J. Polym. Sci. Polym. Chem. Ed.* **1976**, *14*, 2819–2830.

91. Lundberg, R. D.; Makowski, H. S. *J. Polym. Sci. Polym. Phys. Ed.* **1980**, *18*, 1821–1836.

92. Weiss, R. A. *J. Appl. Polym. Sci.* **1983**, *28*, 3321–3332.

93. Peiffer, D. G.; Duvdevani, I., Agarwal, P. K.; Lundberg, R. D. *J. Polym. Sci. Polym. Lett. Ed.* **1986,** *24*, 581–586.

94. Agarwal, P. K.; Duvdevani, I.; Peiffer, D. G.; Lundberg, R. D. *J. Polym. Sci. B Polym. Phys.* **1987,** *25*, 839–854.

95. Sakurai, K.; Douglas, E.; MacKnight, W. J. *Macromolecules* **1993,** *26*, 208–212.

96. Grady, B. P.; O'Connell, E. M.; Yang, C. Z.; Cooper, S. L. *J. Polym. Sci. B Polym. Phys.* **1994,** *32*, 2357–2366.

97. Peiffer, D. G. *J. Polym. Sci. B Polym. Phys.* **1992**, *30*, 1045–1053.

98. Ng, C.-W. A.; Lindway, M. J.; MacKnight, W. J. *Macromolecules* **1994,** *27*, 3027–3032.

99. Ng, C.-W. A.; MacKnight, W. J. *Macromolecules* **1994**, *27*, 3033–3038.

100. Lu, X.; Weiss, R. A. *Macromolecules* **1991**, *24*, 4381–4385.

101. Fox, T. G. *Bull. Am. Phys. Soc.* **1956,** *1*, 123.

102. Couchman, P. R. *Macromolecules* **1978,** *11*, 1156–1161.

103. Kwei, T. K.; Dai, Y. K.; Lu, X.; Weiss, R. A. *Macromolecules* **1993,** *26*, 6583–6588.

104. Weiss, R. A.; Lu, X. *Polymer* **1994,** *35*, 1963–1969.

105. Cimecioglu, A. L.; Weiss, R. A. *Macromolecules* **1995,** *28*, 6343–6346.

106. Belfiore, L. A.; Pires, A. T. N.; Wang, Y.; Graham, H.; Ueda, E. *Macromolecules* **1992,** *25*, 1411–1419.

107. Pearson, R. G. In *Survey of Progress in Chemistry*, Vol. 6; Scott, A., Ed.; Academic: New York, 1969; Chapter 1.

108. Horrion, J.; Jérôme, R.; Teyssie, P. *J. Polym. Sci. C. Polym. Lett.* **1986,** *24*, 69–76.

109. Russell, T. P.; Jérôme, R.; Charlier, P.; Foucart, M. *Macromolecules* **1988,** *21*, 1709–1717.

110. Charlier, P., Jérôme, R., Teyssié, P., Utracki, L. A. *Macromolecules* **1992,** *25*, 2651–2656.

111. Horrion, J.; Jérôme, R., Teyssié, P. *J. Polym. Sci. A Polym. Chem.* **1990,** *28*, 153–171.

112. Horrion, J.; Agarwal, P. K. *Polym. Commun.* **1989,** *30*, 264–267.

113. Cho, K.; Jeon, H. K.; Moon, T. J. *J. Mater. Sci.* **1993,** *28*, 6650–6656.

114. Ha, C. S.; Cho, W. J.; Hur, Y. S.; Kim, S. C. *Polym. Adv. Technol.* **1991,** *2*, 31–40.

115. Xu, X.; Zeng, X.; Li, H. *J. Appl. Polym. Sci.* **1992,** *44*, 2225–2231.

116. Kyu, T.; Yang, J.-C. *Macromolecules* **1990,** *23*, 176–182.

117. Iwakura, K.; Fujimura, T. *J. Appl. Polym. Sci.* **1975,** *19*, 1427–1437.

118. Hara, M.; Bellinger, M.; Sauer, J. A. *Polym. Int.* **1991,** *26*, 137–141.

119. Hara, M.; Bellinger, M.; Sauer, J. A. *Colloid Polym. Sci.* **1992,** *270*, 652–658.

120. Bellinger, M. A.; Sauer, J. A.; Hara, M. *Macromolecules* **1994,** *27*, 6147–6155.

121. Beck Tan, N. C.; Liu, X.; Briber, R. M.; Peiffer, D. G. *Polymer* **1995,** *36*, 1969–1973.

122. Hseih, D.-T.; Peiffer; D. G. *Polymer* **1992,** *33*, 1210–1217.

123. Hseih, D.-T.; Peiffer, D. G. *J. Appl. Polym. Sci.* **1992,** *44*, 2003–2007.

124. Register, R. A.; Bell, T. R. *J. Polym. Sci. B Polym. Phys.* **1992,** *30*, 569 575.

125. Tomita, H.; Register, R. A. *Macromolecules* **1993,** *26*, 2796–2801.

126. Bazuin, C. G.; Rancourt, L.; Villeneuve, S.; Soldera, A. *J. Polym. Sci. B Polym. Phys.* **1993,** *31*, 1431–1440.

CHAPTER 10

APPLICATIONS

As we showed in Figure 1.1, the total number of papers in the field of ionomers now approaches 650 per year; approximately one-third of that number is patents. This level of activity suggests that ionomers are of commercial interest. There is ample justification for this conclusion, because, even at present, ionomers are used in a wide range of applications, judging from the number of patents alone.

The aim of this chapter is to introduce the reader to some of the more interesting applications of ionomers. No attempt is made to provide encyclopedic coverage or even a catalog of applications, but merely to illustrate the wide range of uses for these materials. Therefore, the examples should be considered as illustration, and the space devoted to a particular application is no way related to its commercial usefulness or prospects.

The coverage is divided into four sections. In the first section, membranes and thin films are discussed. The second section is devoted to a discussion of ionomers in bulk. The third covers liquid state applications, specifically drilling fluids. The final section discusses ionomers both as catalysts by themselves and as carriers for catalysts.

10.1. MEMBRANES AND THIN FILMS

10.1.1. Membranes

One of the most useful applications of ionomers has been as membranes exhibiting superpermselectivity (1,2). Perfluorosulfonates, the chemical structure of which was shown in Section 6.2, and perfluorocarboxylates were initially prepared by du Pont (Nafion) and subsequently by several other companies, e.g., Asahi Glass (Flemion), and Asahi Chemical (Aciplex). These materials have more recently been joined by a membrane with a shorter side chain prepared by Dow Chemical. The materials show exceptional stability, resembling that of poly(tetrafluoroethylene), in both thermal and electrochemical applications. The electrochemical applications depend on the superpermselectivity, i.e., the film allowing cations to pass through the membrane relatively rapidly while limiting the transport of anions to much slower rates. Superpermselectivity of this type is thought to rely on the presence of a narrow conductive path or narrow channels, and it is thus intimately related to the morphology of the materials (see Chapter 3).

The largest application of perfluorinated ionomer membranes is undoubtedly in the chlor-alkali industry (1,2). In the chlor-alkali process, the starting material is a NaCl solution, and chlorine and hydrogen are produced at the anode and cathode, respectively; the membrane acts as a separator, keeping the gases apart and allowing sodium hydroxide in high purity to accumulate on the cathode side. The high purity of the NaOH is maintained because the diffusion coefficient for the Cl^- anion is low. Chlorine is one of the large-volume chemicals produced and used in industry, and thus even a moderate saving in production costs produces enormous total savings. In this context the perfluorinated ionomer membranes have been most successful. In chlorine production they are used at elevated temperatures (near 100°C) in strongly oxidizing (or reducing) media and are capable of withstanding high current densities, ~50 A/dm^2. The function of the membrane is thus to separate anode and cathode compartments, allowing the passage of sodium ions but blocking chloride and hydroxide anions. The membrane can maintain permselectivity at high values of water permeability.

In contrast, in fuel cells, the perfluorinated ionomer, usually in the sulfonated form, is used as a solid polymer electrolyte that keeps oxygen and hydrogen apart. In the cell these gases are converted to water and produce electricity. This process is the inverse of that of the chlor-alkali process, i.e., the production of power from a chemical reaction rather than the use of power to produce a particular product; but the general usefulness of the membranes is equally high in both applications. A number of excellent reviews are available on the perfluorinated ionomers, and several books have collected papers and reviews dealing with early work on this system (1,2). More recently, Schlick (3) edited a volume that contains six articles dealing with various aspects of perfluorinated ionomers, including a brief discussion of electrochemical applications that is much more extensive than the one presented here.

10.1.2. Packaging

One of the early applications of ionomers in packaging was based on their high melt strength. It was demonstrated that even the exposure of the film to a sharp tip, e.g., of a fishing hook, does not result in the perforation of the film under heat-seal conditions (4). The applications of ionomers in packaging are currently extensive. The advantages of ionomers, especially ethylene ionomers, as packaging materials are their oil and chemical resistance, flex resistance, resistance to impact, the ability to be heat sealed under a broad range of conditions (including in the presence of grease), adhesion to a wide range of substrates as a result of the presence of both hydrophobic segments and hydrophilic segments (e.g., the carboxylic acid or salt groups), excellent optical properties (as a result of the destruction of the spherulites in the presence of ionic groups), and their high melt strength. For example, a laminar structure, consisting of a layer of an ionomer and a layer of nylon to which the ionomer adheres well, is used in meat packaging because the nylon provides the appropriate barrier properties, and the ionomer suitable heat-seal properties. Similarly, the ionomers can be used in conjunction with metal foils in the packaging of pharmaceuticals, again, because a sandwich combines good barrier properties and good heat-seal properties. The ethylene ionomers are also useful as an interlayer between paper or metal films and polymers, again, because of their broad adhesive range.

The extensive use of ionomers in coatings has been well documented (2). One application involves the direct coating onto glass objects to reduce the danger of breakage or to contain the glass fragments. On occasion, a sandwich construction between two glass layers is used, which leads to a shatterproof material. A wide range of applications in coating involve polyurethane emulsions. The advantage of this material in the coating industry is that the coat can be deposited from an aqueous solution; and after the film dries, the ionic groups improve the properties of the final polymer relative to those of the nonionic systems.

10.1.3. Fertilizer

An application of ionomers as coatings is being explored in the fertilizer industry, in which zinc-neutralized ethylene-propylene diene terpolymer (EPDM) sulfonate is used as a coat for microspheres of urea (5). The thin film of the ionomer is deposited onto the urea sphere by a spray process from solution to which a small amount of a polar cosolvent is added to weaken the ionic interaction and reduce the viscosity. This technique permits easy spraying of the ionomer solution onto the urea particle and the consequent formation of a uniform thin ionomer film. As the solvents evaporate, the strong ionic interactions yield a strong film with high melt strength, which minimizes pin hole formation in the coat. A wide range of release profiles can be achieved, depending on the sphere size and the film thickness. The release can be rapid (hyperbolic), sigmoidal, or linear as a function of time. The sigmoidal profiles are obtained from osmotic swelling followed by membrane rupture. Under the conditions of use, when the ionomer coating the sphere of urea is exposed to water (rainfall, dew), the water diffuses through the membrane, and dissolves the urea. Initially the membrane remains intact. However, as the urea is dissolved progressively and the degree of swelling increases, the pressure on the wall of the membrane increases. At a certain point, the pressure causes surface cracking or pin hole formation, and all of the urea in solution is released. Because of the variation in the sizes of the urea spheres, and presumably some variation in the film thickness, the release can take place over long times. For example, the sigmoidal release profile can extend over a period of ~8 weeks. Thus deposition of the ionomer on the fertilizer particle can be arranged to produce maximum release at the height of the growing season, when it is most needed.

10.1.4. Floor Polish

A successful series of industrial floor polishes was developed based on the idea of ionic cross-linking of water-soluble polymers or emulsions (6). For floor polishes, it is desirable to deposit the film from an aqueous solution. Once deposited, properties such as durability are expected, even when wet footwear comes into contact with the floor. Therefore, once deposited, the film must be water resistant. However, once the film is partly abraded, it becomes necessary to remove it. This removal should naturally be accomplished by aqueous washing. These properties can be achieved with a polish consisting of an emulsion of polymers containing carboxylic acid group along with an amine complex of a zinc ion. On drying, the zinc loses its complexing group, at least in part, and yields a zinc-cross-linked carboxylate copolymer, which provides a stable and durable floor wax. Extensive optimization of the materials was involved; in some formulations, ethylenimine was added to the emulsion polymers containing carboxylic acid group so that the final polymers contain both the carboxylate and amine groups. Removal of the coating can be accomplished in an aqueous solution in the presence of ammonia, which complexes the zinc, softens the coating layer, and allows removal.

10.1.5. Imaging Systems

Tan (7) reviewed ionomers in imaging systems. Ionomers found applications as components of toners in electrophotography. For example, styrene–n-butyl methacrylate–potassium methacrylate terpolymer was used as a binder in toners involving carbon black. More recently, ionomers have found applications as additives to toners, in which they act as compounds that determine the charge of the toner during contact charging. The mechanism of the charging essentially involves removal of the most mobile species (8). A number of polymers have been developed in which light converts a material into its ionic form, which is then soluble in water, or conversely, in which light exposure cross-links the polymers (9). One example of such a material is an ionic polystyrene containing the cyclopropene unit, the structure of which is given in Scheme 10.1, where X^- is a counteranion, such as tetrachloroaluminate,

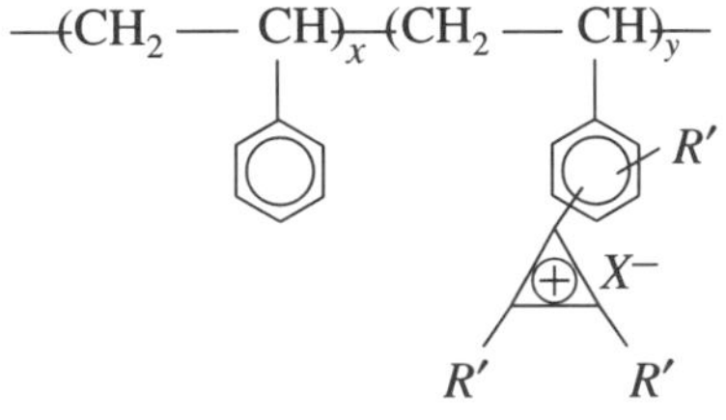

Scheme 10.1

and R' is an aryl group. When the polymer is exposed to light, it becomes cross-linked, which renders it insoluble, in contrast to the unexposed region from which it can be removed with a solvent.

Ionomers have also found extensive applications in silver halide photography as binders between a polyolefin overcoat and a paper base. They have also been important in color image transfer photography, in which the various ionic components serve a number of different functions. This field is extensive, and the reader is referred to the original literature (10–12). The applications include use as mordant polymers for anionic dye molecules, e.g., styrene containing quaternary salts, as acid layer polymers in which polymers containing a carboxylic acid are used to control the pH of the polymer after development, and as timing layers used to control the diffusion of hydroxide ions within the film structure during development.

10.1.6. Magnetic Recording Media

Ionic polyurethanes have found extensive use in magnetic recording media (7). Ionic urethanes can be prepared as aqueous dispersions; they serve both as a binder and as a dispersing agent in magnetic tape formation. The ionic polyurethanes are prepared from polyisocyanate and polyhydroxy compounds in the presence of organic solvents. However, these materials are water dispersible. Other carboxylated ionomers can be added to improve the properties, e.g., ethylene ionomers, which can also interact with polar pigment particles.

10.1.7. Adhesives

Because of their high melt strength and their amphiphilic nature, in addition to their other desirable physical properties, ionomers, such as carboxylated and sulfonated systems, have been used as adhesives (13). Both solvent-based and pressure-sensitive formulations have been developed. The butylene ionomers and EPDM-based systems are most frequently used.

10.2. IONOMERS IN BULK

10.2.1. Plastics

Ionomers have found extensive applications as bulk plastic materials (2,4). Examples of applications include golf ball covers, bowling pin covers, bumper guards, side molding strips, and shoe parts, especially for athletic footwear. In all of these applications, two of the required properties are moldability and toughness. In golf ball covers, the effect of ionic aggregation on the properties is most spectacular, and thus we cite it as an example. The toughness of the ionomers makes the ionomer-covered ball highly cut resistant so that an off-center full power swing, which would result in a cut by the edge of the club in a conventional ball made of materials other than ionomers, does not damage the ionomer-coated ball significantly. Because the material is easily moldable, injection molding can be used to manufacture the golf balls, thus reducing costs (14,15). Furthermore, by appropriate use of both zinc and sodium ionomers in a blend form, the cold crack problem encountered in the sodium ionomer and the low rebound problem encountered in the zinc ionomers have been effectively eliminated (16). The use of the blend gives the best properties of both systems. Therefore, it is not surprising that since the early 1970s golf balls with ethylene ionomer covers have been used extensively in the amateur market.

10.2.2. Elastomers

Sulfonated EPDM terpolymer with a zinc counterion can be used as a self-vulcanizing rubber. At high temperatures, the ionic interactions within the multiplets are loosened; the addition of zinc stearate reduces the viscosity in the melt even further and allows the material to be injection molded. On cooling, the ionic interactions become strong enough to provide a strong physical cross-link and to give the material desirable rubber properties (13). Thus injection molding technology can be used to prepare rubbery materials. The materials can be extended with the usual additives, e.g., carbon black, and oil (17). The use of these materials in roofing membranes has been proposed, and a large number of potential applications have been explored (2,9). Because the cross-links are physical, a certain amount of creep is expected, which prevents the material from being used in rubber tire applications. However, the presence of physical cross-links avoids the problem of rupture, which is encountered with chemical cross-links, where the shortest chains break under large strain. For physical cross-links, large strain induces ion hopping rather than breakage of chains. These materials, therefore, have advantages over chemically vulcanized rubbers for some applications. In addition, reprocessing by simply reheating the material and remolding it is easy.

10.3. DRILLING FLUIDS

Sulfonated ionomers have been employed in oil-based drilling fluids (13,18). The function of the sulfonated polystyrene is to enhance the suspension of the various agents and by-products in the drilling operation. It should be remembered that in the drilling operation the material must work at high temperatures, and the advantage of the styrene backbone is the stability of the polymer up to ~330°C. The interaction of the functional groups with solid particles produced during drilling also appears to be advantageous in preventing settling of the material, which might result in damage to the drill bit.

10.4. CATALYSTS AND CATALYTIC SUPPORTS

Ionomers can be used catalytically. Olah's group (19,20) found that the acid form of the perfluorosulfonic acid can be used as a superacid catalyst. For example, perfluorosulfonic acid can act as a catalyst in some organic reactions, such as pinacol rearrangement, deacetylation, and decarboxylation of aromatic rings. When the acid sample is neutralized, the ionic aggregates can be considered as microreactors, which catalyze chemical reactions owing to the presence, after reduction, of metal particles, metal ions, and activated metal sites. It was found that several different metal ions can react with gases, such as H_2, CO, NO, C_2H_2, and N_2H_4, in perfluorosulfonates or sulfonated polystyrene systems (2,21,22). The metal cations in the ionic aggregates can also be reduced to form metal particles. Chemistry in polystyrene- and poly(tetrafluoroethylene)-based ionomers was discussed in Sections 5.11.2 and 6.2.10, respectively.

10.5. REFERENCES

1. *Perfluorinated Ionomer Membranes*; Eisenberg, A.; Yeager, H. L., Eds.; ACS Symposium Series 180; American Chemical Society: Washington, DC, 1982.

2. Risen, W. M. Jr. In *Ionomers: Characterization, Theory, and Applications*; Schlick, S., Ed.; CRC; Boca Raton, FL, 1996; Chapter 12.

3. *Ionomers: Characterization, Theory, and Applications*; Schlick, S., Ed.; CRC: Boca Raton, FL, 1996.

4. Statz, R. J. *Polym. Prepr. Am. Chem. Soc. Div. Polym. Chem.* **1989,** *29,* 435–437.

5. Drake, E. N. *Polym. Prepr. Am. Chem. Soc. Div. Polym. Chem.* **1994,** *35,* 14–15.

6. Rogers, J. R.; Randall, F. J. *Polym. Prepr. Am. Chem. Soc. Div. Polym. Chem.* **1989,** *29,* 432.

7. Tan, J. S. In *Structure and Properties of Ionomers*; Pineri M.; Eisenberg, A., Eds.; NATO ASI Series C: Mathematical and Physical Sciences 198; Reidel: Dordrecht, 1987; pp. 439–451.

8. Diaz, A.; Fenzel-Alexander, D.; Miller, D. C.; Wollmann, D.; Eisenberg, A. *J. Polym. Sci. C Polym. Lett.* **1990,** *28,* 75–80.

9. Wadsworth, D. H. and co-workers. U.S. Patent 4 3779 989, 1973.

10. Hanson, W. T. Jr. *Photogr. Sci. Eng.* **1976,** *20,* 155–164.

11. Hanson, W. T. Jr. *J. Photogr. Sci.* **1977,** *25,* 189–196.

12. Fleckenstein, L. J. In *The Theory of Photographic Process*; James, T. H., Ed.; Macmillan: New York, 1977; p. 366.

13. Lundberg, R. D. In *Structure and Properties of Ionomers*; Pineri M.; Eisenberg, A., Eds.; NATO ASI Series C: Mathematical and Physical Sciences 198; Reidel: Dordrecht, 1987; pp. 429–438.

14. Statz, R. J. Presented at the *First World Scientific Congress of Golf*, St. Andrews, 1990.

15. Sullivan, M. J.; Melvin, T. In *Science and Golf II*; Cochran, A. J.; Farrally, M. R., Eds., Span: London, 1994; p. 334.

16. Molitor, R. P. U.S. Patent 3 819 768, 1974.

17. Paeglis, A. U.; O'Shea, F. X. *Rubber Chem. Technol.* **1988,** *61,* 223–237.

18. Portnoy, R. C.; Lundberg, R. D.; Peiffer, D. G. *Polym. Prepr. Am. Chem. Soc. Div. Polym. Chem.* **1989,** *29,* 433–434.

19. Olah, G. A.; Meidar, D. *Synthesis* **1978,** 358.

20. Olah, G. A.; Laali, K.; Mehrota, A. K. *J. Org. Chem.* **1977,** *42,* 4187–4191.

21. Risen, W. M. Jr. In *Structure and Properties of Ionomers*; Pineri, M.; Eisenberg, A., Eds.; NATO ASI Series C: Mathematical and Physical Sciences 198; Reidel: Dordrecht, 1987; pp 87–96.

22. Barnes, D. M.; Chaudhuri, S. N.; Chryssikos, G. D., Mattera, V. D. Jr.; Peluso, S. L.; Shim, I. W.; Tsatsas, A. T.; Risen, W. M. Jr. In *Coulombic Interactions in Macromolecular Systems*; Eisenberg, A.; Bailey, F. E., Eds.; ACS Symposium Series 302; American Chemical Society: Washington, DC, 1986; Chapter 5.

INDEX